Introduction to Biochemistry and Metabolism

D. Anandhi
Department of Biochemistry
Dwaraka Doss Goverdhan Doss Vaishnav College
Chennai, Tamil Nadu

PEARSON
Chennai • Delhi

Associate Editor—Acquisitions: R. P. Mathivathani
Editor—Production: M. R. Ramesh

ISBN 978-81-317-7485-4

First Impression

Published by Dorling Kindersley (India) Pvt. Ltd, licensees of Pearson Education in South Asia.

Head Office: 7th Floor, Knowledge Boulevard, A-8(A), Sector 62, Noida 201 309, UP, India.
Registered Office: 11 Community Centre, Panchsheel Park, New Delhi 110 017, India.

Compositor: Cameo Corporate Services Limited, Coimbatore.
Printer: Pushp Print Services

Contents

Preface

Introduction to Biochemistry and Metabolism has been written with the aim of introducing graduate students to this exciting field of science. It is based on the syllabus prescribed by major Indian universities for undergraduate students of zoology, botany, microbiology, biochemistry and biotechnology. It also caters to the needs of students who study this subject as an ancillary course, as also those of biomedical students who pursue the subject from the standpoint of research, laboratory and formulations work.

Chapters 1–5 give a brief account on the structure of biomolecules, their sources, classification, biomedical importance, and properties with colour reaction principle. These chapters give students an insight into the practical applications of biomolecules as well as the theoretical principles behind their functional utility.

To understand the concepts better, Chapters 6–13 are focused on the chemistry that occurs at the site of synthesis and give an overview of the metabolism, mechanism and regulation that influence these reactions. Each chapter concludes with a sufficient number of review questions to help the student recapitulate on the topics covered therein.

I am sure that this book will help beginner students in various disciplines of life sciences to understand and appreciate the basic concepts of biochemistry and metabolism. If any misrepresentation of facts or concepts is detected in this book, it may please be brought to my notice or that of the publisher. I welcome constructive criticisms and suggestions for further improvement of this book.

D. Anandhi

Acknowledgements

FEAR OF GOD IS THE BEGINNING OF WISDOM

> "The mediocre teacher tells. The good teacher explains. The superior teacher demonstrates. The great teacher inspires."
>
> —*William Arthur Ward*

With this inspiration and profound influence of my great teachers, colleagues and admirable students, I had set out to write a book on biochemistry and metabolism and believe that I have succeeded in it to a fair extent. The book focuses on biochemistry, metabolism in particular, and enables students gain better coherence on the subject by providing a one-point source that elaborates upon all the major metabolic pathways in a simple language.

As Jacques Barzun rightly said, "In teaching, you cannot see the fruit of a day's work. It is invisible and remains so, maybe for twenty years." I am sure that such collective wisdom and guidance will create scholars and researchers in the field of biochemistry from our mother nation, India, in the days to come.

I bow and thank the Lord for giving me this wisdom and good health.

"What you sow, so shall you reap." I remember my teachers, Dr A. Geetha, Associate Professor, Bharathi Women's College, Chennai, and Dr Niranjali, Professor and Head, Department of Biochemistry, University of Madras, Chennai, at this juncture. Yes, genetic engineering is a field in which you reap what you sow—intricate knowledge is passed on from a teacher to a student to another student and so on. The wisdom continues…

I am indebted to my guide Dr K. Revathi, Associate Professor, Department of Zoology, Ethiraj College for Women, Chennai, for her enormous and consistent support in all my efforts. I thank Dr S. Narasimhan, Principal, DG Vaishnav College, and Shri. P. Haridas, Secretary of DG Vaishnav College, for their support and motivation. I am obliged to Mrs M. Kavitha, who helped me in writing one of the chapters of the book.

Introduction to Biochemistry and Metabolism has been written with a single-point objective of presenting the subject in a lucid language with emphasis on coherence while covering all the major metabolic pathways in a single book. The idea of writing this book, though deep-rooted in me for quite some time, was kindled by my former students from SRM, present students of DG Vaishnav College, and my Department Head, Dr P. T. Srinivasan.

I am grateful to my mother, husband and son for their motivation, patience and encouragement. I am obliged to all my colleagues who helped me to bring this book to fruition. Finally, I thank R. P. Mathivathani of Pearson Education for efficiently seeing the book through the press.

D. Anandhi

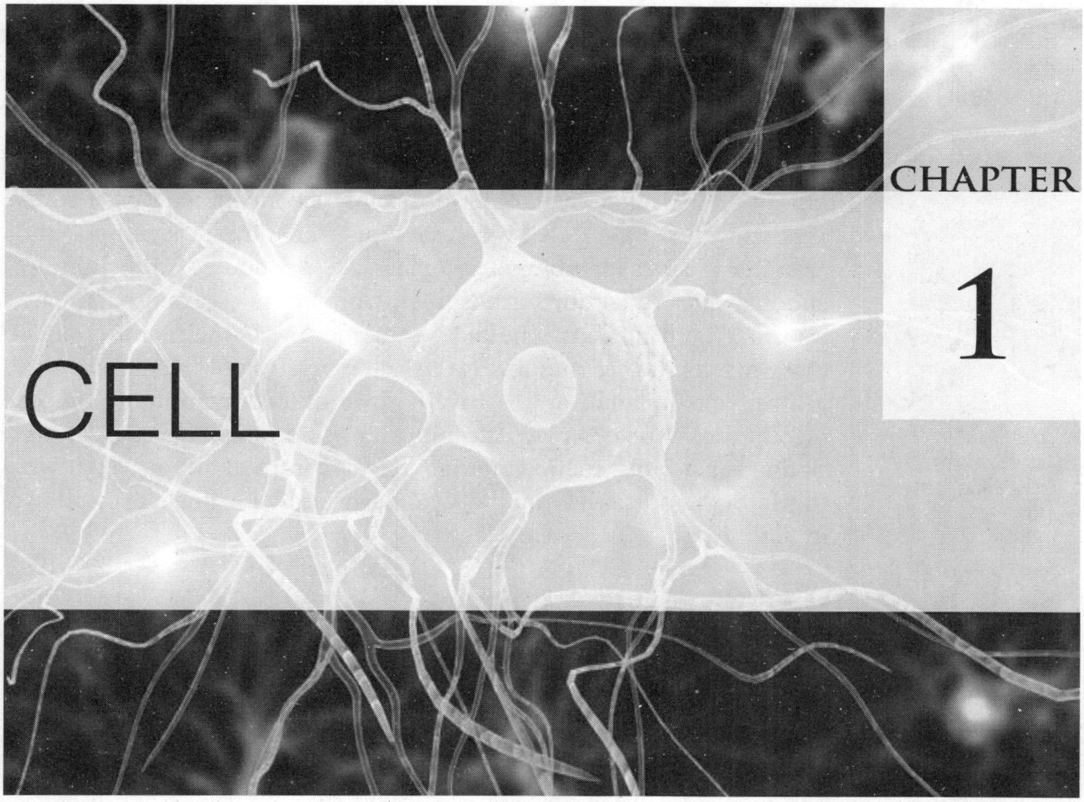

CHAPTER 1

CELL

Concept Outline

- Introduction
- Cell and its Structure
- Cell Components and Their Functions
- Nucleus
- Protoplasm

INTRODUCTION

Cell is considered to be the smallest unit of life. It is self-sufficient such that all kinds of cells can survive independently. There are several differences between an animal cell and a plant cell. Most cells have organelles, which are sub-cellular parts that perform specific functions. These sub-cellular parts (organelles) help the cell to efficiently perform all its activities.

CELL AND ITS STRUCTURE

Cell is the fundamental unit of living things. It is defined as the basic unit of biological activity. The word cell was contributed by Schieiden and Schwann (1838). Cell is microscopic in nature. It is composed of an outer plasma membrane, cytoplasm, Golgi bodies, mitochondira, endoplasmic reticulum, ribosomes, microtubules, microfilaments, lysosome, and nucleus. The plasma membrane is the main organelle that protects the cell and forms the outer covering of the cell. Nucleus is a cellular organelle and is covered by a double membrane (nuclear membrane). Nucleus plays an important role in cell division as well as the transfer of genetic material. The structure of a typical cell with the internal organelles in the cytoplasm and the nucleus is shown in Figure 1.1.

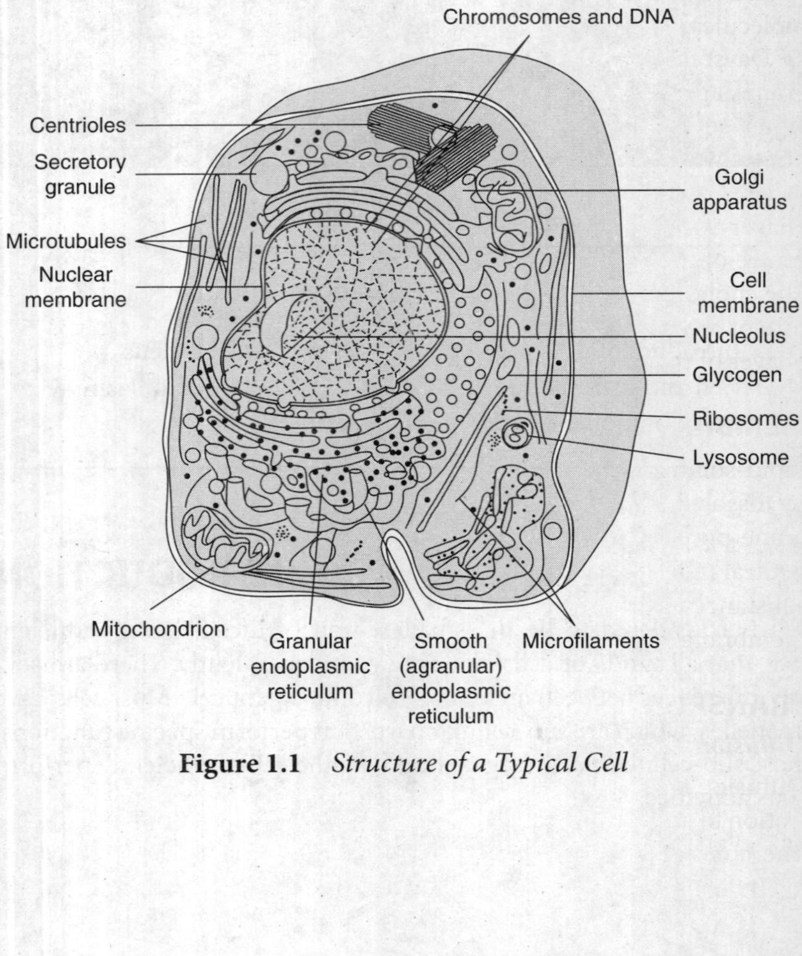

Figure 1.1 *Structure of a Typical Cell*

BIOSYNTHESIS OF FATTY ACID

INTRODUCTION

The dietary carbohydrates and amino acids, when consumed in excess, can be converted to fatty acids and stored as triglycerols. De novo (new) synthesis of fatty acids occurs predominantly in liver, kidney, adipose tissue, and lactating mammary glands. The enzyme machinery for the fatty acid production is located in cytosomal fraction of the cell.

Site of synthesis: Cytosol

Source: Acetyl-CoA – Source of carbon atom
NADPH – Provides the reducing equivalents
ATP – Supply energy for fatty acid formation

FATTY ACID SYNTHASE COMPLEX

Fatty acid synthase is a dimmer composed of two identical subunits (monomers), each with a molecular weight of 240,000. Each subunit contains the activities of seven enzymes of FAS an ACP (acyl-carrier protein) with 4'-phosphopantothiene —SH group. The two subunits lie in anti-parallel (head to tail) orientation. The SH group of phosphopantetheine of one subunit is in close proximity to the SH group of cysteine residue (of the enzyme ketoacyl synthase) of the other subunit.

Each monomer of FAS contains all the enzyme activities of fatty acid synthesis. But only dimer is functionally active. This is because the functional unit consists of half of each subunit interacting with the complementary half of the other. Thus, the FAS structures have both functional division and subunit division. The two functional subunits of FAS independently operate and synthesise two fatty acids simultaneously.

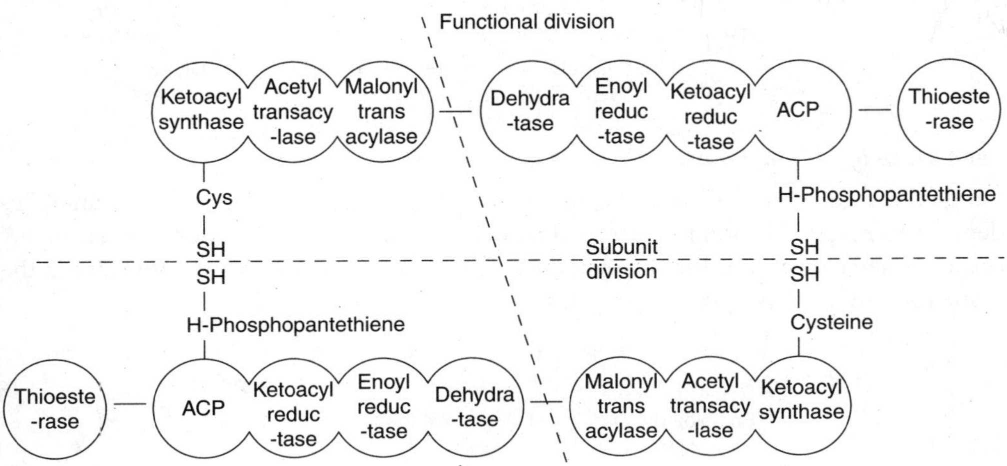

FATTY ACID SYNTHESIS OCCURS IN THREE STAGES

1. Production of Acetyl-CoA and NADPH
2. Conversion of Acetyl-CoA to malonyl-CoA
3. Reactions of fatty acid synthase complex

Production of Acetyl-CoA and NADPH

Acetyl-CoA and NADPH are pre-resquisites for fatty acid synthesis. Acetyl-CoA is produced in the mitochondria by the oxidation of pyruvate and fatty acids and degradation of carbon skeletons of amino acids and from ketone bodies. Mitochondrion is not permeable to acetyl-CoA, so Acetyl-CoA condenses with oxaloacetate in mitochondria to form citrate. Citrate lyase liberates Acetyl-CoA and oxaloacetate. Oxaloacetate in the cytosol is converted to malate.

Malic Enzyme

It converts malate to pyruvate. NADPH and CO_2 are regenerated in this reaction. Both of them are utilised for fatty acid synthesis.

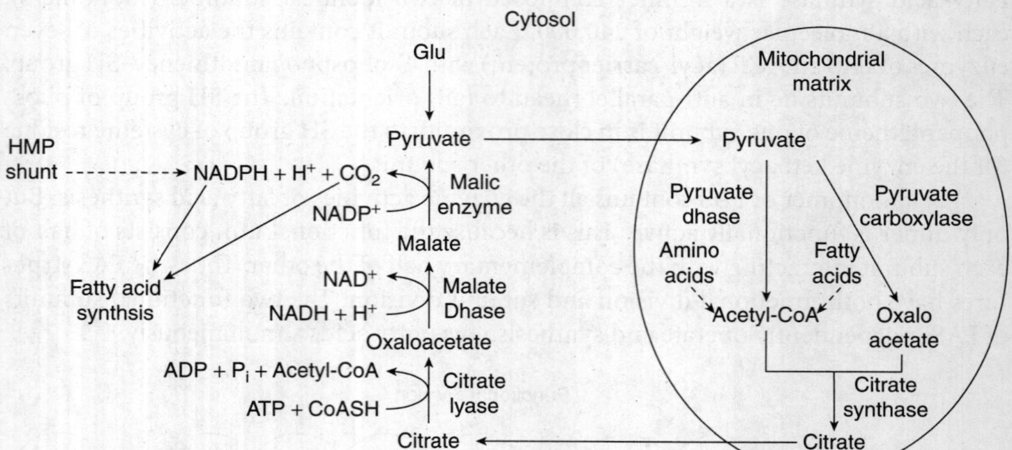

Formation of Malonyl-CoA

Acetyl-CoA carboxylase has a requirement for the vitamin biotin. This is an ATP-dependent reaction biotin for carbon fixation. The mechanism of action of acetyl-CoA carboxylase is similar to that of pyruvate carboxylase. Acetyl-CoA carboxylase is the regulatory enzyme in fatty acid synthesis.

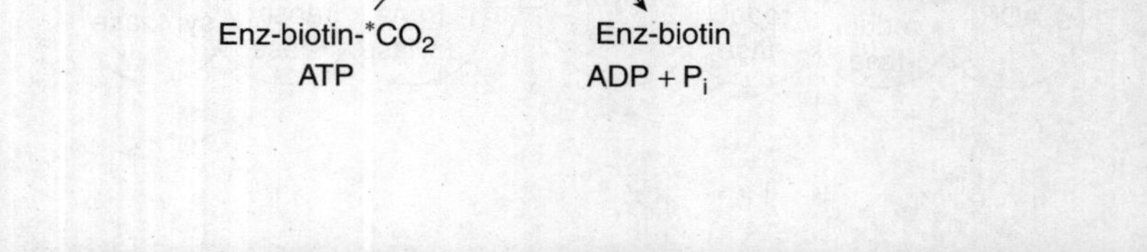

Reaction of Fatty Acid Synthase Complex

The sequence of reactions for the synthesis of fatty acid (palmitate) is shown below. The carbon chain attached to ACP is transferred to cysteine residue, and reactions 1 to 4 are repeated

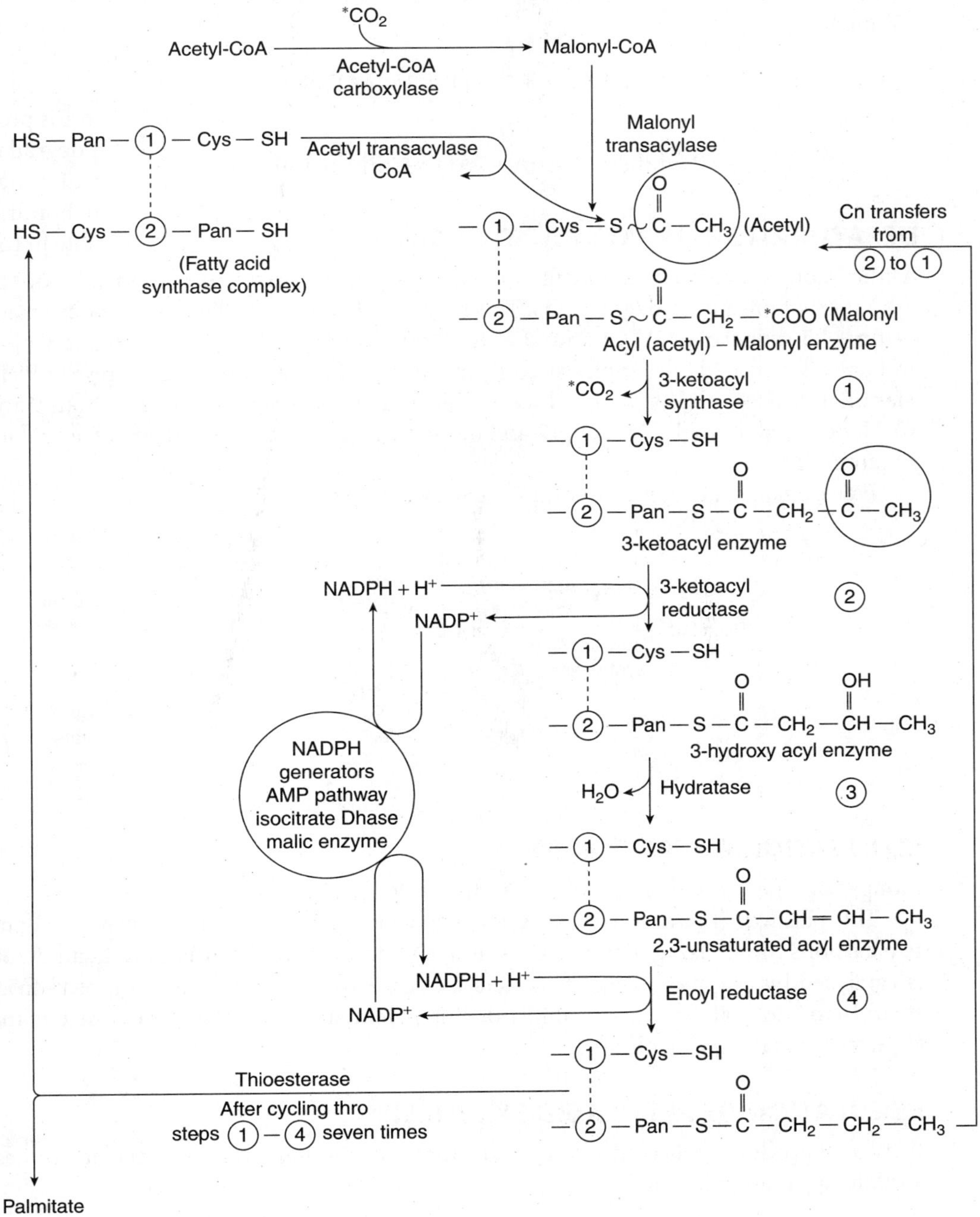

6 more times. Each time, the fatty acid chain is lengthened by a two-carbon unit. At the end of the seventh cycle, the fatty acid synthesis is completed, and a 16-carbon fully saturated fatty acid (palmitate) bound to ACP is produced. The enzyme thioesterase separates palmitate from fatty acid synthase. This completes the synthesis of palmitate.

$$\text{8Acetyl-CoA} + 7\text{ ATP} + 14\text{ NADPH} + 14\text{ H}^+$$
$$\downarrow$$
$$\text{Palmitate} + 8\text{ CoA} + 7\text{ADP} + 7\text{Pi} + 6\text{H}_2\text{O} + 7\text{CO}_2$$

DESATURATION OF FATTY ACID CHAINS

A microsomal enzyme system called fatty acyl-CoA desaturase is responsible for the formation of unsaturated fatty acids. This reaction also involves flavin-dependent cytochrome b5 reductase NADH and O_2. The monounsaturated fatty acids, namely oleic and palmitoleic acids, are synthesised from stearic acid and palmitic acid, respectively. Mammals lack the ability to introduce double bond in fatty acid between carbons 9 and 10. Hence linoleic acid – 18: 2, 9, 12 and linolenic acid – 18: 3; 9, 12, 15 are essential for a man in diet.

Fate of palmitate after biosynthesis is shown below

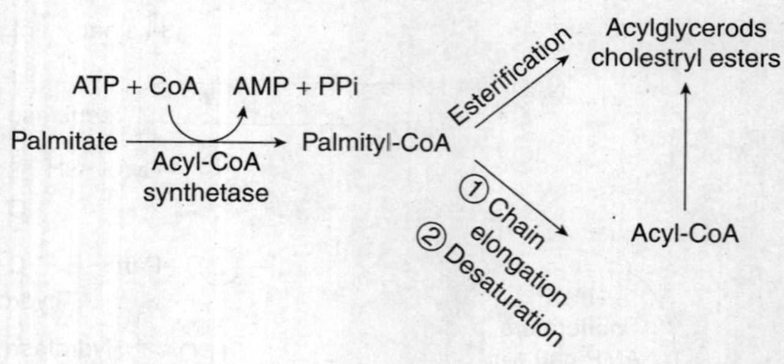

ELONGATION OF FATTY ACID

Elongation of fatty acid chains occurs in the endoplasmic reticulum.

This pathway (the microsomal system) converts fatty Acyl-CoA derivative having two carbons more, using malonyl-CoA as acetyl donor and NADPH as reductant and is catalysed by the microsomal fatty acid elongase system of enzymes. Elongation of stearyl-CoA in brain increases rapidly during myelination in order to provide C_{22} and C_{24} fatty acids for sphingolipids.

REGULATION OF FATTY ACID SYNTHESIS

Fatty acid production is controlled by enzymes, metabolites, end products, hormones, and dietary manipulation.

ACETYL-CoA CARBOXYLASE

This enzyme controls the committed step in fatty acid synthesis. Acetyl-CoA carboxylase exists as an inactive promoter (monomer) or an active polymer. Citrate promotes polymer formation and hence increases fatty acid synthesis. Palmityl-CoA and malonyl-CoA cause dephosphorylation of the enzyme and, therefore, inhibits fatty acid synthesis.

HORMONAL INFLUENCE

Hormones regulate acetyl-CoA carboxylase by a separate mechanism – phosphorylation (inactive form) and dephosphorylation (active form) of the enzyme. Glucagon, epinephrine, and norepinephrine inactivate the enzyme by cAMP-dependent phosphorylation. Insulin, on the other hand, dephosphorylates and activates the enzyme. Thus, insulin promotes fatty acid synthesis while glucagons inhibits. Insulin stimulates tissues' uptake of glucose and conversion of pyruvate to acetyl-CoA. This also facilitates fatty acid formation.

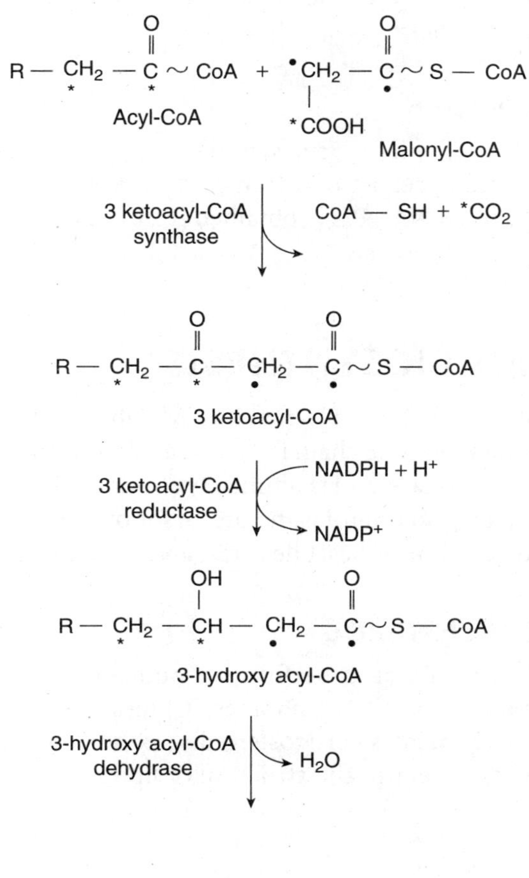

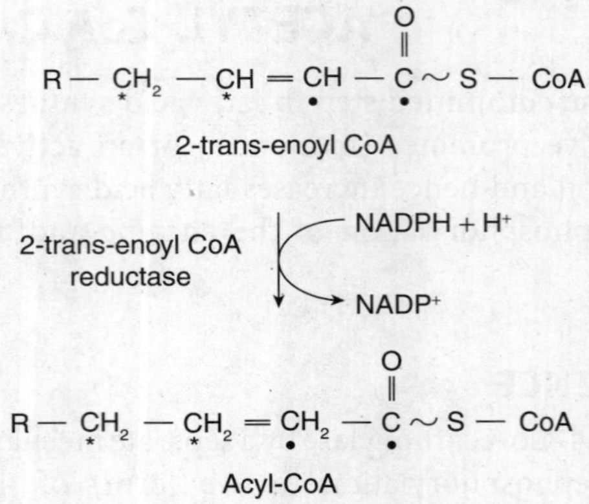

DIETARY REGULATION

Consumption of high carbohydrate of fat-free diet increases the synthesis of acetyl-CoA carboxylase and fatty acid synthase, which promotes fatty acid formation. On the other hand, fasting or high fat diets decreases fatty acid production by reducing the synthesis of these two enzymes.

AVAILABILITY OF NADPH

The reducing equivalents for fatty acid synthesis are provided by NADPH, which comes either from citrate (Acetyl-CoA) transport or hexose monophosphate shunt. About 50%–60% of required NADPH is obtained from hexose monophosphate shunt, which influences fatty acid synthesis.

CHOLESTEROL BIOSYNTHESIS

Cholesterol is present in tissues and plasma. It is available either as free cholesterol or as a storage form attached to a long chain fatty acid as cholesteryl ester. It is synthesised in many tissues from Acetyl-CoA and is ultimately eliminated from the body in the bile as cholesterol or bile salts. Cholesterol is the precursor of all other steroids in the body such as corticosteroids, sex hormones, bile acids, and vitamin D.

FUNCTIONS OF CHOLESTEROL

Cholesterol is essential to life as it performs a number of important functions. It is a structural component of all membranes. Cholesterol is the precursor of all other steroids in the body such as corticosteroids, sex hormones, bile acids, and vitamin D. It is essential ingredient in the structure of lipoproteins from which lipids in the body are transported. Fatty acids are transported to liver through cholesteryl ester oxidation.

LOCATION

About 1 g of cholesterol is synthesised per day in adults. Almost all the tissues of the body participate in cholesterol biosynthesis. The largest contribution is made by liver (50%), intestine (15%), skin, adrenal cortex, reproductive tissue, and so on.

SITE OF SYNTHESIS

Cytosol and mitochondrial fraction (ER) – enzymes involved in cholesterol synthesis are formed in these organelles.

BIOSYNTHESIS

The biosynthesis of cholesterol may be divided into five steps:

1. Synthesis of HMG-CoA (3 hydroxy 3-methyl glutaryl CoA)
2. Biosynthesis of mevalonate
3. Formation of isoprenoid units
4. Synthesis of squalene
5. Conversion of squalene to cholesterol

Synthesis of HMG-CoA

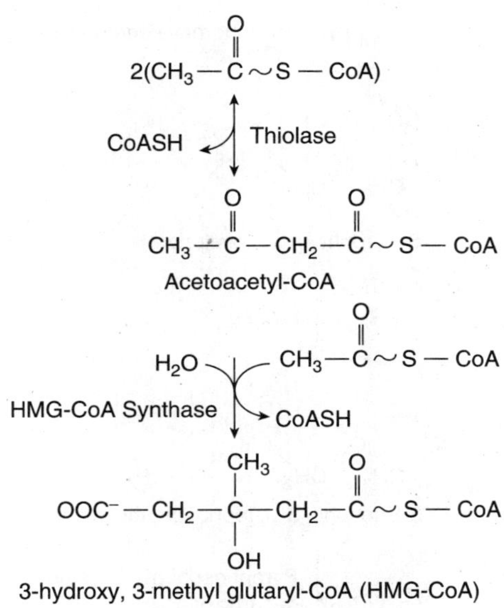

Synthesis of Mevalonate

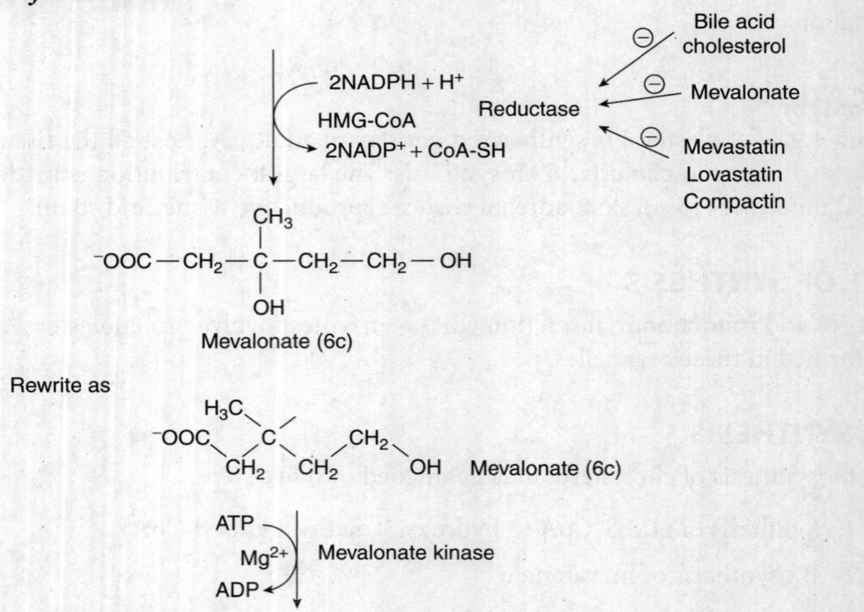

Formation of Isoprenoid Units

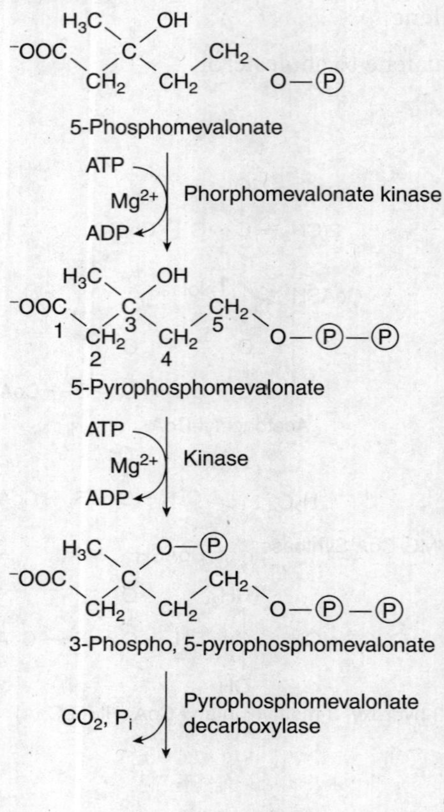

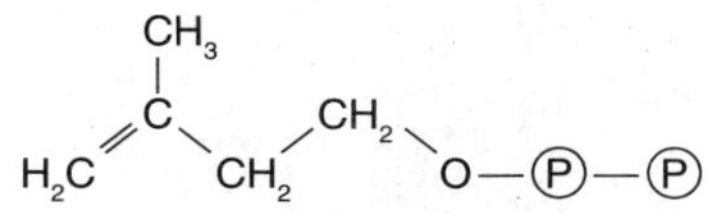

Synthesis of Squalene

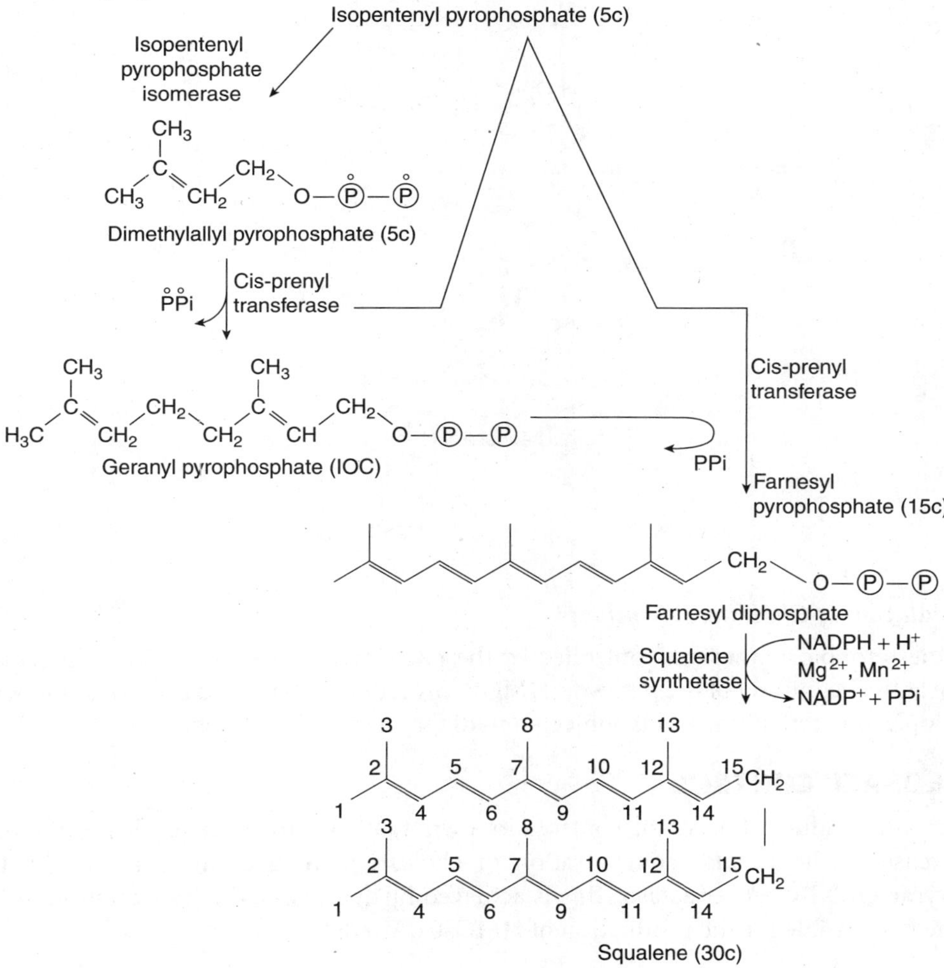

Conversion of Squalene to Cholesterol

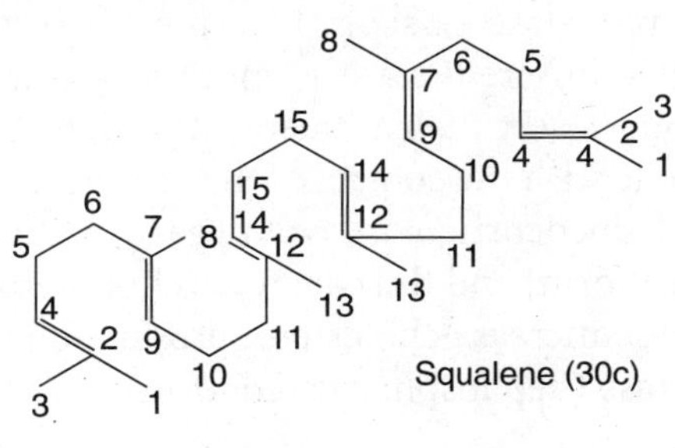

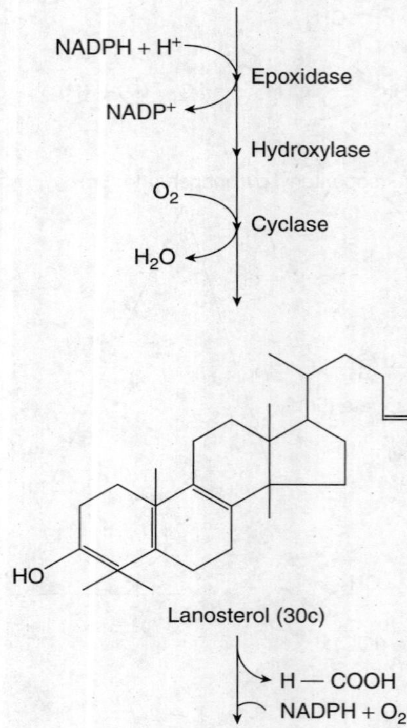

Regulation of Cholesterol Synthesis

Cholesterol biosynthesis is controlled by the rate-limiting enzyme HMG-CoA reductase at the beginning of the pathway. HMG-CoA reductase is found in association with endoplasmic reticulum and is subjected to different metabolic controls.

FEEDBACK CONTROL

The end product cholesterol controls its own synthesis by a feedback mechanism. Increase in the cellular concentration of cholesterol reduces the synthesis of the enzyme HMG-CoA reductase. This is achieved by decreasing the transcription of the gene responsible for the production of HMG-CoA reductase.

HORMONAL CONTROL

The enzyme HMG-CoA reductase exists in two interconvertible forms. The dephosphorylated form of HMG-CoA reductase is more active while the phosphorylated is less active. The hormones exert their influence through cAMP by a series of reactions, which are comparable with the control of the enzyme glycogen synthase. The net effect is that glucagon and glucocorticoids favour the formation of inactive HMG-CoA reductase (phosphorylated form) and thus decrease cholesterol synthesis. On the other hand, insulin and thyroxine increase cholesterol production by enhancing the formation of HMG-CoA reductase (dephosphorylated form).

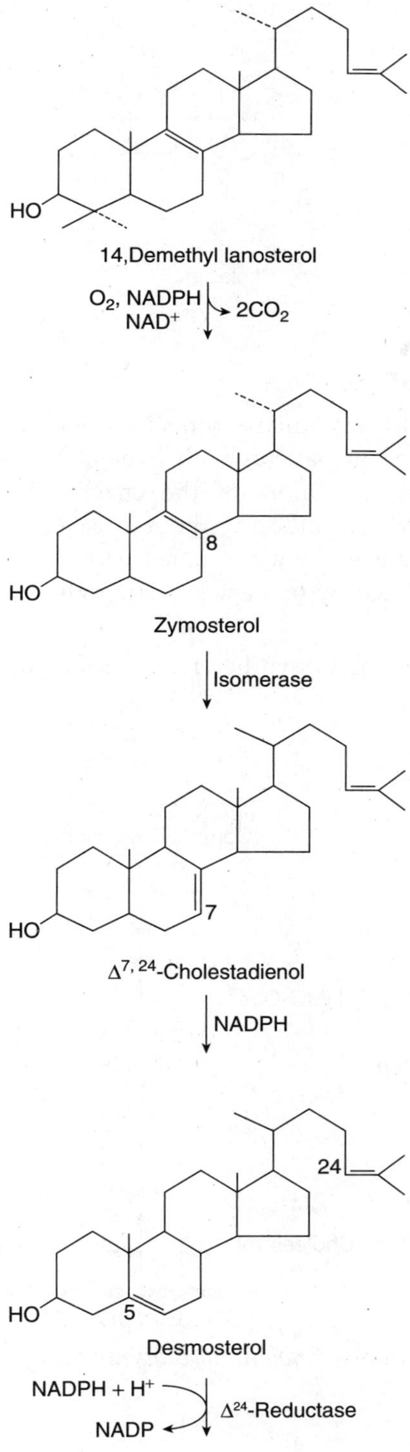
HO
14,Demethyl lanosterol
O$_2$, NADPH
NAD$^+$
2CO$_2$
8
HO
Zymosterol
Isomerase
7
HO
$\Delta^{7,24}$-Cholestadienol
NADPH
24
HO
5
Desmosterol
NADPH + H$^+$
NADP
Δ^{24}-Reductase

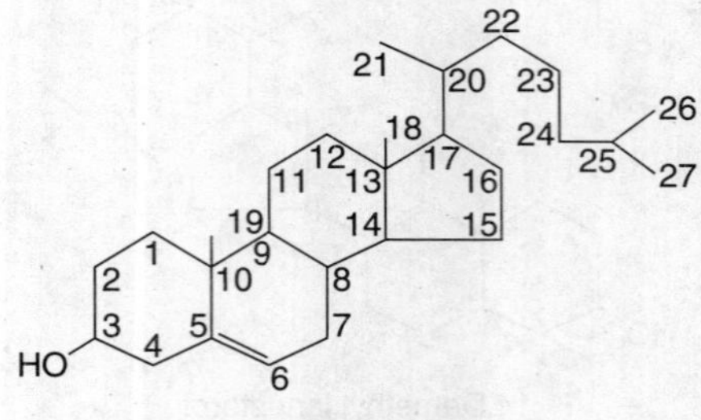

INHIBITION OF DRUGS

The drugs compactin and lovastatin are fungal products. They are used to decrease the serum cholesterol level in patients with hypercholesterolemia. Compactin and lovastatin are competitive inhibitors of the enzyme HMG-CoA reductase and, therefore, reduce cholesterol synthesis. About 50%–60% decrease in serum cholesterol level has been reported by a combined use of these two drugs. HMG-CoA reductase activity is inhibited by bile acids; fasting also reduced the activity of this enzyme.

Regulation of cholesterol biosynthesis by HMG-CoA reductase is shown in Figure 10.4

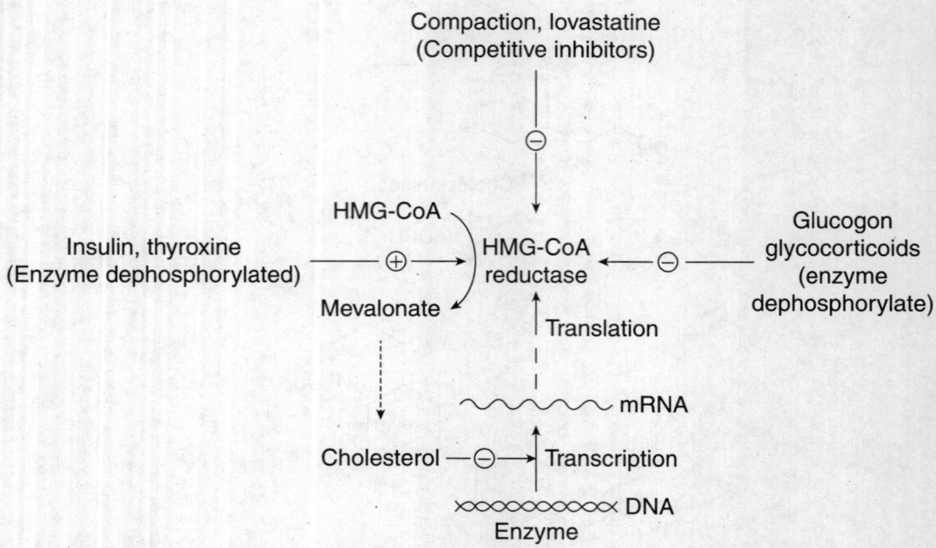

Figure 10.4 *Regulation of Cholesterol Biosynthesis by HMG-CoA Reductase*

DEGRADATION OF CHOLESTEROL

The ring structure of cholesterol cannot be metabolised to CO_2 and H_2O in humans. The steroid ring is eliminated from the body by the following ways:

1. Conversion to bile acids (bile salts)
2. Conversion to steroid hormones
3. Conversion to vitamin D

CONVERSION TO BILE ACIDS

The bile acids contain twenty-four carbons, with two or three hydroxyl groups and a side chain that terminates in a carboxyl group.

Site of Synthesis

Primary bile acids are synthesised in liver from cholesterol.

Degradation of Cholesterol

Cholesterol is degraded to bile acids by a multi-step pathway, in which the hydroxyl groups are inserted at specific positions on the steroid structure of cholesterol; the double bond of B-ring is reduced. The hydrocarbon chain is shortened by three carbons, introducing a carboxyl group at the end of the chain. The resulting compounds, cholic and chenodeoxy cholic acids, are called primary bile acids.

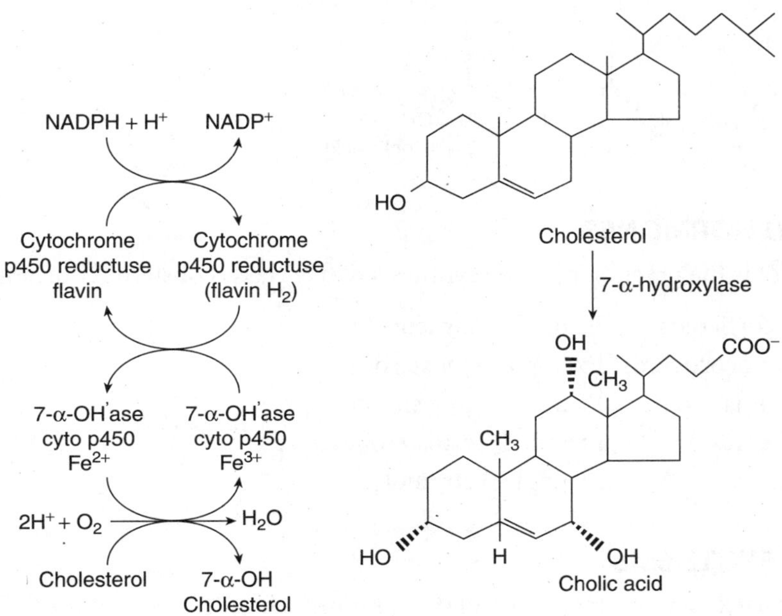

Bile Salts

Bile salts leave the liver as a molecule conjugated to either glycine or taurine by an amide linkage between the carboxyl group of the bile acid and the amino group on the added compounds. These modified structures are called bile acids, which include glycocholic acid, chenodeoxycholic acid, taurocholic acid, and taurochenodeoxycholic acid.

Structure of Bile Salts

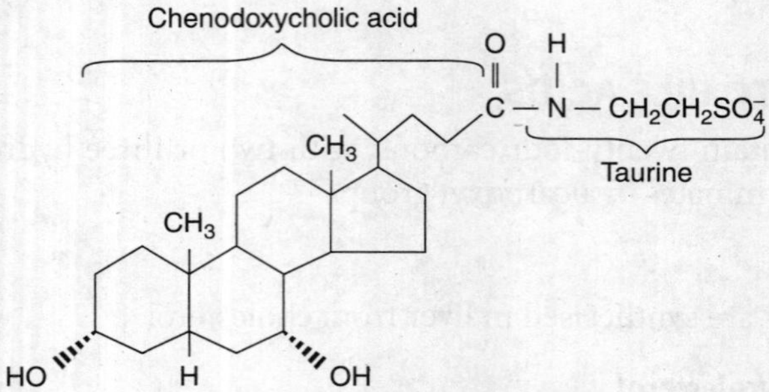

Taurochenodeoxycholic acid

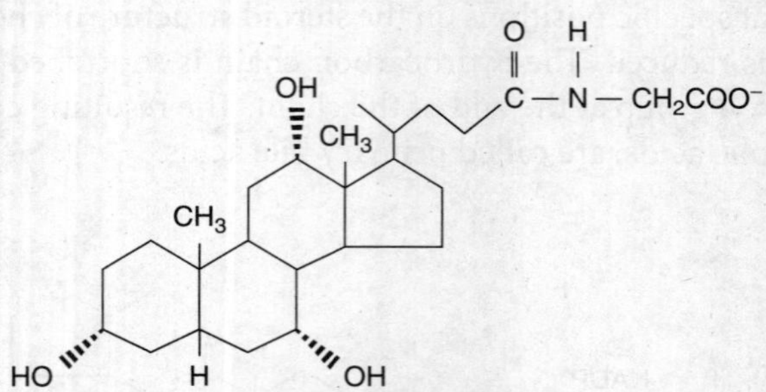

Glycocholic acid

STEROID HORMONES

Cholesterol is the precursor for the synthesis of five classes of steroid hormones:

Glucocorticoids Example: Cortisol
Mineralocorticoids Example: Aldosterone
Progestins Example: Progesterone
Androgens Example: Testosterone
Estrogens Example: Estradiol

SITE OF SYNTHESIS

Adrenal cortex, ovaries, testis, cytoplasm, and mitochondria are the sites of synthesis of bile salts.

STRUCTURE OF STEROID HORMONES

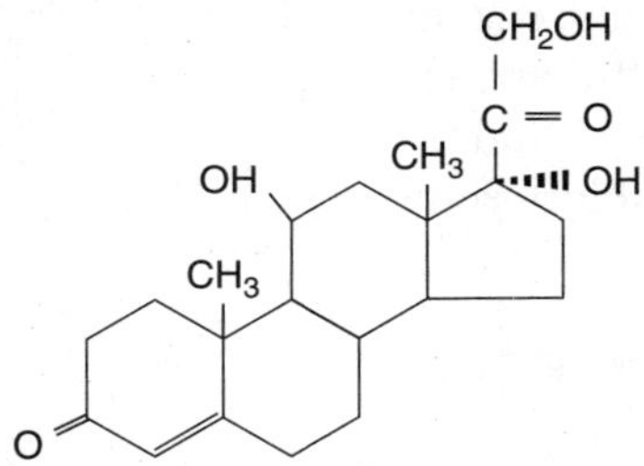

Cortisol
(c-21)

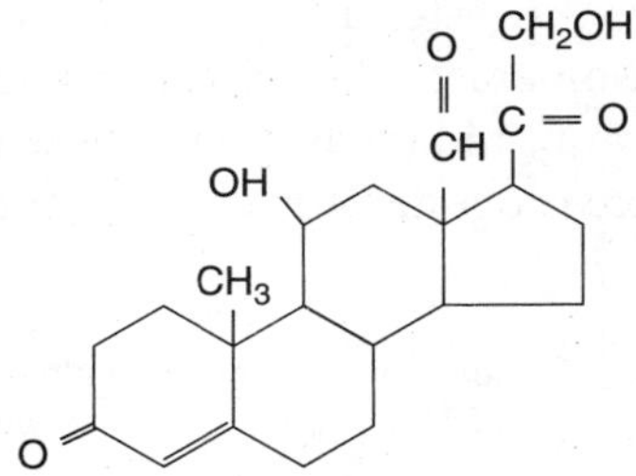

Aldosterone
(c-21)

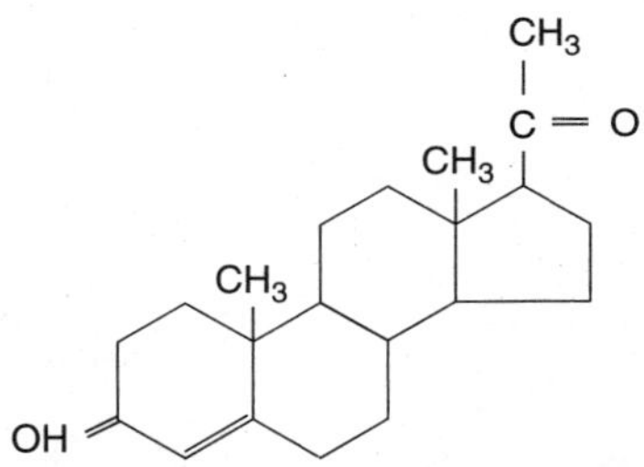

Progesterone
(c-21)

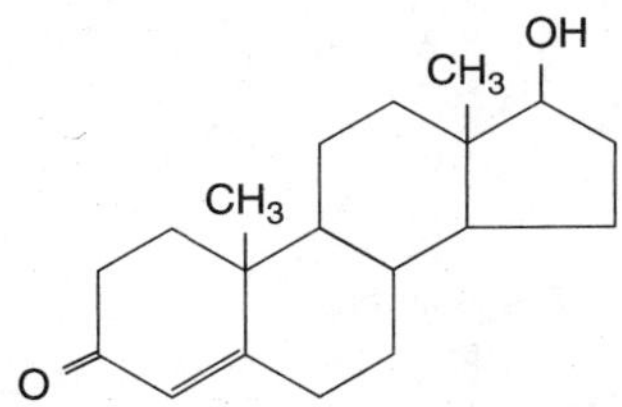

Testosterone
(c-19)

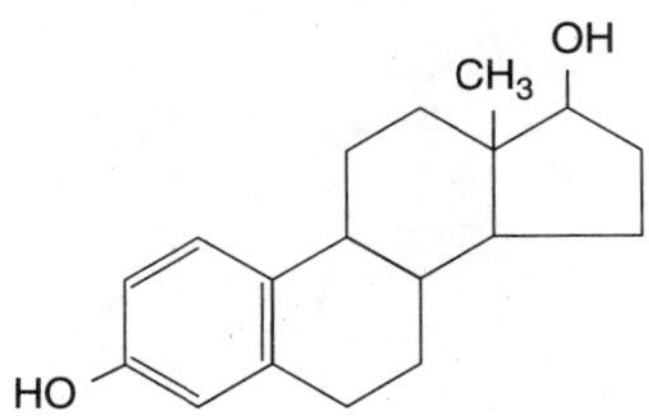

Estradiol
(c-18)

TYPES OF STEROID HORMONES

The steroid hormones are majorly synthesised by two ways:

1. Adrenal cortical hormones
2. Sex hormones

BioSynthesis of Adrenal Cortical Hormones

Conversion of cholesterol to adrenal cortical hormones

Cholesterol is the major precursor and undergoes side chain cleavage to form Δ^5–pregnenolone, releasing a C_6 aldehyde, isocaproaldehyde, Δ^5-pregnenolone is mandatory in the synthesis of all steroid hormones. Pregnenolone can be converted directly to progesterone, which requires two plasmic enzymes, 3-β-OH dehydrogenase and $\Delta^{4.5}$ isomerase. The dehydrogenase converts the 3—OH group of pregnenolone to a 3-keto group, and the isomerase moves the double bond from the B-ring to the A-ring to produce progesterone, which is converted to aldosterone (or) cortisol.

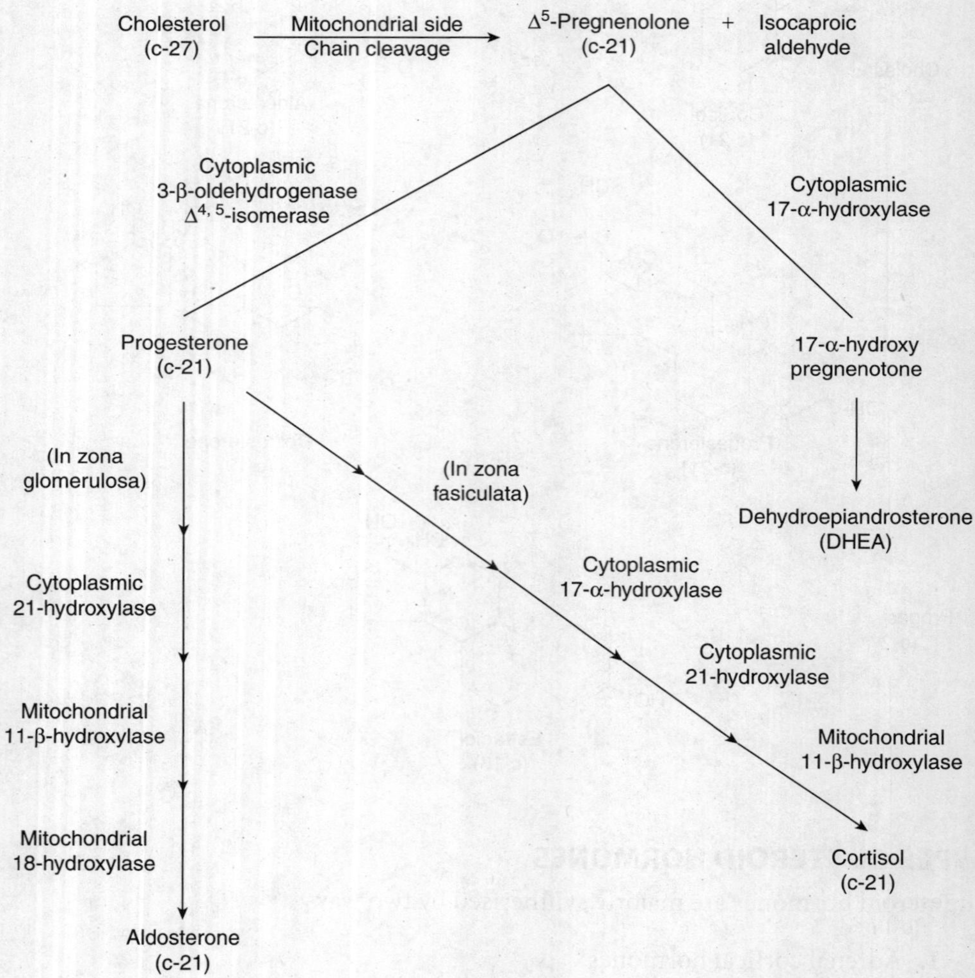

Conversion of Cholesterol to Sex Hormones

Cholesterol is also converted to sex hormones by the way of Δ^5-pregnenolone. Progesterone can be formed and further converted to testosterone. Progesterone can enter an alternate pathway to form dehydroxyepiandrosterone. This compound can be converted to 17-β estradiol action of 17-reductase. Also, estradiol can be formed from testosterone by the action of the aromatase system. Steroid hormones are generally converted into metabolic excretion production in the liver. Approximately, 20% to 30% of these metabolites are secreted into the bile and then excreted in the faeces, whereas the remainder are released into the blood and are filtered from the plasma in the kidney, passing into urine.

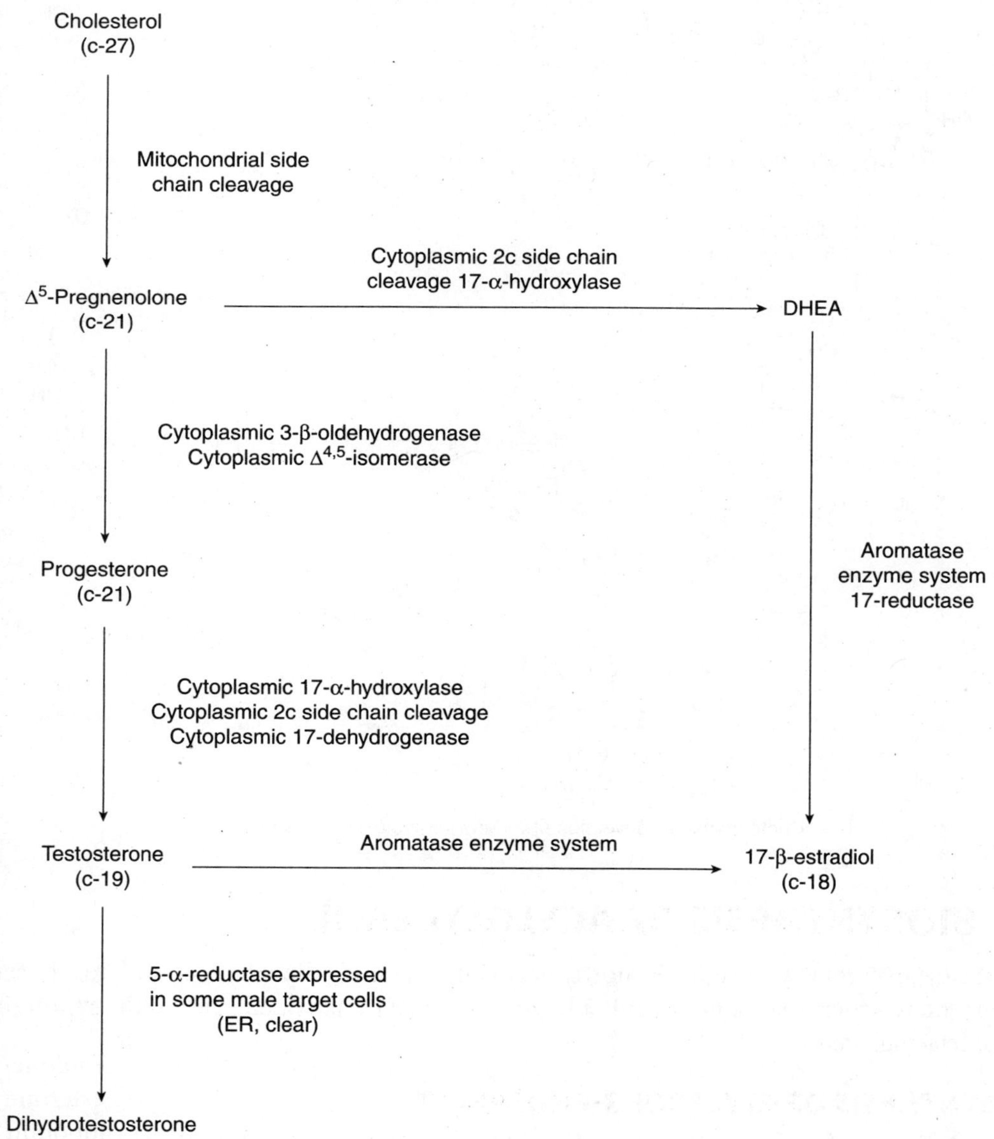

CONVERSION OF CHOLESTEROL TO VITAMIN D

7-Dehydro cholesterol is an intermediate in the pathway of cholesterol biosynthesis and is converted in the skin to previtamin-D_3 by irradiation with UV rays of the sun. Previtamin-D_3 is converted thermally and slowly to the double bond isomers by a non-enzymatic reaction to the biologically active vitamin, cholecalciferol (vitamin D_3).

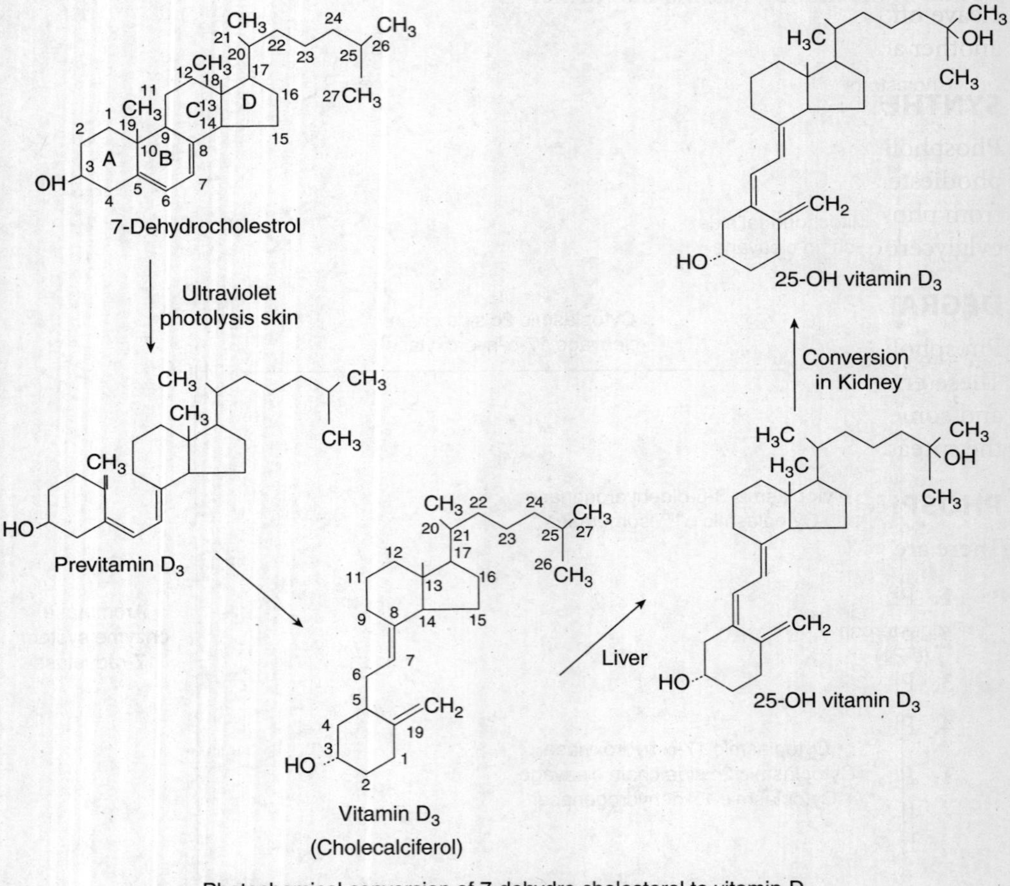

Photochemical conversion of 7-dehydro cholesterol to vitamin-D_3

BIOSYNTHESIS OF ACYLGLYCEROL

Triacylglycerol (TG) synthesis mostly occurs in liver and adipose tissue and, to a lesser extent, in other tissues. Fatty acids and glycerol must be activated prior to the synthesis of triacylglycerol.

SYNTHESIS OF GLYCEROL 3-PHOSPHATE

In the liver, glycerol is activated by glycerol kinase. This enzyme is absent in adipose tissue. In both adipose tissues, glucose serves for glycerol 3-phosphate. Dihydroxyacetone phosphate (DHAP) produced in glycolysis is reduced by glycerol 3-phosphate dehydrogenase to glycerol 3-phosphate.

Glycerol 3-phosphate acyltransferase catalyses the transfer of an acyl group to produce lysophosphatidic acid. DHAP can also accept acyl group, ultimately resulting in the formation of lysophosphatidic acid. Another acyl group is added to lysophosphatidic acid to form phosphatidic acid (1, 2-diacylglycerol phosphate). The enzyme phosphates cleave off phosphate of phosphatidic acid to produce diacylglycerol. Incorporation of another acyl group finally results in synthesis of triacylglycerol as shown in Figure 10.5.

SYNTHESIS OF PHOSPHOLIPIDS

Phospholipids are polar, ionic compounds made up of an alcohol and linked by a phosphodiester bond to either diacylglycerol or sphingosine. Phospholipids are synthesised from phosphatidic acid and 1,2-diacyl glycerol, intermediate in the production of triacylglycerol. Phospholipid synthesis occurs in the smooth endoplasmic reticulum.

DEGRADATION OF PHOSPHOLIPIDS

Phospholipids are degraded by phospholipase, which cleaves phosphodiester bonds. These enzymes are found in mammalian tissues, pancreatic juice, and snake venom and some toxins. Certain pathogenic bacteria produce phospholipase, which helps in the spread of infection by dissolving cell membranes.

PHOSPHOLIPASE

There are four kinds of phospholipase:

1. Phospholipase A_1
2. Phospholipase A_2
3. Phospholipase C
4. Phospholipase D

1. *Phospholipase A_1*
   It cleaves the fatty acid at C_1 position of phospholipids, resulting in lysophospholipids.
   Lysophospholipid can be further acted by lysophopholipase to remove second acyl group at C_2 position.
2. *Phospholipase A_2*
   It hydrolyses the fatty acid at C_2 position of phospholipid. This enzyme is found in snake venom and in many tissues and pancreatic juice. Phospholipase A_2 acts on phosphptidyl inositol; it liberates arachidonic acid, the substrate for the synthesis of prostaglandins.
3. *Phospholipase C*
   It cleaves the bond between phosphate and glycerol of phospholipids. This enzyme is present in lysosomes of hepatocytes. The toxin is isolated from clostridia, and other bacillus contains phospholipase C.

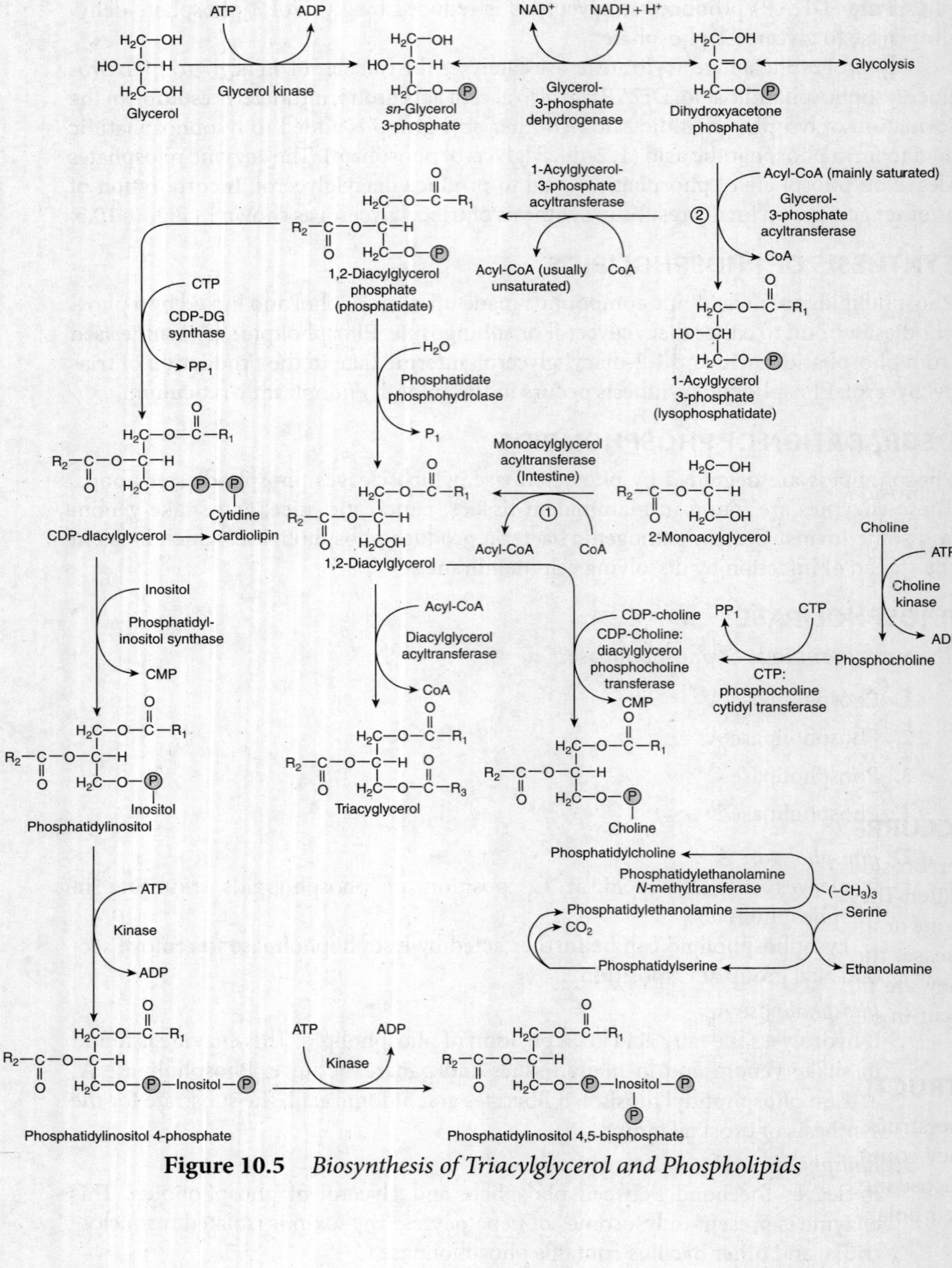

Figure 10.5 *Biosynthesis of Triacylglycerol and Phospholipids*

4. *Phospholipase D*

   It hydrolyses and removes nitrogenous bases from phospholipids. It is present in plant sources (cabbage, cotton seed, and so on).

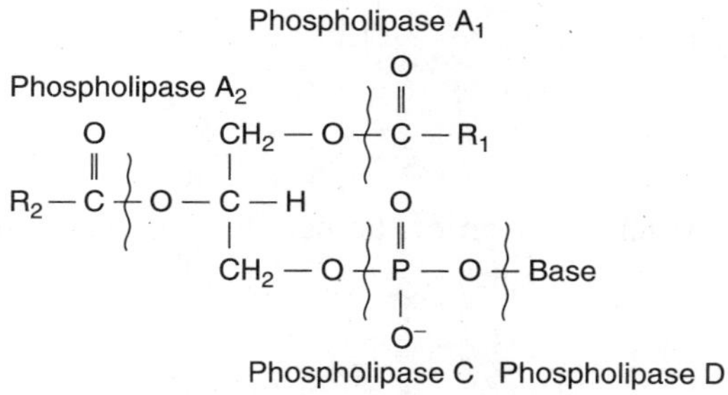

COMPLEX LIPIDS

Cerebrosides are also known as galactose lipids because of their galactose content. They have no phosphoric acid group. They are compounds of fatty acid of high molecular weight, sphingosine and galactose.

Four cerebrosides have been identified according to their fatty acid component. They are as follows:

- Kerasin contains lignoceric acid
- Cerebron contains cerebronic acid
- Neuron contains neuronic acid
- Oxynervon contains oxynervonic acid

OCCURRENCE

Cerebrosides are important constituents of brain, where they amount to 8% of the solid matter. These may also occur in tissues other than brain. Since the head group consists of one or more sugar units, cerebrosides are often called glycosphingosides. In Gaucher disease, the cerebrosides appear in relatively large amount in the liver and spleen. They are present in large amount in the brain in Niemann–Pick disease. They also occur in some plant organs.

STRUCTURE

The structure of cerebrosides is somewhat similar to that of phosphosphingosides. They contain a high molecular weight fatty acid, sphingosine, and either galactose or glucose instead of choline but no phosphoric acid. They have no electric charges since their polar groups are neutral. In general properties, they resemble sphingomyelin.

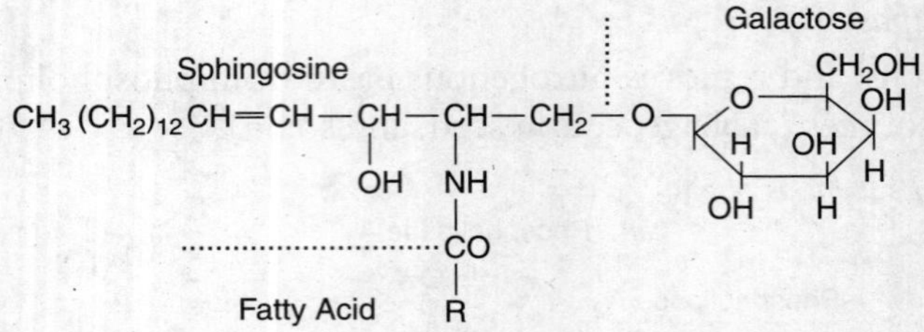

The sphingosine – fatty acid portion of the cerebrosides is commonly referred to as ceramide.

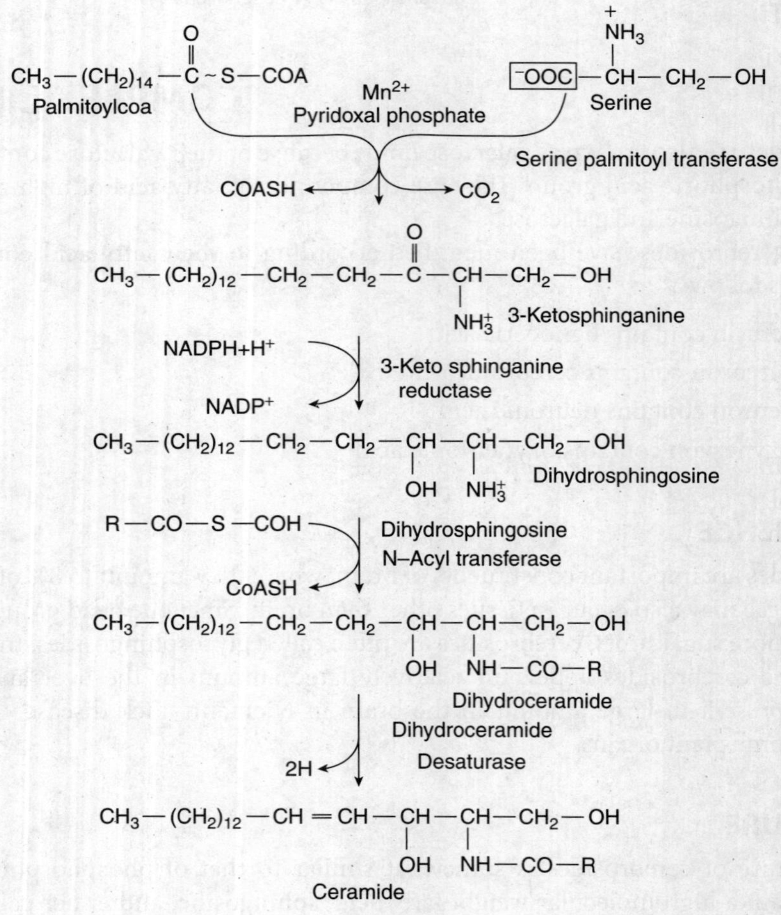

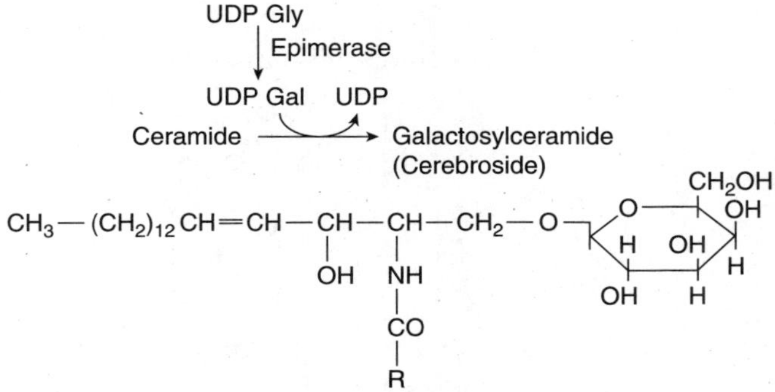

Cerebroside

FORMATION OF CERAMIDE

Ceramide is synthesised in endoplasmic reticulum. First, the amino acid serine following activation by combination with pyridoxal phosphate combines with palmitoyl-CoA to form 3-keto sphingosine. This is converted to dihydrosphingosine in a reductive step utilising NADPH. Dihydro ceramide is formed by combination with acyl-CoA, followed by desaturation to form ceramide.

CONVERSION OF CERAMIDE TO CEREBROSIDE

Ceramide is converted to cerebroside with the help of the enzyme epimerase, which converts UDP. Thus, cerebroside is synthesised.

CARDIOLIPIN

Cardiolipin is made up of two molecules of phosphatidic acid combined to a molecule of glycerol. It is the only phosphoglyceride that is antigenic in nature. It is synthesised by phosphotidylglycerol. Phosphatidic acid is a precursor of phosphptidylglycerol, which, in turn, gives rise to cardiolipin in mitochondria.

STRUCTURE

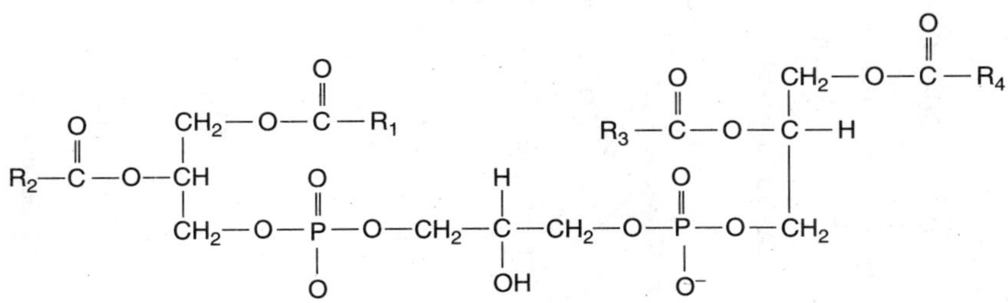

Cardiolipin

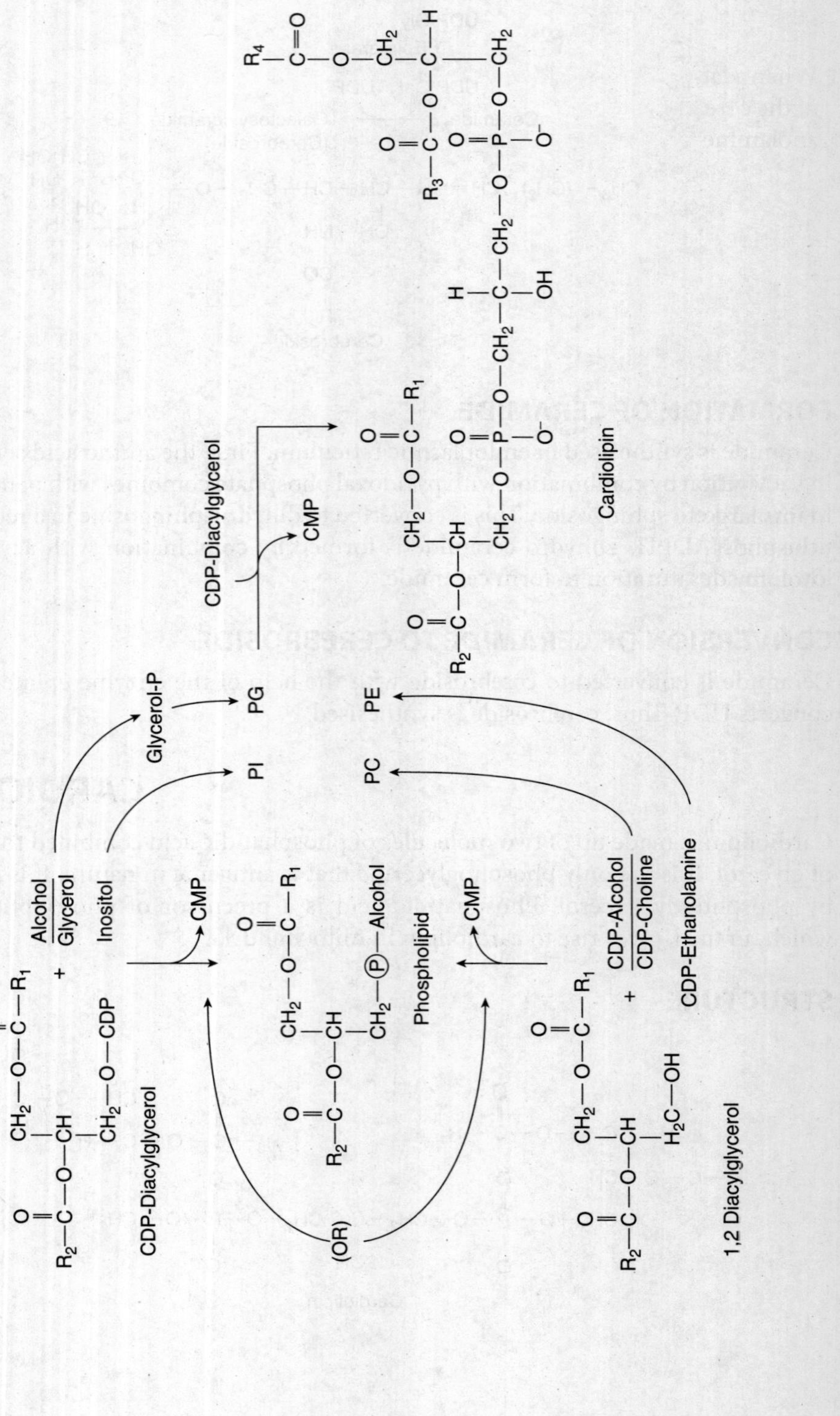
CDP-Diacylglycerol
Alcohol
Glycerol
Inositol
CMP
Glycerol-P
PI
PG
CDP-Diacylglycerol
CMP
Cardiolipin
(OR)
Phospholipid
CMP
PC
PE
1,2 Diacylglycerol
CDP-Alcohol
CDP-Choline
CDP-Ethanolamine

PLASMALOGENS

When a fatty acid is attached by an ether linkage rather than by an ester linkage, at carbon 1 of the core glycerol molecule, a plasmalogen is produced. For example, phosphotidyl ethanolamine is the plasmalogen that is similar in structure to phosphotidyl ethanolamine.

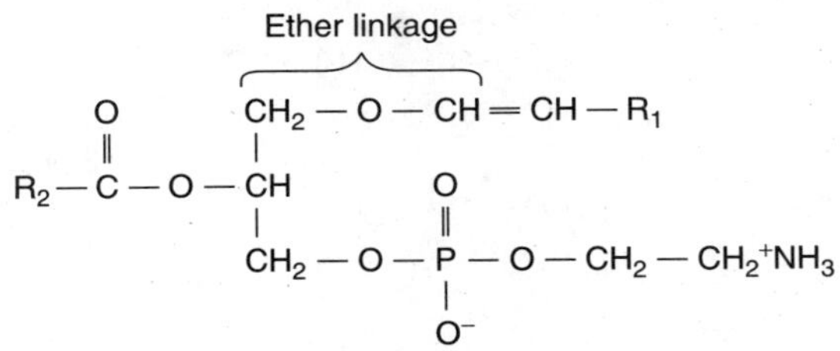

Myelin contains large amount of ethanolamine plasmalogen, and heart muscle contains large amounts of choline plasmalogen. Plasmalogen constitutes as much as 10% of the phospholipids of brain and muscle. Typically, the alkyl radical is an unsaturated alcohol. In some instances, choline, serine, or inositol may be substituted for ethanolamine.

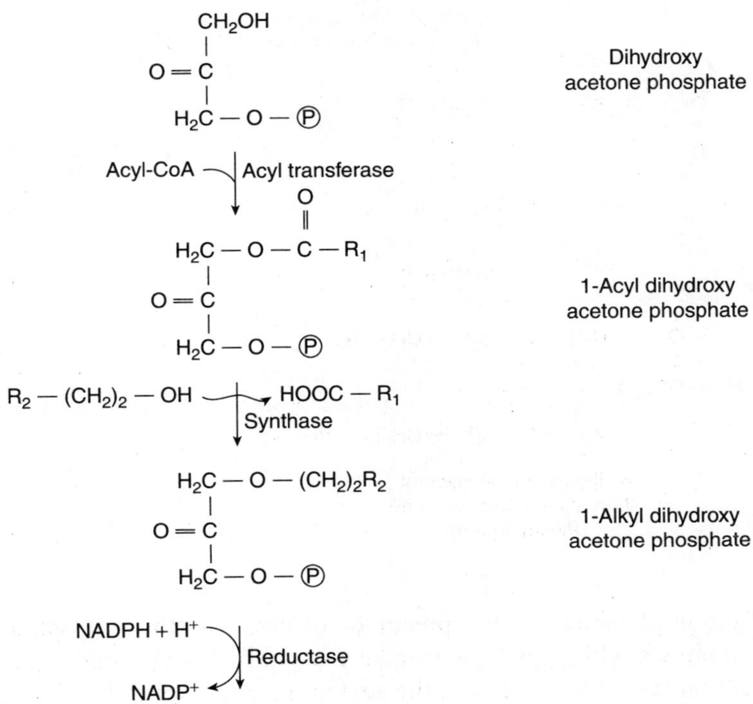

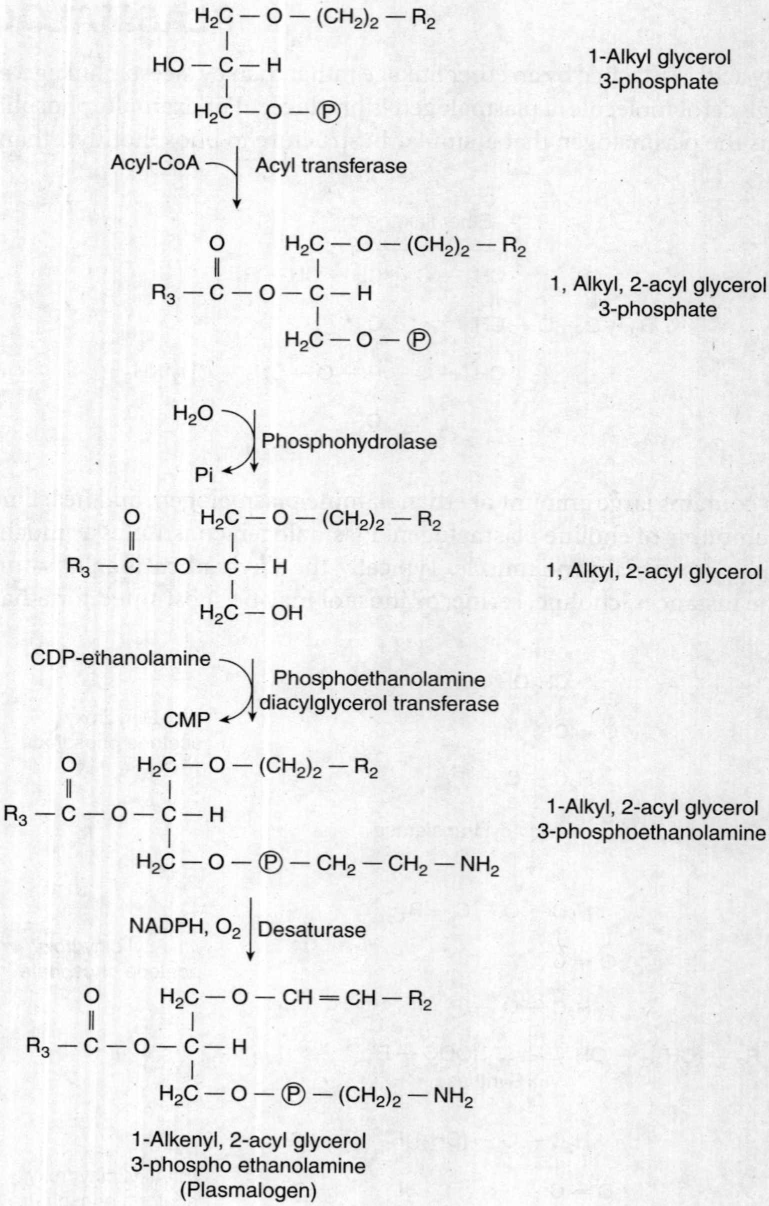

Dihydroxyacetone phosphate is the precursor of glycerol ether phospholipids. This compound combines with acyl-CoA to give 1-acyldihydroxyacetone phosphate. An exchange reaction takes place between the acyl group and a long chain alcohol to give a 1-alkyldihydroxyacetone phosphate, which, in the presence of NADPH, is converted to 1-alkyl glycerol 3-phosphate. After further acylation in the 2 position, the resulting 1-alkyl, 2-acyl glycerol 3-phosphate is hydrolysed to give the free glycerol derivatives.

Plasmalogens are formed by the desaturation of the analogues 3-phosphoethanolamine derivative. Most of the phospholipid in mitochondria consist of plasmalogen.

DEGRADATION OF SPHINGOMYELIN

Sphingomyelin is degraded by lysosomal enzyme sphingomyelinase to form ceramide and phosphorylcholine.

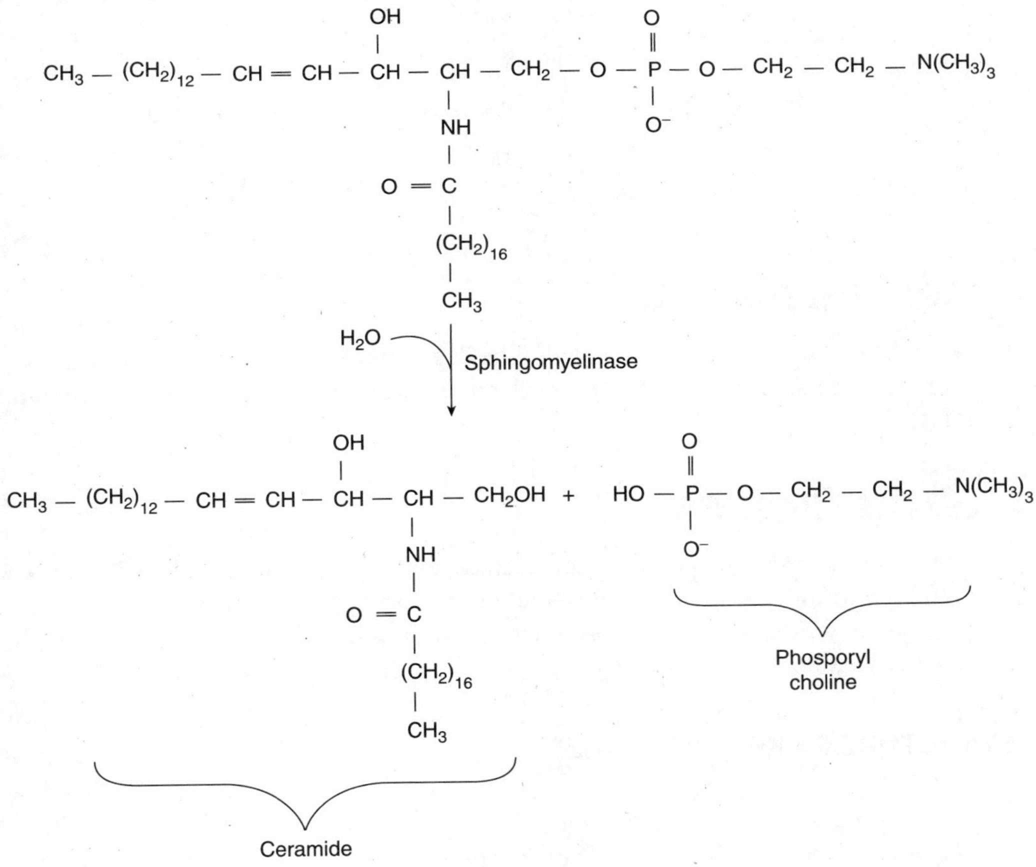

NIEMANN–PICK DISEASE

Deficiency of sphingomyeliase enzyme leads to Niemann–pick disease. Sphingomyelin cannot be degraded in the absence of the enzyme sphingomyelinase. In severe form, liver and spleen are the sites of lipid deposits, and they are enlarged. There is a large amount of sphingosine and a fatty acid.

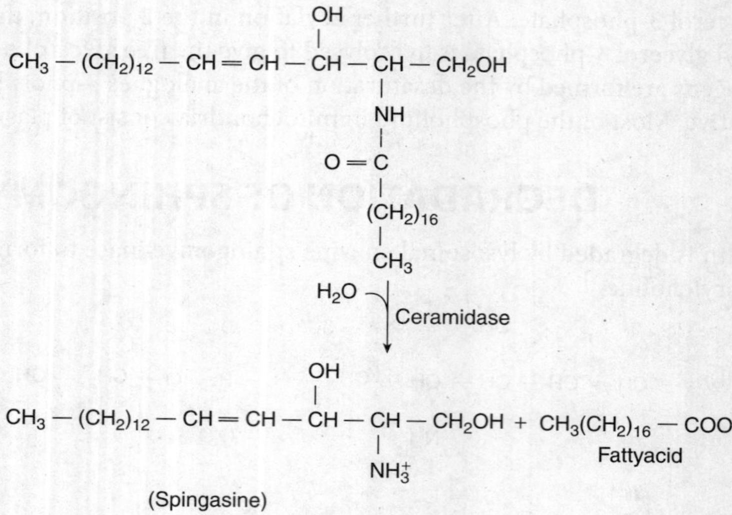

FARBER'S DISEASE

If ceramidase is deficient, ceramide cannot be degraded. This is called Farber's disease.

This disease is characterised by skeletal deformation, dermatitis, and mental retardation.

KETONE BODIES

The compounds namely acetone, acetoacetate, and β-hydroxy butyrate (3-hydroxy butyrate) are known as ketone bodies. Only the first two are true ketone bodies, and β-hydroxy butyrate does not possess a keto group (C=O). Ketones are water soluble and energy yielding.

STRUCTURE OF KETONE BODIES

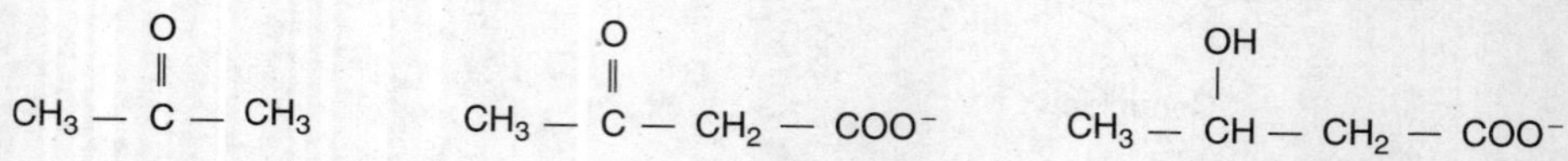

Acetone Acetoacetate β-hydroxybutyrate

KETOGENESIS – SITE OF SYNTHESIS

The synthesis of ketone bodies occurs in the liver. The enzymes for ketone bodies synthesis are located in the mitochondrial matrix. Acetyl-CoA formed by oxidation of fatty acids, pyruvate, or some amino acids is the precursor for ketone bodies; ketogenesis occurs through the following reactions:

1. Two moles of acetyl-CoA condense to form acetoacetyl-CoA. This reaction is catalysed by thiolase, an enzyme involved in the final step of β-oxidation.
2. Acetoacetyl-CoA combines with water molecule of acetyl-CoA to produce β-hyroxy β-methyl glutaryl-CoA (HMG-CoA). HMG-CoA synthase catalyses this reaction and regulates the synthesis of ketone bodies.
3. HMG-CoA lyase cleaves HMG-CoA to produce acetoacetate and acetyl-CoA.
4. Acetoacetate can undergo spontaneous decarboxylation to form acetone.
5. Acetoacetate can be reduced by a dehydrogenase to hydroxybuyurate.

REGULATION OF KETOGENESIS

The ketone body formation (particularly overproduction) occurs primarily due to non-availability of carbohydrates to the tissues. This is an outcome of excessive utilisation of fatty acids to meet the energy requirements of cells. Hormones glucagon stimulates ketogenesis, whereas insulin inhibits. The increased ratio of glucagon/insulin in diabetes mellitus promotes ketone body formation. This is due to the disturbance caused in carbohydrate and lipid metabolism in diabetes.

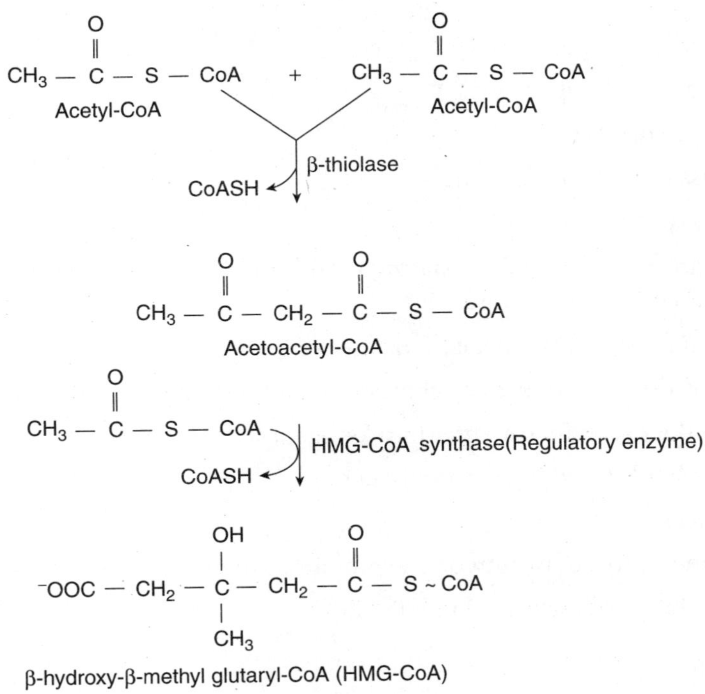

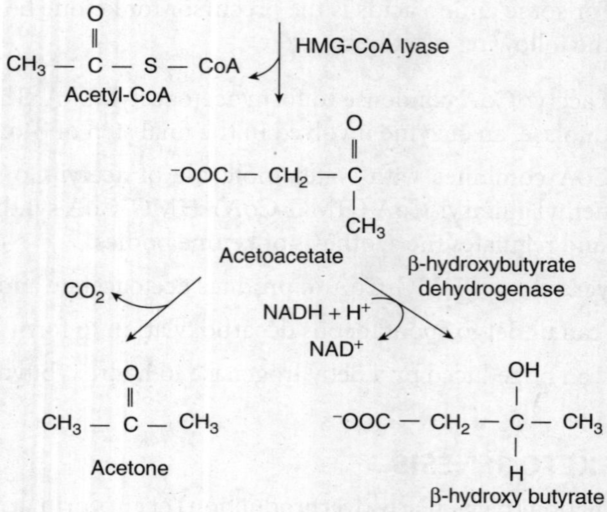

QUESTION BANK

SECTION A

1. What are sphingomyelins?
2. What are apolipoproteins?
3. What are plasmalogens?
4. Discuss the role of carnitine.
5. What are chylomicrons?
6. How are water-insoluble triacylglycerol and cholesterol transported in the aqueous medium of blood?
7. Explain fatty acid synthetase enzyme complex components.
8. What is the effect of dietary cholesterol on the cholesterol biosynthesis?
9. When does acetyl-CoA enter into ketogenesis?
10. How is lipid stored in adipose tissue?
11. How is fat synthesised?
12. Explain the role of insulin on the mobilisation of fat.
13. How is bile acid converted to bile salt?

SECTION B

1. How are phosphoglycerides synthesised?
2. Comment on the degradation of odd chain fatty acids.
3. Discus about omega oxidation of fatty acids.
4. How are phospholipids degraded?
5. How is bile salt synthesised?
6. Explain the action of lipase and the effect of hormone on its action.
7. How are phospholipids interconverted?
8. How is oleic acid degraded?

SECTION C

1. What are ketone bodies? How are they synthesised?
2. What is β-oxidation? How are fatty acids oxidised by β-oxidation?
3. Discus the biosynthesis of cholesterol regulation.
4. How is fatty acid synthesised?

CHAPTER 11

NUCLEOTIDE METABOLISM

Concept Outline

- Introduction
- Fate of Dietary Nucleic Acids
- Biosynthesis of Purine Ribonucleotides
- Biosynthesis of Pyrimidine Nucleotides
- Coenzymes

INTRODUCTION

Nucleotides consist of a nitrogenous base, a pentose and a phosphate. The pentose sugar is deoxyribonucleotides of DNA. The sugar is 2-deoxy D-ribose. Nucleotides participate in almost all the biochemical processes, either directly or indirectly. They are the structural components of nucleic acid (DNA and RNA) coenzymes involved in the regulation of several metabolic reactions.

FATE OF DIETARY NUCLEIC ACIDS

Ribonuclease and deoxyribonuclease, which are produced in the pancreatic juice, convert RNA and DNA primarily to oligonucleotides. This is further degraded to 3′- and 5′-mononucleotides by the enzyme pancreatic phosphodiesterase. A group of nucleotides degrade phosphate and release the bases before uptake (dietary purines and pyrimidines are not utilised for the synthesis of uric acid). The purine bases are further catabolised to uric acid by the intestinal mucosal cells and are finally excreted in the urine. The remnants of purine bases are metabolised by intestinal flora. The summary of this pathway is shown in Figure 11.1.

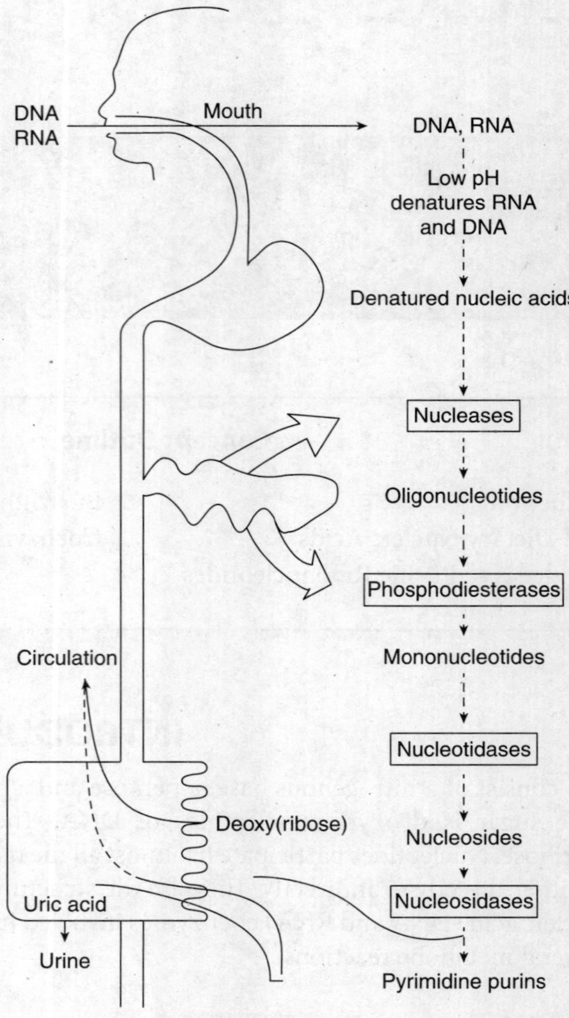

Figure 11.1 *Digestion of Dietary Nucleic Acids*

BIOSYNTHESIS OF PURINE RIBONUCLEOTIDES

Human and most other vertebrates can synthesis some amount of purine and pyrimidine nucleotides de novo (i.e. from amphibolic intermediates)

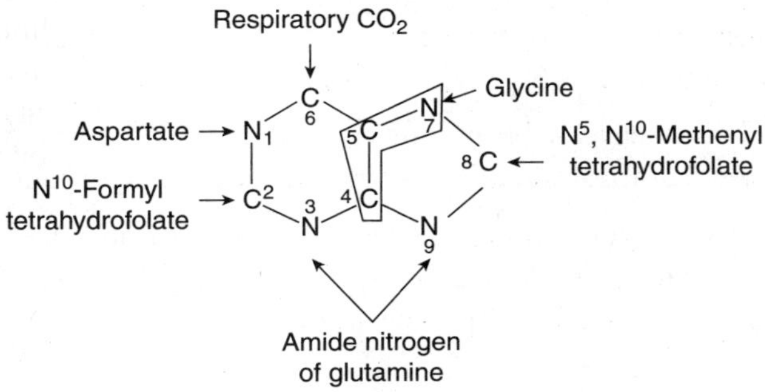

THE SOURCES OF THE NITROGEN AND CARBON ATOMS OF PURINE RING

Site of Synthesis

Liver is the major site for purine nucleotide synthesis. Erythrocytes, polymorphonuclear leukocytes, and brain cannot produce purines.

De Novo Synthesis of Purines

The pathway for synthesis of inosine monophosphate (IMP), the parent purine nucleotide, is described below:

1. Ribose 5-phosphate produced in the hexose monophosphate shunt is the starting material of purine nucleotide synthesis. It reacts with ATP to form phosphoribosyl pyrophosphate (PRPP).
2. Glutamine transfers its amide nitrogen to phosphoribosyl pyrophosphate to replace pyrophosphate and produce 5-phosphoribosylamine. The enzyme phosphoribosyl pyrophosphate glutamyl amidotransferase is controlled by feedback inhibition of nucleotides (IMP, AMP, and GMP). This reaction is the committed step in purine nucleotide biosynthesis.
3. Phosphoribosylamine reacts with glycine in the presence of ATP to form glycinamide ribosyl 5-phosphate or glycinamide ribotide (GAR).
4. N^{10}-formyltetrahydrofolate donates the formyl group, and the product formed is formyl glycinamide ribosyl 5-phosphate.
5. Glutamine transfers the second amido amino group to produce formyl glycinamidine ribosyl 5-phosphate.

6. The imidazole ring of the purine is closed; an ATP-dependent reaction yields 5-amino-imidazole ribosyl 5-phosphate.
7. Incorporation of CO_2 (carboxylation) occurs yield aminoimidazole carboxylate ribosyl 5-phosphate. The reaction does not require vitamin biotin and ATP; this is the case with most of the carboxylation reactions.
8. Aspartate condenses with the product in reaction 7 to form amino imidazole 4-succinyl carboxamide ribosyl 5-phosphate.
9. Adenylosuccinate lyase cleaves off fumarate and only the amino group of aspartate is retained to yield aminoimidazole 4-carboxamide ribosyl 5-phosphate.
10. N^{10}-formyltetrahydrofolate donates a one-carbon moiety to produce formaminoimidazole 4-carboxamide ribosyl 5-phosphate, with this reaction; all the carbon and nitrogen atoms of purine ring are contributed by the respective sources.
11. The final reaction catalysed by cyclohydrase leads to ring closure with an elimination of water molecule. The product obtained is inosine monophosphate (IMP), from which other purine nucleotides can be synthesised. Synthesis of purine nucleotides in Figure 11.2.

CONVERSION OF IMP TO AMP AND GMP

Purine and their nucleosides may be converted directly to mononucleotides by salvage reactions.

Conversion of purines and deoxyribonucleotides to mononucleotides is carried out by the so-called salvage pathway, which requires far less energy than the de novo synthesis. The most important step of the salvage pathway is phosphoribosylation by PRPP of a free purine to form a purine 5′-mononucleotide (Pu-RP).

$$\text{Pu} + \text{PP} - \text{RP} \longrightarrow \text{PPi} + \text{Pu} - \text{RP}$$

Adenine phosphoribosyltransferase catalyses the formation of AMP from adenine. Hypoxanthine-guanine phosphoribosyltransferase (HGPRT) converts guanine and hypoxanthine, respectively to GMP is IMP. Phosphoribosyl pyrophosphate (PRPP) is the salvage pathway. A second salvage mechanism involves phosphoryl transfer from ATP to a purine ribonucleoside (PuR).

$$\text{PuR} + \text{ATP} \longrightarrow \text{ADP} + \text{PuR} - \text{P}$$

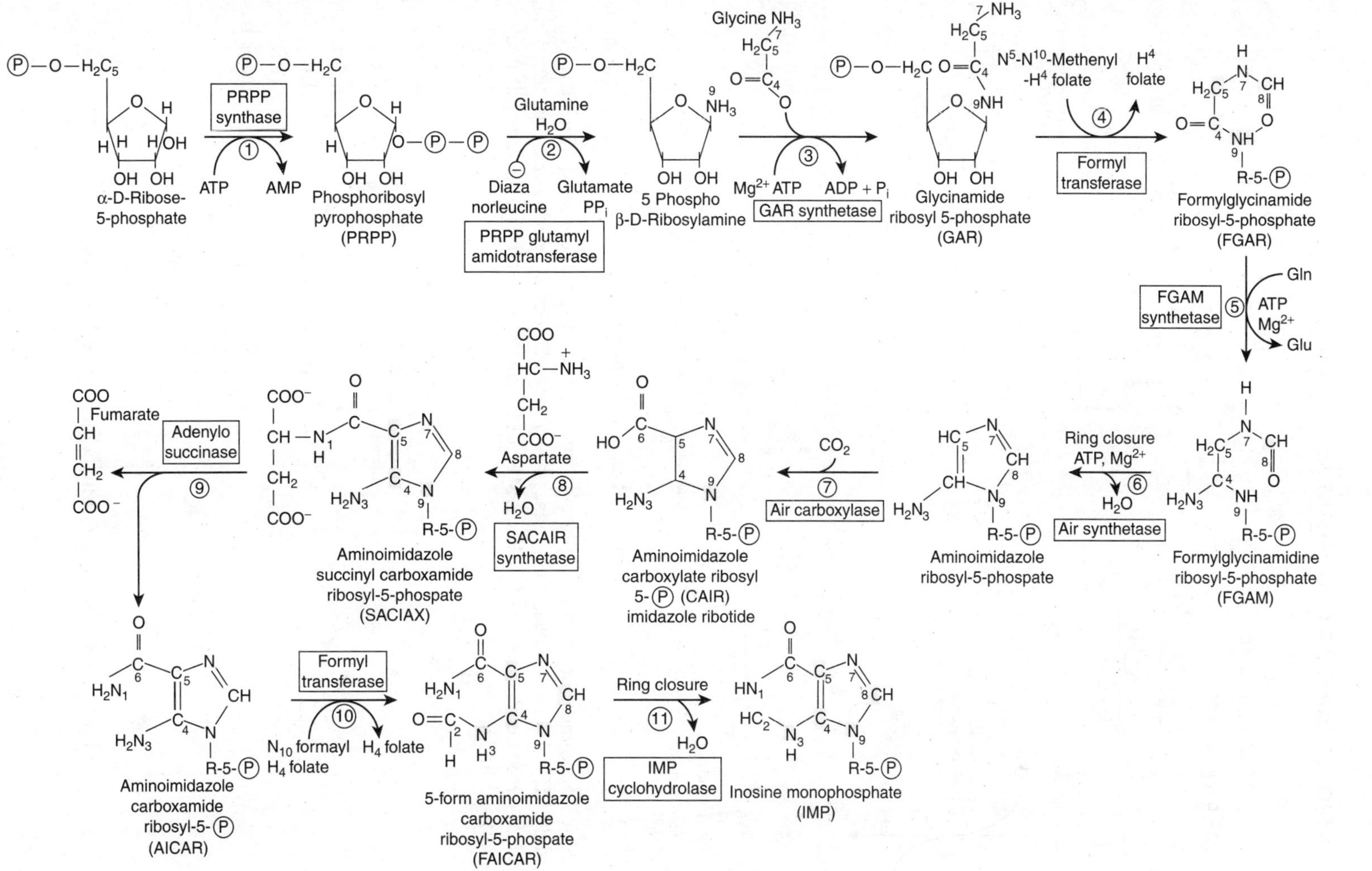

Figure 11.2 *Synthesis of Purine Nucleotides*

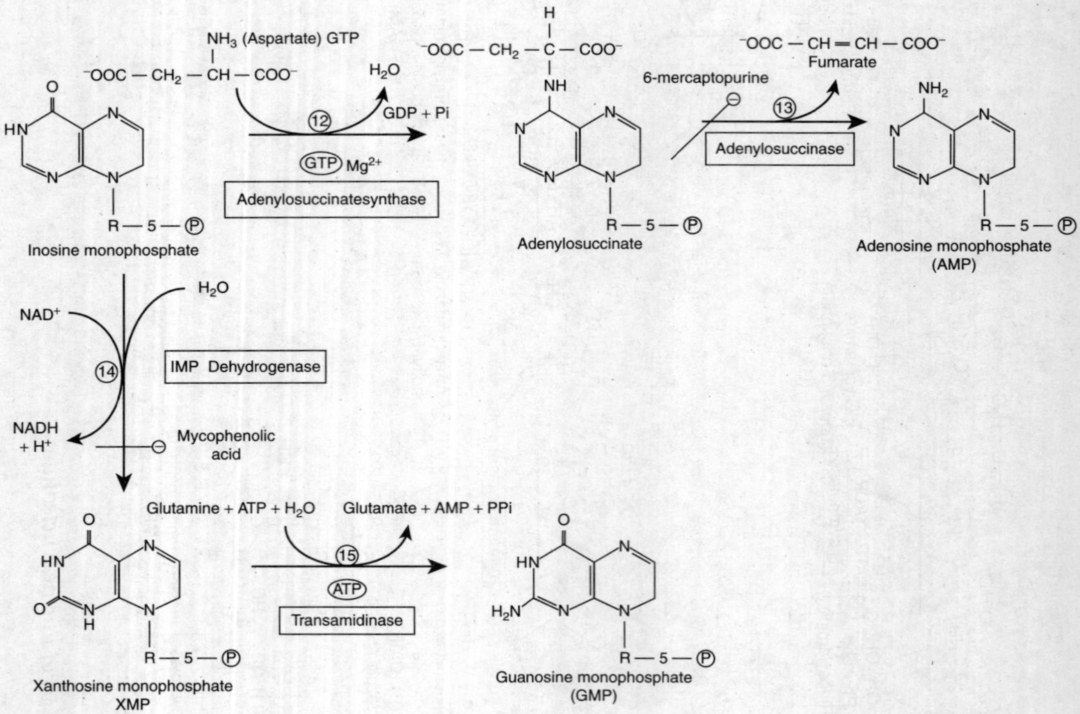

Figure 11.2 *(Continued)*

Adenosine kinase catalyses phosphorylation of adenosine and deoxyadenosine to AMP and dAMP, while deoxycytidine kinase phosphorylates deoxycytidine, deoxyadenosine, and 2′-deoxyguanosine to dCMP, dAMP, and dGMP, respectively. Salvage pathway is described in Figure 11.3.

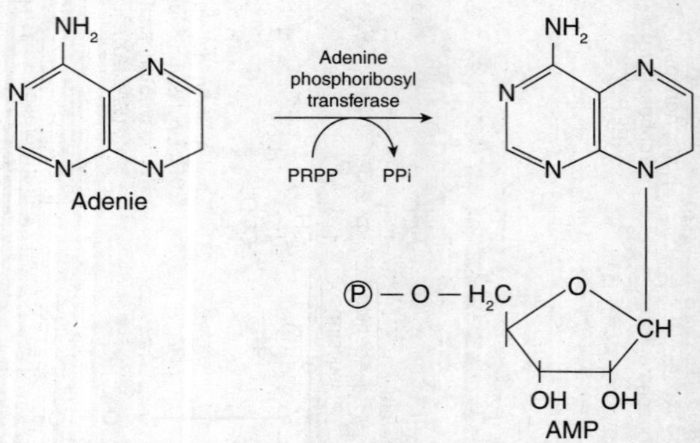

Figure 11.3 *Salvage Pathway for Purine Nucleotides Synthesis*

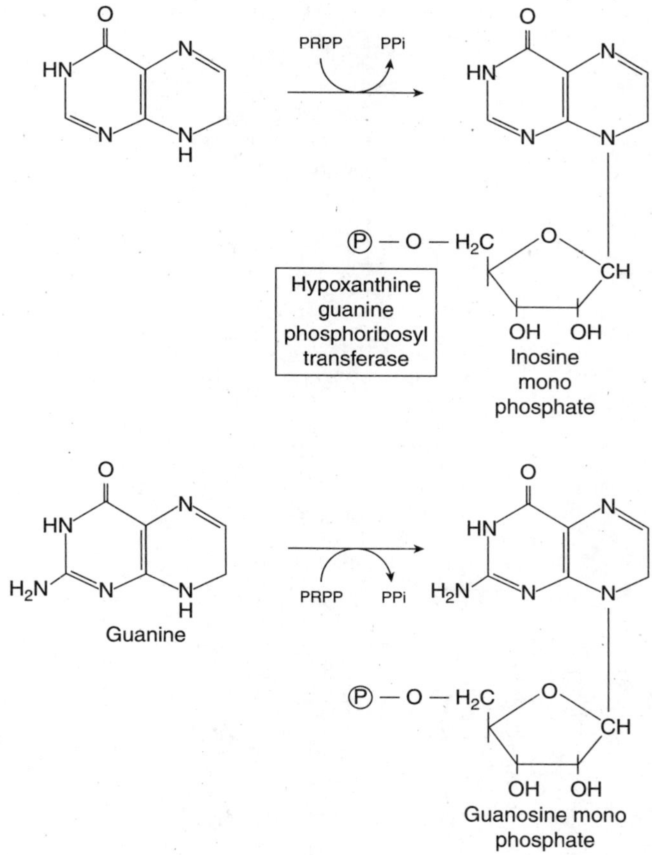

Figure 11.3 *(Continued)*

REGULATION OF PURINE NUCLEOTIDE BIOSYNTHESIS

The purine nucleotide synthesis is well coordinated to meet the cellular demands. The intracellular concentration of PRPP regulates purine synthesis to a large extent. This increase depends on the availability of ribose 5′-phosphate and the enzyme PRPP synthase. PRPP glutamyl amidotransferase is controlled by a feedback mechanism by purine nucleotides. That is, if AMP and GMP are available in adequate amounts to meet the cellular requirements, their synthesis is turned off at the amidotransferase reaction. Another important stage of regulation is in the conversion of IMP to GMP and AMP. AMP inhibits adenylsuccinate synthase, while GMP inhibits IMP DHase. Thus, AMP and GMP control their respective synthesis from IMP by feedback mechanism as described in Figure 11.4.

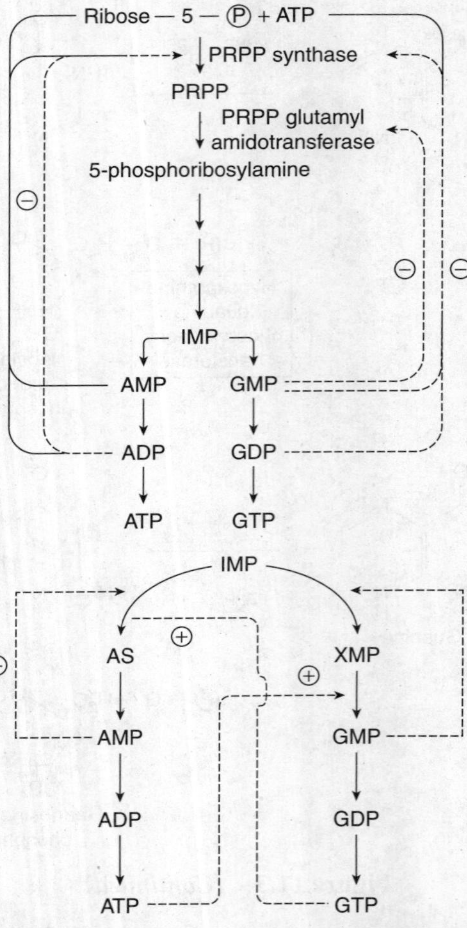

Figure 11.4 *Regulation of Purine Nucleotides Synthesis*

CONVERSION OF RIBONUCLEOTIDES TO DEOXYRIBONUCLEOTIDES

The synthesis of purine and pyrimidine deoxyribonucleotides occurs from ribonucleotides by a reduction of the C_2 of ribose moiety as shown in Figure 11.5. This reaction is catalysed by a multi-subunit (two B_1 and two B_2 subunits) enzyme ribonucleotide reductase.

The ribonucleotide reductase itself provides the hydrogen atoms needed for the reduction from its sulfhydryl groups. The reducing equivalents, in turn, are supplied by thioredoxin, a monomeric proteins with two cysteine residues. NADPH-dependent thioredoxin reductase converts the oxidised thioredoxin to reduced form, which can be recycled again and again. Thioredoxin thus serves as a cofactor in an enzymatic reaction.

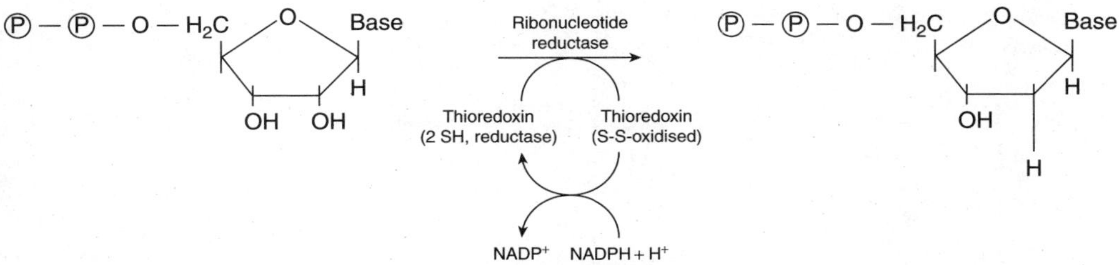

Figure 11.5 *Conversion of Ribonucleotides to Deoxyribonucleotides*

DEGRADATION OF PURINE NUCLEOTIDES

The end product of purine metabolism in human is uric acid. The sequence of reactions in purine nucleotides degradation is given below:

1. The nucleotide monophosphates (AMP, IMP, and GMP) are converted to their respective nucleoside forms, adenosine inosine and guanosine by the action of nucleotidase.
2. The amino group either from AMP or adenosine can be removed to produce IMP or inosine, respectively.
3. Inosine and guanosine are respectively converted to hypoxanthine and xanthine (purine bases) by purine nucleoside phosphorylase. Adenosine is not degraded by this enzyme; hence, it has to be converted to inosine.
4. Guanine undergoes deamination by guanase to form xanthine.
5. Xanthine oxidase is an important enzyme that converts hypoxanthine to xanthine to uric acid. This enzyme contains fad, molybdenum, and iron, and is exclusively found in liver and small intestine. Xanthine oxidase liberates H_2O_2, which is harmful to the tissues. Catalase cleaves H_2O_2 to H_2O and O_2.
6. Uric acid is the final excretory product of purine metabolism in humans as described in Figure 11.6. Uric acid can serve as an important antioxidant by getting itself converted to allantoin. It is believed that the antioxidant role of ascorbic acid in primates is replaced by uric acid since their animals have lost their ability to synthesis ascorbic acid.
7. Most animals (other than primates), however, oxidise uric acid by the enzyme uricase to allantoin, where the purine ring is cleaved.
8. Allantoin is then converted to allantoic acid and excreted in some fishes. Further degradation of allantoic acid may occur to produce urea in amphibians (most fishes and molluscs)and alter to ammonia (in marine vertebrates).

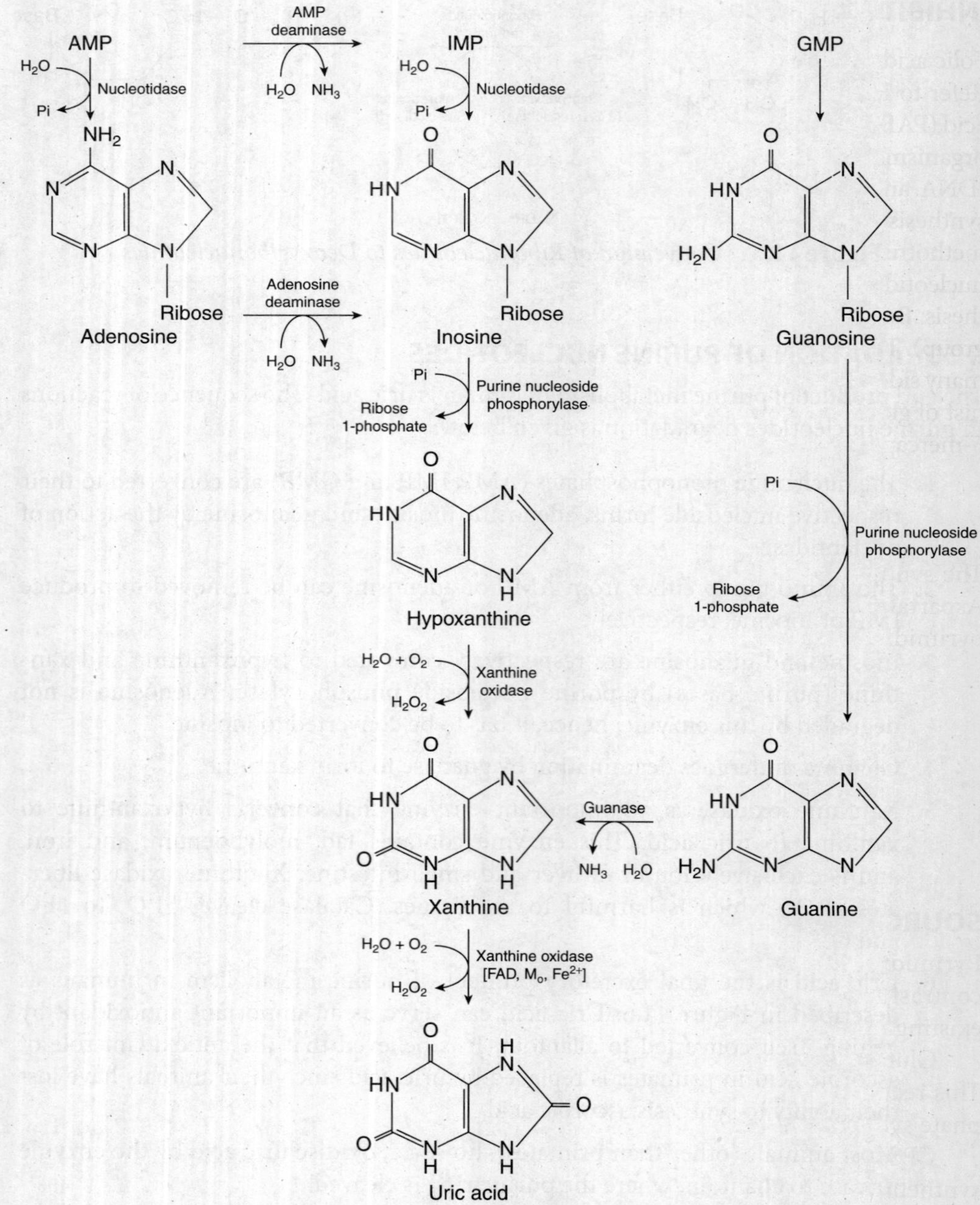

Figure 11.6 *Degradation of Purine Nucleotides*

INHIBITORS OF PURINE SYNTHESIS

Folic acid (THF) is essential for the synthesis of purine nucleotides (reactions 4 and 10)- Refer to Figure 11.2. Sulphonamides are the structural analogues of para aminobenzoic acid (PABA). These sulpha drugs can be used to inhibit the synthesis of folic acid by micro-organisms. This indirectly reduces the synthesis of purines and therefore the nucleic acids (DNA and RNA). Sulphonamides have no influence on humans since folic acid is not synthesised and is supplied through diet. The structural analogues of folic acid (example: methotrexate) are widely used to control cancer. They inhibit the synthesis of purine nucleotide viz reaction 4 and 10 refer Figure 11.2 and thus inhibits the nucleic acid synthesis. Both these reactions are concerned with the transfer of one carbon moiety (formyl group). These inhibitors also affect the proliferation of normally growing cells. This causes many side effects including anemia, baldness, scaly skin, and so on. Azaserine is an antagonist of glutamine, and this can inhibit the reaction 5. Diaz Norleucine inhibits reaction 2. 6-mercaptopurine inhibits reactions 13 and 14. Mycophenolic acid inhibits reaction 14.

BIOSYNTHESIS OF PYRIMIDINE NUCLEOTIDES

The synthesis of pyrimidine is a much simpler process compared to that of purines. Aspartate, glutamine (amide group), and CO_2 contribute the atoms in the formation of pyrimidine ring.

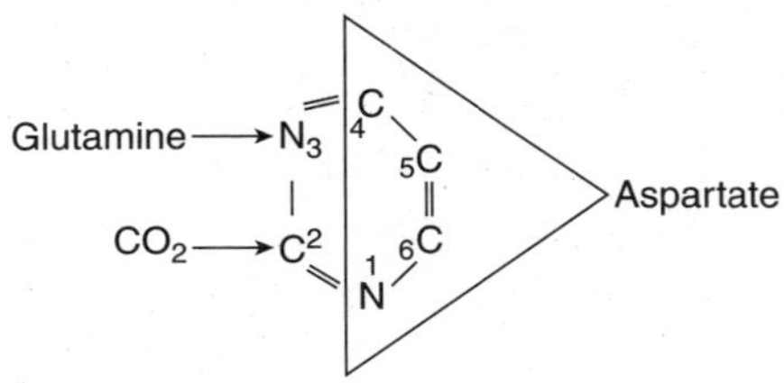

SOURCES OF INDIVIDUAL ATOMS IN PYRIMIDINE RING

Pyrimidine ring is first synthesised and then attached to ribose 5-phosphate. This is in contrast to purine nucleotide synthesis, where in purine, the ring is built upon a pre-existing ribose 5-phosphate.

Glutamine transfers its amido nitrogen to CO_2 to produce carbamoyl phosphate. This reaction is ATP dependent and is catalysed by cytosol enzyme carbamoyl phosphate synthetase II (CPS II).

CPS II is activated by ATP and PRPP and inhibited by UTP. Carbamoyl phosphate synthetase I (CPS I) is a mitochondrial enzyme which synthesises carbamoyl phosphate from ammonia and CO_2 and in turn urea. Carbamoyl phosphate condenses with aspartate to form carbamoyl aspartate. This reaction is catalysed by aspartate transcarbamoylase. Dihydroorotase catalyses the pyrimidine ring closure with a loss of water. The three enzymes (CPS II, aspartate transcarbamoylase, and dihydroorotase) are the domains (functional units) of the same portion. This is a multifunctional enzyme.

The next step in pyrimidine synthesis is an NAD$^+$-dependent dehydrogenation, leading to the formation of orotate. Ribose 5-phosphate is now added to orotate to produce uridine monophosphate (OMP). This reaction is catalysed by orotate phosphoribosyl transferase, an enzyme comparable with HGPRT in its function. OMP undergoes decarboxylation to uridine monophosphate (UMP). Orotate phosphoribosyl transferase and OMP decarboxylase are domains of a single protein. By an ATP-dependent kinase reaction, UMP is condensed to UDP, which serves as a precursor for the synthesis of dUMP, dTMP, UTP, and CTP. Ribonucleotide reductase converts UDP to dUDP by a thioredoxin-dependent recrimination; thymidylate synthetase catalyses the transfer of a methyl group from N^5, N^{10} methylenetetrahydrofolate to produce deoxythymidine monophosphate (dTMP). UDP undergoes an ATP-dependent kinase reaction to produce UTP. Cytidine triphospahte (CTP) is synthesised from UTP by amination. CTP synthetase is the enzyme and glutamine provides the nitrogen. Thus, the pyrimidine nucleotides are synthesised from UMP as described in Figure 11.7.

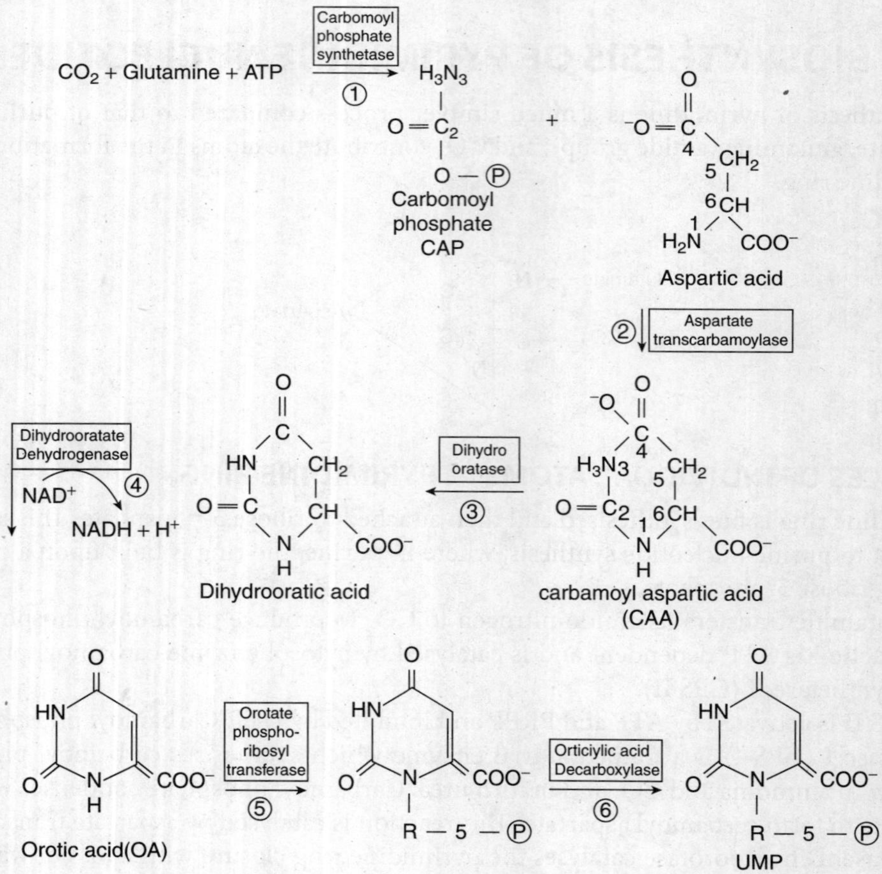

Figure 11.7 *Synthesis of Pyrimidine Nucleotides*

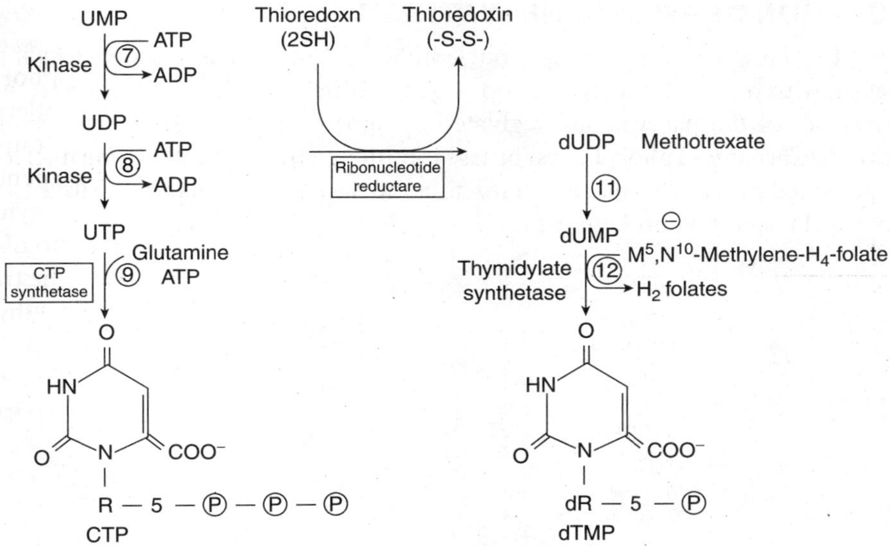

Figure 11.7 *(Continued)*

SALVAGE PATHWAY

Mammalian cells lack efficient means of salvaging pathway for pyrimidine bases. Salvage reactions convert the pyrimidine ribonucleotides uridine and cytidine and the deoxyribonucleosides thymidine and deoxycytidine to their respective nucleotides. Deoxycytidine is phosphorylated by deoxycytidine kinase, an enzyme that also phosphorylates deoxyguanosine and deoxyadenosine. Orotate phosphoribosyl transferase, an enzyme of de novo synthesis of pyrimidine nucleotides, can salvage orotic acid to OMP.

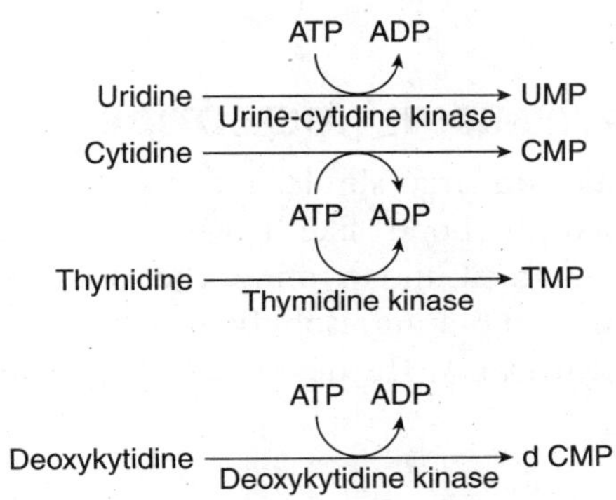

REGULATION OF PYRIMIDINE SYNTHESIS

The first two enzymes of pyrimidine nucleotides biosynthesis is sensitive to allosteric regulation with respect to the regulation of gene expression. The first three and the last two enzymes of the pathway are regulated by apparently coordinate repression and depression. Carbamoyl phosphate synthase is inhibited by UTP and purine nucleotides but is activated by PRPP. Aspartate transcarbamoylase is particularly sensitive to inhibition by CTP as shown in Figure 11.8.

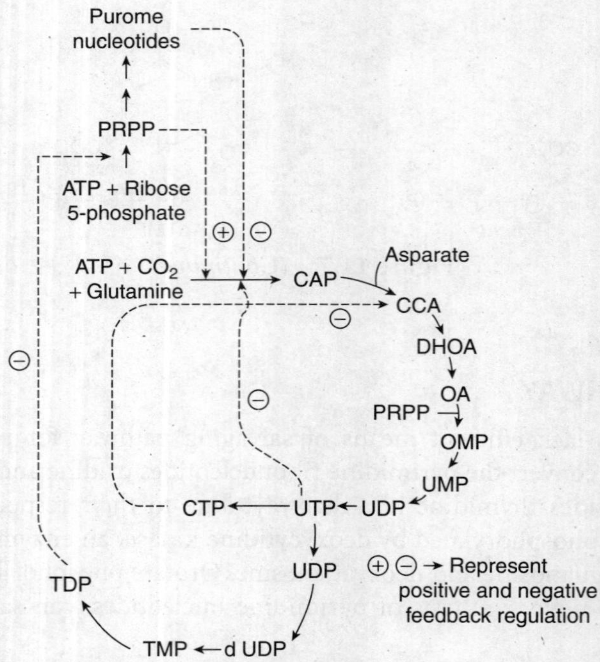

Figure 11.8 *Regulation of Pyrimidine Synthesis*

DEGRADATION OF PYRIMIDINE NUCLEOTIDES

The pyrimidine nucleotides undergo similar reaction (dephosphorylation, deamination, and cleavage of glycosidic bond) like that of purine nucleotides to liberate the nitrogenous bases cytosine, uracil, and thymine. The bases are then degraded to highly soluble products β-alanine and β-aminoisobutyrate and other reactions to finally produce acetyl CoA and succinyl CoA. The degradation of pyrimidine nucleotides is represented in Figure 11.9.

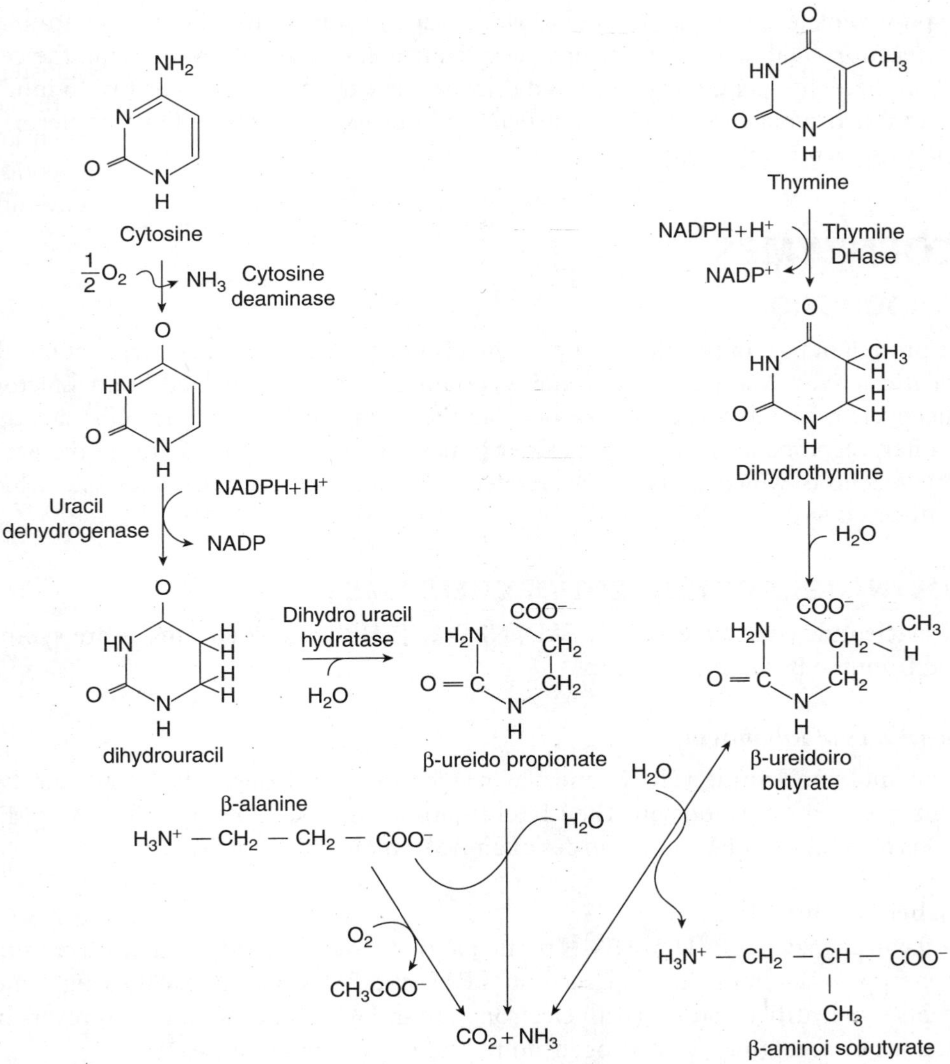

Figure 11.9 *Degradation of Pyrimidine Nucleotides*

INHIBITORS OF PYRIMIDINE SYNTHESIS

Reaction 12 (Refer Figure 11.7) is the sole reaction of pyrimidine nucleotide biosynthesis, which requires a tetrahydrofolate derivative. During the transfer process, the methylene group of N^5 and N^{10} methylene THF is reduced to a methyl group, and the tetra hydrofolate carrier oxidised to dihydrofolate must be reduced to tetrahydrofolate in a reaction catalysed by tetrahydrofolate reductase. Consequently, dividing the cells that are necessity-generating TMP and dihydrofolate are especially sensitive to inhibitors of dihydrofolate reductase. One such inhibitor is methotrexate (amethopterin), a widely used anti-cancer drug.

COENZYMES

INTRODUCTION

The protein of part of the enzyme on its own is not always adequate to bring about the catalytic activity. Many enzymes require certain non-protein small additional factors, collectively referred to as cofactors for catalysis. Coenzymes are second substrates and are often regarded as co-substrates. Coenzymes undergo alterations during the enzymatic reactions, which are layer regenerated. This is in contrast to the substrate, which is converted to the product.

BIOSYNTHESIS OF NUCLEOTIDE COENZYMES

The nucleotide coenzymes FMN, FAD, NAD^+, $NADP^+$, and coenzyme-A are synthesised from the B-complex vitamins.

Coenzymes of Riboflavin

Flavin mononucleotide (FMN) and flavin adenine dinucleotide (FAD) are the two coenzyme forms of riboflavin. The ribitol (5-arbon) is linked to a phosphate in FMN. FAD is formed from FMN by transfer of an AMP moiety from ATP.

Biochemical function

The flavin coenzymes FAD and FMN participate in many of the redox reactions required for energy production. The catalytic unit of FAD and FMN is isoalloxazine ring, which accepts two hydrogen atoms (with electrons). Both FAD and FMN undergo reversible reactions by accepting two hydrogen atoms and form $FMNH_2$ or $FADH_2$.

Enzymes that contain FMN and FAD as coenzymes are said to be flavoproteins. The coenzymes (prosthetic group) can bind tightly to the protein (apoenzyme) either by non-covalent bonds or covalent bonds in the holoenzyme. In non-covalent linkage, the flavoproteins contain metals such as Fe and Mb, and they are called as metallo flavoproteins.

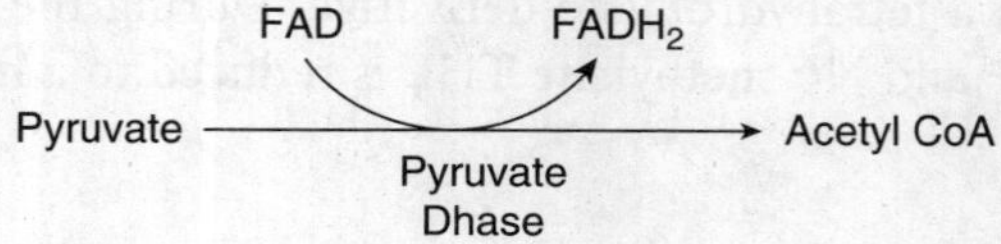

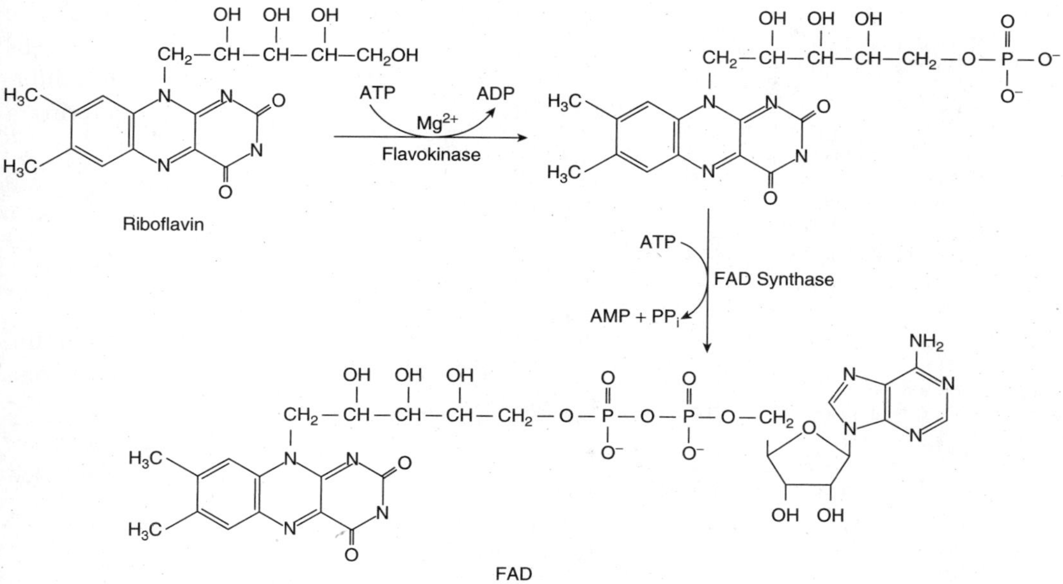

Coenzymes of Niacin

Nicotinamide adenine dinucleotide (NAD^+) and nicotinamide adenine dinucleotide phosphate ($NADP^+$) are the coenzymes derived from dietary nicotinamide, niacin, and tryptophan. Nicotinamide is deaminated in the body as niacin. Niacin is further converted through a series of reactions to NAD^+ and $NADP^+$. Finally, tryptophan produces quinolonate, which then produces nicotinamide and ultimately NAD^+ and $NADP^+$.

For the synthesis of NAD^+, phosphoribosyl pyrophosphate and ATP provide ribose phosphate and AMP moiety. Glutamine provides the amino group. In the structure of coenzymes, nitrogen atom of nicotinamide possesses a positive charge because it forms an extra bond. Nicotinamide, which is formed from the degradation of NAD^+ and $NADP^+$, is excreted in urine as N-methyl nicotinamide.

Biochemical functions

The coenzymes NAD^+ and $NADP^+$ involve in a variety of oxidation-reduction reactions. They accept hydride ion (hydrogen atom and one electron H^-) and undergo reduction in the pyridine ring. This results in the neutralisation of positive charges. The nitrogen atom and the fourth carbon atom of nicotinamide ring participate in the reaction. While the atom of hydrogen is accepted by the coenzyme, the other ion is released into the surrounding medium. This reaction is reversed when NADH is oxidised to NAD^+. $NAPD^+$ also functions like NAD^+ in oxidation-reduction reactions.

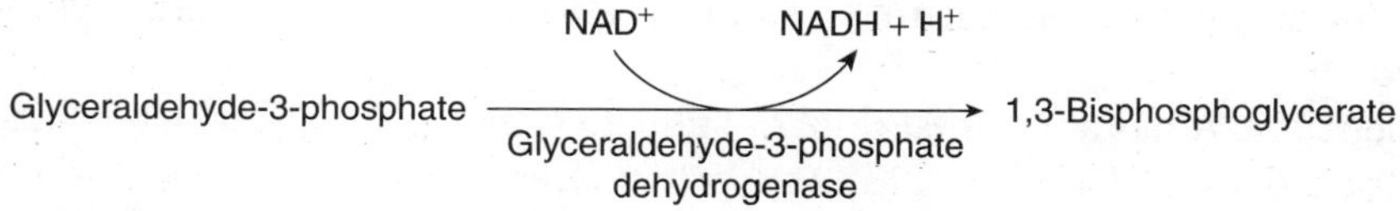

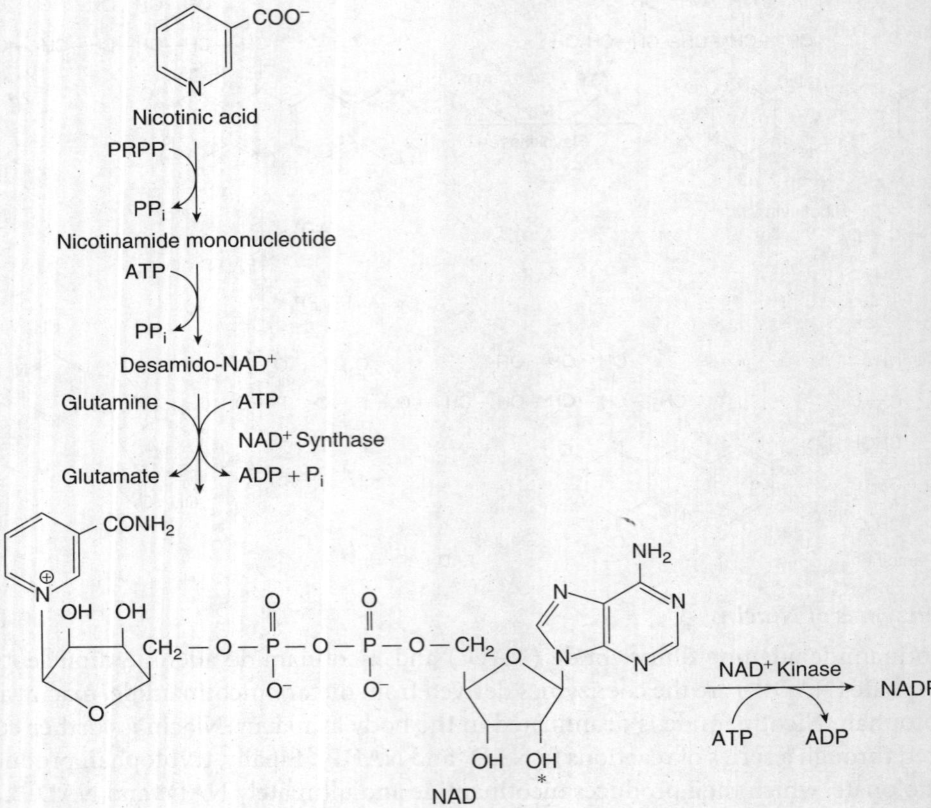

Coenzyme of Pantothenic Acid

Pantothenic acid is made up of two units—pantoic acid and β-alanine—which are joined together by a peptide linkage. A-form pantothenate occurs in a series of reactions. Pantothenate is first phosphorylated, to which cysteine is added. Decarboxylation followed by addition of AMP moiety and a phosphate results in coenzyme A. The structure of coenzyme A or CoA (A for acetylation) consists of pantothenic acid joined to β-mercaptoethylamine on one end. On the other side, pantothenic acid is held by a phosphate bridge to adenylic acid, which is made up of adenine, and a phosphate linked to carbon-3 of ribose.

Biochemical function

The functions of pantothenic acid are excreted through coenzyme A or CoA.

CoA is a central molecule involved in the entire metabolism (carbohydrate, lipid, and protein).

Coenzyme A has a terminal thiol or sulfhydryl group (—SH), which is the reactive site; hence, CoA—SH is also used. Acyl group (free fatty acid) is linked to coenzyme

A by a thioester bond to give acyl CoA. When bound to acetyl unit, it is called acetyl CoA, and with succinate, succinyl CoA is formed.

$$\text{Pyruvate} \xrightarrow[\text{Pyruvate Dhase}]{\text{CoA}} \text{Acetyl CoA}$$

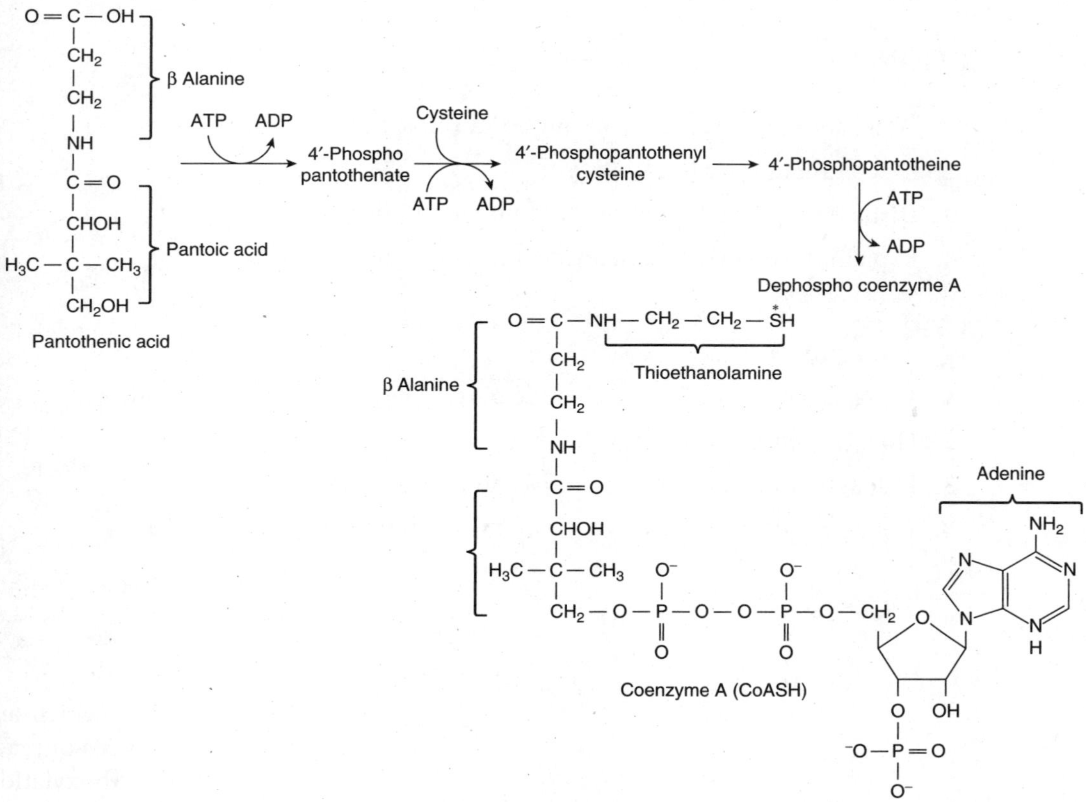

QUESTION BANK

SECTION A

1. What are the excretory products of purine catabolism?
2. Draw the nuclear skeleton of purine and indicate the sources of various atoms in it.
3. Give any two inhibitors of nucleotide biosynthesis.

4. How is purine base converted to purine nucleotide?
5. Name any two inhibitors of pyrimidine biosynthesis and enumerate their importance.
6. Define NAD.
7. What is the role of FAD in metabolism?
8. Draw the nuclear skeleton of pyrimidine.

SECTION B

1. Write about the inhibitors of nucleotide biosynthesis.
2. Discus about salvage pathway.
3. Discuss the regulation of pyrimidine nucleotides.
4. Explain the degradative pathway of pyrimidine nucleotide.

SECTION C

1. Write about pyrimidine nucleotide biosynthesis. Explain its regulation.
2. How is adenine synthesised?
3. Discuss the catabolism of purines and pyrimidines.
4. Define the coenzyme derived from nucleotides.

CHAPTER

12

DETOXICATION MECHANISM

Concept Outline

- Introduction
- Oxidation
- Reduction
- Hydrolysis
- Conjugation
- Detoxification of Drugs
- Role of Cytochrome p450

INTRODUCTION

The term *detoxication* or *detoxification* refers to the series of biochemical reactions occurring in the body to convert the foreign compounds to non-toxic or less toxic and more easily executable forms.
Examples: pollutants and pesticides.

TOXIC SUBSTANCES

Toxic substances may include the following:

1. Products which are formed in the intestines due to normal metabolic reactions.
   Examples
   (i) Foreign organic compounds which are not ordinarily ingested or taken once in a way.
   (ii) Drugs used for therapeutic purposes and chemicals used for therapeutic purposes.
2. *Putrefaction:* Products produced in the intestines through bacterial enzymatic reaction or normal digestion products.
3. Sex hormones or bile pigments which are normal physiological compounds and are eliminated to some extent as sulphate and glucuronide.

SITE OF DETOXICATION

1. The detoxification reactions are carried out mainly in the liver, which is equipped with the enzyme machinery.
2. Kidney and other organs may sometimes involve.
3. The products formed by detoxification are mostly excreted by the kidney and less frequently excreted via faeces or expired air.

MECHANISM OF DETOXIFICATION

The metabolism of xenobiotics may be divided into two phases:

1. Phase I
2. Phase II

Phase I

The reactions of phase I are oxidation, reduction, and hydrolysis.

Phase II

The reaction of phase II is conjugation.

Generally, detoxification of a compound involves phase I as well as phase II reactions. It is shown in Figure 12.1.

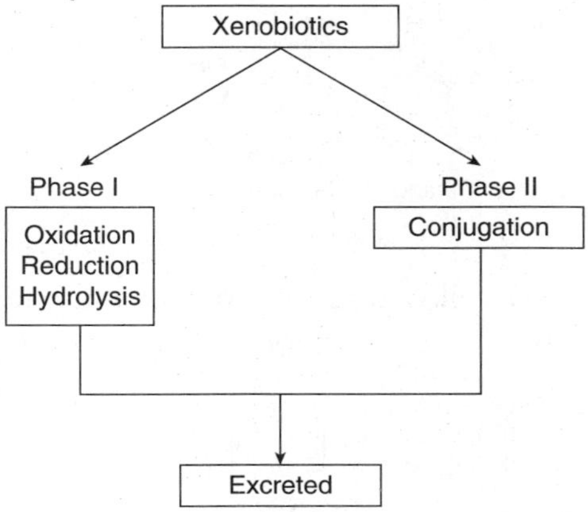

Figure 12.1 *Detoxification of a Compound*

OXIDATION

A large number of foreign compounds, which include alcohols, aldehydes, amines, aromatic hydrocarbons, and sulphur compounds, are detoxified by the process by oxidation. In general, aliphatic compounds are more easily oxidised than aromatic.

ALCOHOL

Aliphatic and aromatic alcohols undergo oxidation to form the corresponding acids.

$$\underset{\text{Methanol}}{CH_3OH} \longrightarrow \underset{\text{Formic acid}}{HCOOH}$$

$$\underset{\text{Ethanol}}{C_2H_5OH} \longrightarrow \underset{\text{Acetic acid}}{CH_3COOH}$$

$$\underset{\text{Benzyl alcohol}}{C_6H_5CH_2OH} \longrightarrow \underset{\text{Benzoic acid}}{C_6H_5COOH}$$

Methanol is a highly toxic substance that when it is consumed, it produces blindness.

AROMATIC HYDROCARBONS

Benzene may be oxidised to mono-, di-, and tri-hydroxy phenols as shown below.

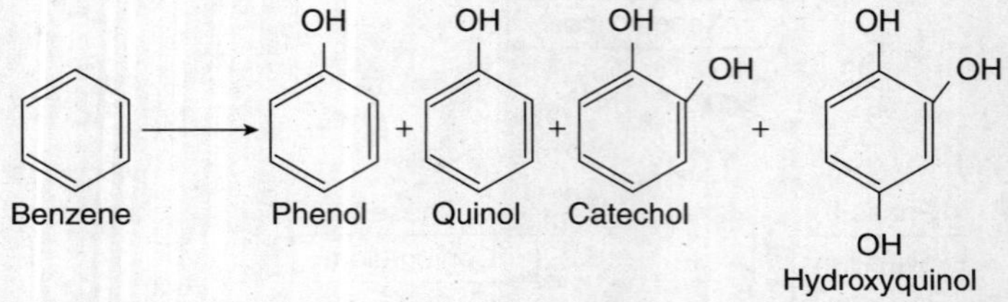

If the aromatic hydrocarbon contains alkyl group, it is oxidised to acid.

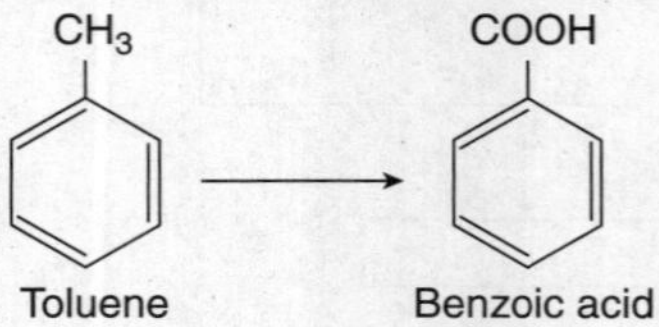

ALDEHYDES

Aldehydes are oxidised to produce the corresponding acids.

$$\underset{\text{Benzaldehyde}}{C_6H_5CHO} \longrightarrow \underset{\text{Benzoic acid}}{C_6H_5COOH}$$

$$\underset{\text{Chloral}}{CCl_3CHO} \longrightarrow \underset{\text{Trichloroacetic acid}}{CCl_3COOH}$$

AMINES

Aliphatic amines are converted to the corresponding acids, liberating urea while aromatic amino acids are oxidised to phenols.

$$\underset{\text{Aliphatic amines}}{RCH_2NH_3} \longrightarrow \underset{\text{Aliphatic acid}}{R-COOH} + \underset{\text{Urea}}{NH_2-CO-NH_2}$$

$$\underset{\text{Aniline}}{C_6H_5NH_2} \longrightarrow \underset{\text{P-amino phenol}}{HO-C_6H_4-NH_2}$$

SULPHUR COMPOUNDS

Organic sulphur is oxidised to sulphuric acid.

$$\text{Sulphur compounds} \longrightarrow \text{Sulphuric acid}$$

DRUGS

Meprobamate is a tranquillizer. It is oxidised to hydroxy meprobamate and excreted in urine.

REDUCTION

In man, reduction is less common than oxidation. It has been observed that certain metabolites, which are formed due to reduction, are more toxic than the original compounds.

$$\underset{\text{Picric acid}}{C_6H_2OH(NO_2)_3} \longrightarrow \underset{\text{Picramic acid}}{C_6H_2OH(NO_2)_2NH_2}$$

$$\underset{\text{Chloral}}{CCl_3CH(OH)_2} \longrightarrow \underset{\text{Trichloroethanol}}{CCl_3CH_2OH}$$

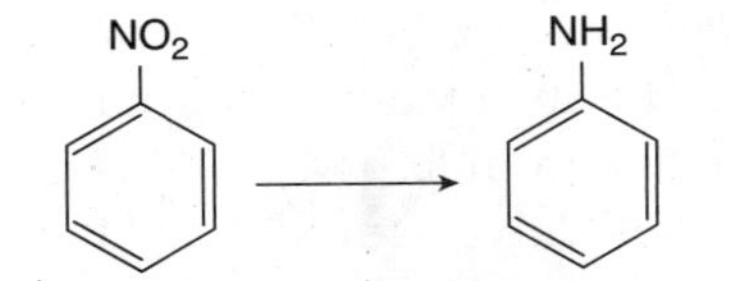

Nitrobenzene Aminobenzene

HYDROLYSIS

Some of the compounds which are ingested for therapeutic purposes undergo hydrolysis in the body. Compounds such as ester, glycoside, and amide are important in the metabolism of xenobiotics.

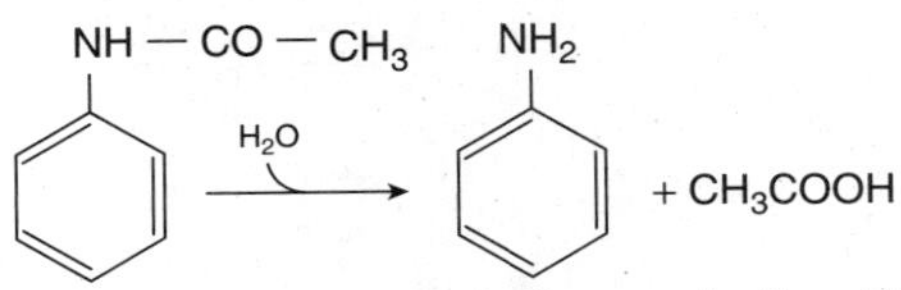

Acetanilide Aniline Acetic acid

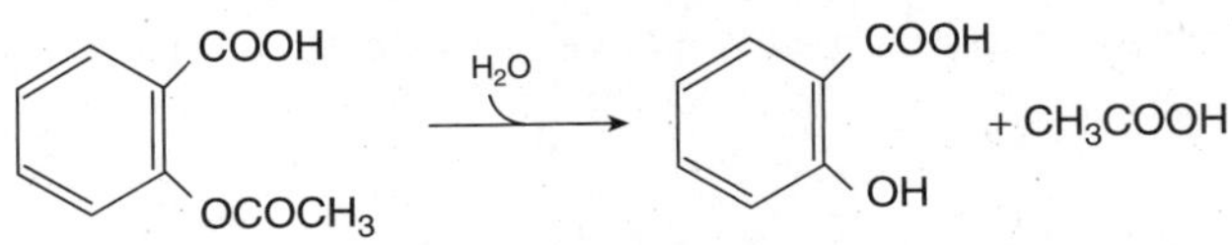

Aspirin Salicylic acid Acetic acid

$$\underset{\text{Diisopropyl fluorophosphate}}{(C_3H_7O)_2\ POF} \xrightarrow{H_2O} \underset{\text{Dialkyl phosphate}}{(C_3H_7O)_2 PO(OH) + HF}$$

CONJUGATION

Several xenobiotics undergo detoxification by conjugation to produce less toxic or more easily executable compounds. Conjugation is the process in which a foreign compound combines with a substance produced in the body. At least, eight different conjugating agents have been identified in the body. They are as follows:

1. Glucuronic acid
2. Glycine
3. Glutamine
4. Methyl group
5. Sulphuric acid
6. Cysteine

7. Acetic acid
8. Thiosulphate

GLUCURONIC ACID

The functional form of glucuronic acid is UDP-glucuronic acid and this is produced in the uronic acid pathway. The microsomal enzyme UDP-glucuronyl transferase is involved in the formation of glucuronide. Two types of glucuronides are known.

Ether linkage: Phenol combines with glucuronic acid through ether linkage and forms glucuronide.

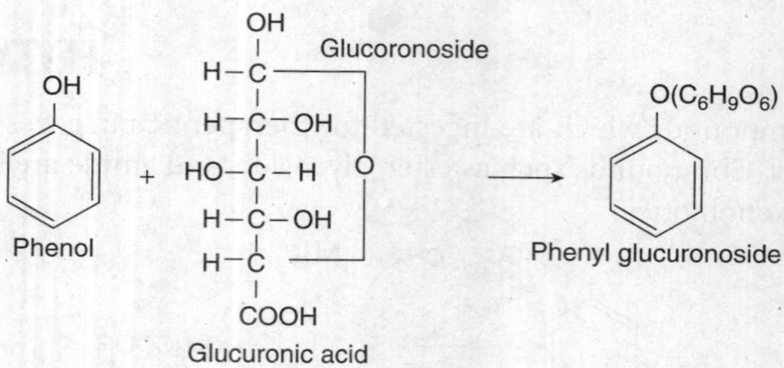

Ester linkage: Aromatic acids from ester linkage with glucuronic acid. In this case, glucuronic acid couples with the carboxyl group of benzoic acid to form an ester linkage.

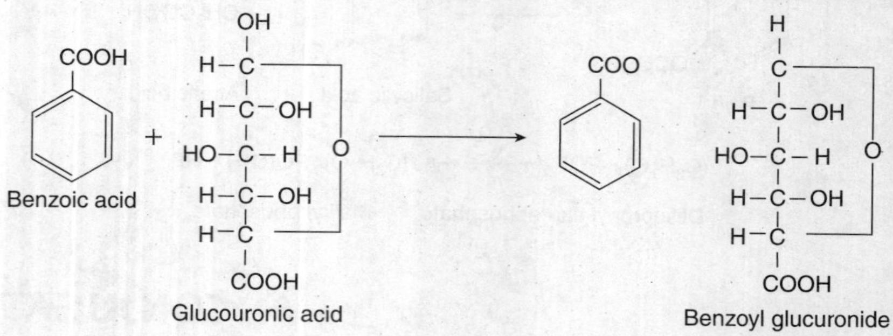

SULPHURIC ACID

Theactiveformofsulphate-3′-phosphoadenosine5′-phosphosulphate(PAPS)participates in conjugation reactions, and the enzyme sulfotransferase is involved in this process.

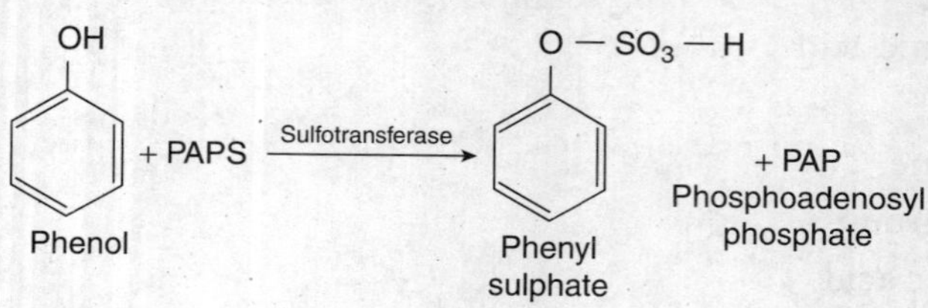

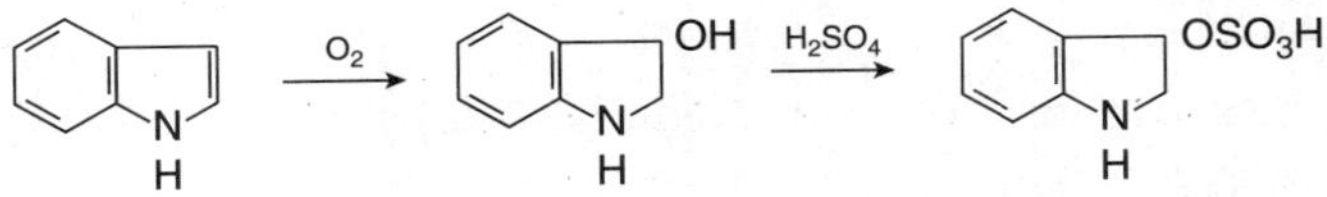

Indole → Indoxyl → Indoxyl sulphuric acid

GLYCINE

Many aromatic carboxylic acids are conjugated with glycine. Hippuric acid is formed when glycine is conjugated with benzoyl-CoA. The excretion of hippuric acid was first reported in 1829 in the urine of cows and horses.

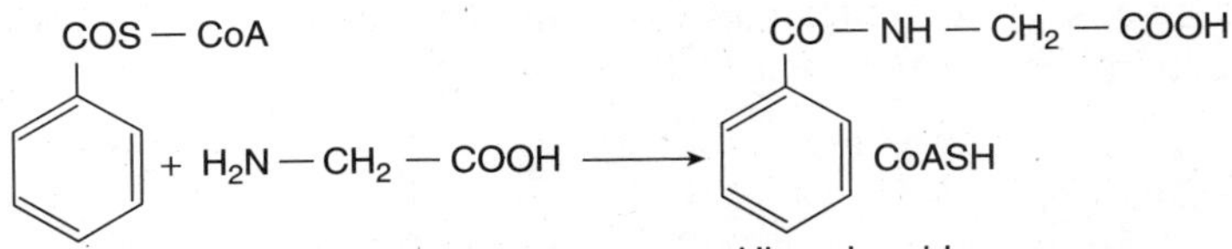

Benzoyl CoA + Glycine → Hippuric acid

$$\text{Cholic acid} + \text{Glycine} \longrightarrow \text{Glycolic acid}$$

CYSTEINE

A wide range of organic compounds such as alkyl or acyl aldehydes, alkenes with compounds, and epoxides get conjugated with cysteine of glutathione.

R — X + GSH

↓ GSH transferase (H — X released)

R — SG

↓ γ-Glutamyl transpeptidase (Glutamatel released)

Cysteinyl glycine

↓ Cysteinyl glycinase (Glycine released)

$R-S-CH_2-CH(NH_2)-COOH$ R-cysteine

↓ N-acetyltransferase (Acetyl CoA → CoASH)

$R-S-CH_2-CH(NH-COCH_3)-COOH$

Mercapturic acid

ACETIC ACID

Acetyl-CoA is the active form of acetic acid that takes part in conjugating reactions.

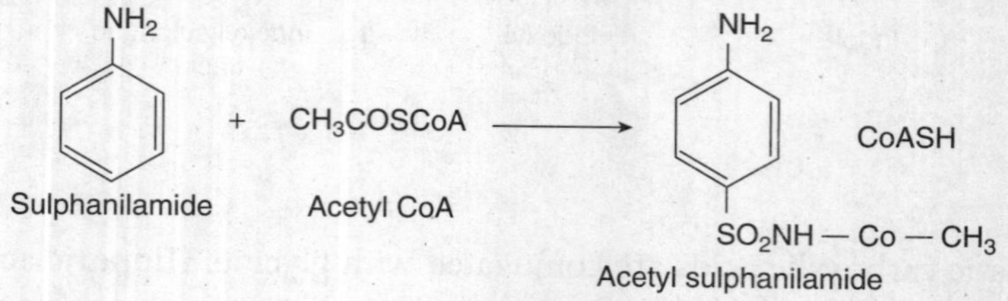

GLUTAMINE

Phenylacetic acid is detoxicated by conjugation with glutamine to form phenylacetylglutamine.

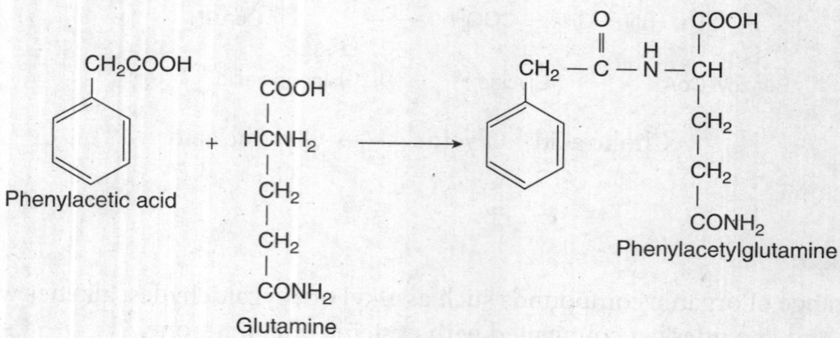

METHYLATION

This is rather limited in the body. The methyl donor for such reactions is s-adenosylmethionine.

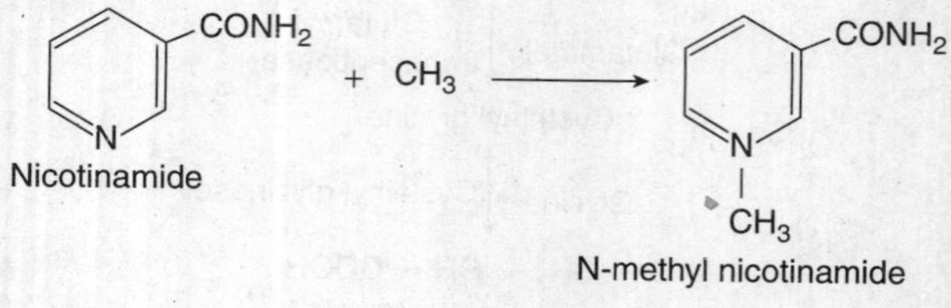

THIOCYANATE FORMATION

The highly toxoid cyanides are conjugated with thiosulphate to form less toxic thiocyanate.

$$\underset{\text{Cyanide}}{HCN} + \underset{\text{Sodium thiosulphate}}{Na_2S_2O_3} \longrightarrow \underset{\text{Thiocyanate}}{HCNS} + \underset{\text{Sodium sulphate}}{Na_2SO_3}$$

DETOXIFICATION OF DRUGS

The toxic effects of certain metals such as arsenic, mercury, and cadmium could be overcome by administering BAL (British Anti-Lewisite). This compound was developed during World War II and was used as a detoxifying agent for certain war poisons. The mechanism of action of BAL was not clearly known, but it combines with metals and gets easily executable into urine.

ROLE OF CYTOCHROME P450

Cytochrome p450 is also known as mixed function oxidase. Almost all the oxidation reactions of detoxification are catalysed by monooxygenase or cytochrome p450. Cytochrome p450 is mostly involved in the addition of a hydroxyl group to aliphatic or aromatic compounds.

$$RH + O_2 + NADPH \longrightarrow ROH + H_2O + NADP^+$$

SALIENT FEATURES OF CYTOCHROME P450

1. Multiple forms of cytochrome p450 are present ranging from 20 to 200. Among this, six species have been isolated and studied in detail.
2. It belongs to a family of haemoprotein, which contain haem as the prosthetic group.
3. It is located in higher concentrations in the microsomes of liver and adrenal gland and also occurs in mitochondria.
4. The mode of reaction of cytochrome p450 is quite complicated and is dependent on NADPH.
5. The main constituent of cytochrome p450 is the phospholipid phosphatidylcholine, which is necessary for the action of this enzyme.
6. It is an inducible enzyme and therefore its synthesis is increased by the administration of drugs such as phenobarbital.
7. It is very specific towards the metabolism of polycyclic aromatic hydrocarbons and hence is also called as aromatic hydrocarbon hydroxylase.

QUESTION BANK

SECTION A

1. Define detoxication.
2. What are the toxic compounds that are detoxified by the system?
3. Define the stages involved in detoxication.

4. Explain the oxidation process involved in detoxication.
5. How is the conjugation process involved in detoxication?
6. Explain the hydrolysis phase of detoxification.
7. How does glucoronic acid favour the conjugation phase of detoxication?

SECTION B

1. Explain the various phases in the detoxication mechanism.
2. What are the conjugation agents that have been identified in the system for detoxication?

SECTION C

1. Define the detoxification mechanism in detail.
2. Give a detailed account on the role of cytochrome p450 in detoxication.

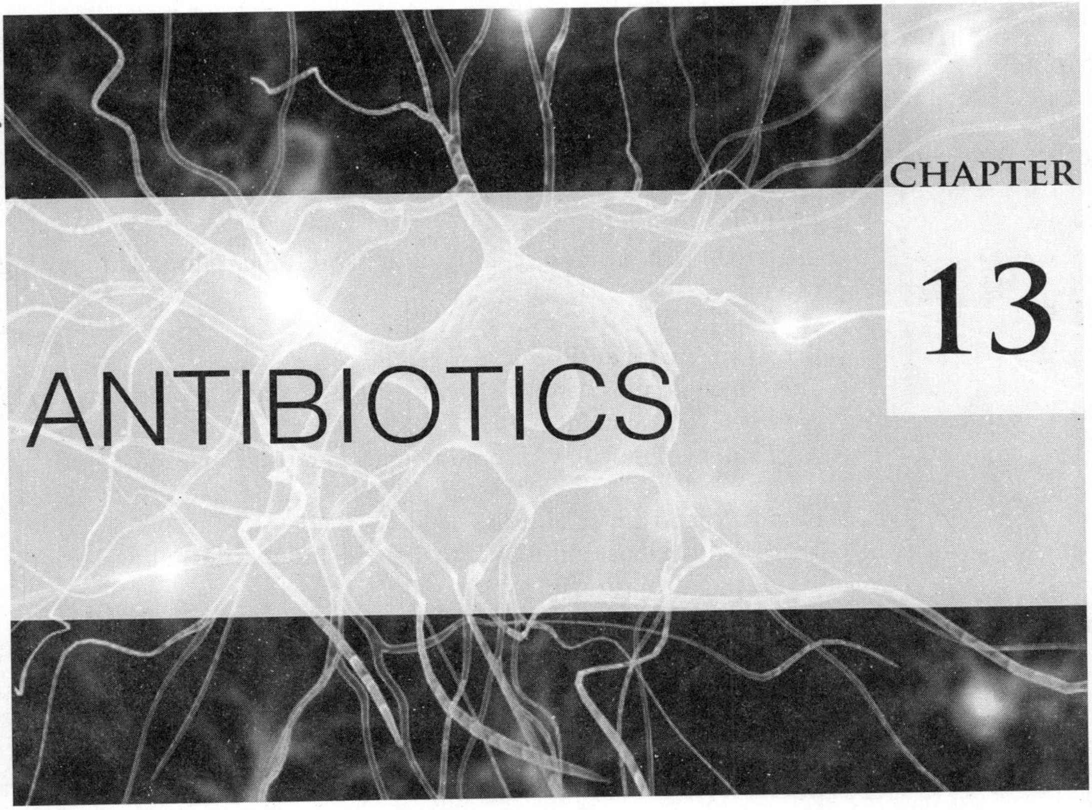

CHAPTER 13

ANTIBIOTICS

Concept Outline

- Introduction
- Penicillin
- Streptomycin
- Chloramphenicol
- Tetracycline

INTRODUCTION

Antibiosis is a process in which one organism may destroy another to survive.

The term *antibiotic* was first of all introduced by Vuillemin in 1889.

'Antibiotic is a chemical substance produced by or derived from living cells which are capable, in small concentrations, of inhibiting the life process or even destroying the microorganisms.'

CLASSIFICATION OF ANTIBIOTICS

Antibiotics can be classified in different ways.

First Classification

The broad classification of antibiotics divides them into the following two types.

Broad spectrum antibiotics

These include those antibiotics which may be used as curative agents against several ailments.

Example: Penicillin, chloramphenicol, tetracycline, and so on.

Narrow spectrum antibiotics

These include those antibiotics that are highly specific in their action.

Example: Bacitracin, nystatin, and so on.

Second Classification

Another classification of antibiotics is available, which depends upon the type of bacteria (Gram positive or Gram negative) the antibiotic can destroy. This classification was originally worked out by Christian Gram and is called Gram staining method (Table 13.1).

Table 13.1 *Classification of bacteria based on Gram staining method*

Gram positive bacteria	Gram negative bacteria
Diphtheria bacillus	Coli and typhoid bacillus
Leprosy bacillus	Gonococcus
Pneumococcus	Meningococcal
Staphylococcus	Plague bacillus
Streptococcus	Vibrios
Tubercle bacillus	

Third Classification

This classification of antibiotics is based on their chemical structure.

Penicillin – derived from amino acids. *Example:* Penicillin

Aminoglycosides – contain a sugar molecule linked to amino compound. *Example:* Streptomycin.

Chloramphenicol – synthetic analogue

Tetracycline – have four six-membered fused ring. *Example:* Tetracycline and terramycin.

PENICILLIN

Penicillin is the name given to the mixture of neutral compounds which have the molecular formula $C_9H_{11}N_2O_48R$ and differ only in R. The general structure of penicillin is given below in Figure 13.1:

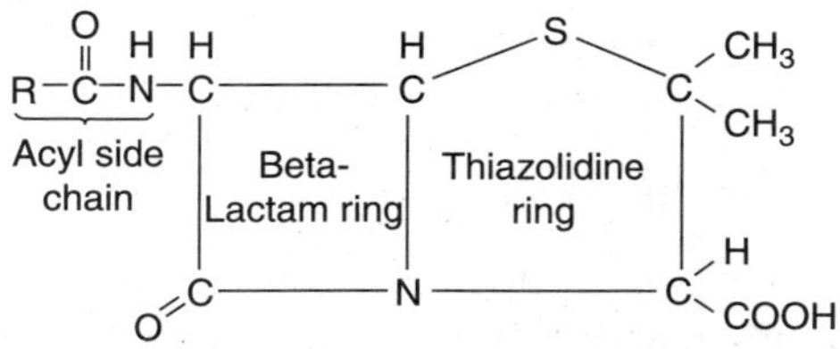

Figure 13.1 *Structure of Penicillin*

R may vary and according to the nature of the functional group, penicillin may be classified as shown in Table 13.2.

Table 13.2

Type	R
Penicillin 1 or F	$-CH_2-CH=CH-CH_2-CH_3$
Penicillin 2 or G	$-CH_2C_6H_5$
Penicillin 3 or X	$-CH_2C_6H_4OH$
Penicillin 4 or K	$-(CH_2)_6CH_3$
Penicillin 5	$-CH_2OC_6H_5$
Dihydro F penicillin	$-(CH_2)_4CH_3$

PROPERTIES OF PENICILLIN

Penicillin is white or slightly yellowish white crystalline powder.

All the natural penicillin is dextrorotatory.

Penicillin is only sparingly soluble in water.

Penicillin is hydrolysed by alkaline medium to yield penicilloic acid, which loses carbon dioxide to form penicilloic acid.

If the hydrolysis of penicillin is carried out in acidic medium, the amide side chain is involved with the opening of β-lactam ring. This causes loss of its activity. Due to this reason, natural penicillins are not very much effective when given orally. Further, the stomach juices contain the enzymes penicillinases. These enzymes also cause hydrolysis of penicillin, resulting in the opening of the lactam ring, followed by the loss of activity.

Mechanism of Action

Bacteria constantly reconstruct their peptidoglycan cell walls and break down portions of the cell walls as they grow and divide. Penicillin inhibits the formation of peptidoglycan cross-links in the bacterial cell wall. This is achieved by the binding of the Lactam ring of penicillin to the enzyme transpeptidase. Hence, peptidase cannot catalyse the formation of the cross-links. This produces imbalanced cell wall production and starts degradation, thereby causing the cell to rapidly die.

Uses

Penicillin is widely used in the treatment of Gram-positive bacterial infections.

Toxicity (side reactions)

Penicillins are found to have low toxicity in comparison to other sulpha drugs. However, in some cases diarrhoea and allergic reactions may occur. Penicillin is hence injected after a test prick is given.

STREPTOMYCIN

Streptomycin is the first antibiotic used in the treatment of tuberculosis. It belongs to aminoglycosides compound. It is a bactericidal antibiotic. Streptomycin cannot be given orally but should be administered by regular intramuscular injections.

Structure of Streptomycin

Streptomycin was first isolated by Albert Schatz. The molecular formula is $C_{21}H_{39}N_7O_{12}$. It consists of three fragments—an N-methyl-L-Glucosamine ring, an L-streptose and a streptidine ring—which are linked together by two glycosidic bonds. It also contains an elaborate network of hydrogen bonds linking selenite and streptomycin oxime ions and water molecules (refer Figure 13.2).

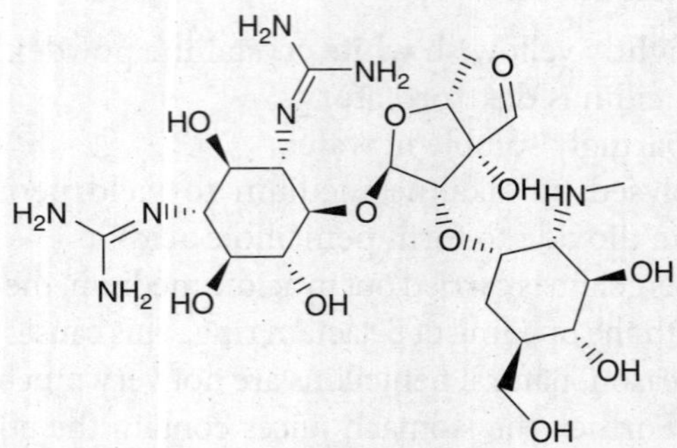

Figure 13.2 *Structure of Streptomycin*

Mechanism of Action

Streptomycin is an inhibitor of protein synthesis. It binds to the small 16S rRNA of the 30S subunit of the bacterial ribosome, and binds to formyl-methionyl-tRNA.

This leads to codon misreading and hence ultimately the death of microbial cells. The human ribosomes are different from that of bacterial ribosomes; this makes the antibiotic selective only for bacteria.

Uses

Streptomycin is used for treating tuberculosis in combination with other anti-TB drugs. It is also used as a pesticide to combat the growth of bacteria, fungi and algae. It controls bacterial and fungal diseases of certain fruits, vegetables, seeds and ornamental crops.

CHLORAMPHENICOL

Chloramphenicol (Chloromycetin) is a bacteriostatic antimicrobial. It is considered as a broad-spectrum antibiotic. Chloramphenicol is active against a wide variety of Gram-positive and Gram-negative bacteria including anaerobic organisms. Chloramphenicol was derived from the bacterium *Streptomyces venezuelae*, and it was isolated by David Gottlieb.

Structure of Chloramphenicol

The molecular formula of chloramphenicol is $C_{11}H_{12}Cl_2N_2O_5$ (refer Figure 13.3). It is soluble in lipid. It remains unbound to protein.

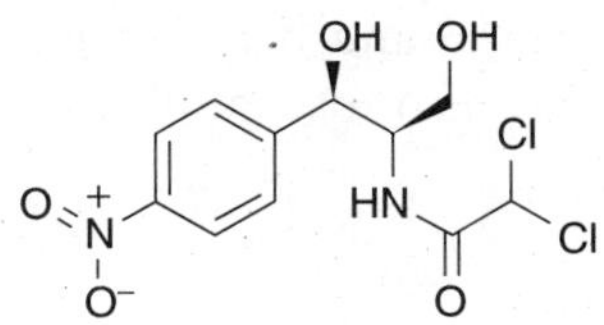

Figure 13.3 *Structure of Chloramphenicol*

Mechanism of Action

Chloramphenicol is a bacteriostatic drug and it inhibits bacterial growth. It controls the bacterial growth by inhibiting protein synthesis. Chloramphenicol controls the protein chain elongation by inhibiting the peptidyl transferase activity of the bacterial ribosome. It specifically binds to the 23S rRNA of the 50S ribosomal subunit, thereby preventing the peptide–bond formation.

Uses

Chloramphenicol is widely used in the treatment of typhoid. It is active against *Neisseria meningitidis, Streptococcus pneumonia* and *Haemophilus influenza.* It is widely used in the treatment of conjunctivitis.

TETRACYCLINE

Tetracycline is a broad-spectrum antibiotic produced by the *Streptomyces* genus of actinobacteria used against various bacterial infections. The tetracyclines were discovered as natural products by Benjamin Minge Duggar in 1945. Refer Figure 13.4 for the structure of tetracycline.

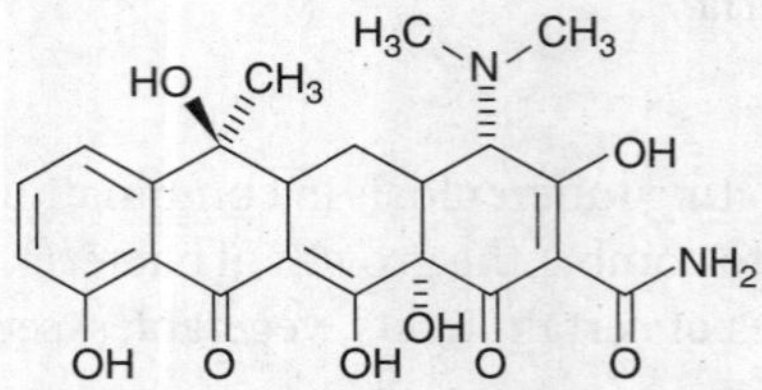

Figure 13.4 *Structure of Tetracycline*

Mechanism of Action

Tetracyclines bind to the 30S subunit of microbial ribosomes. They prevent protein synthesis by blocking the attachment of the charged amino acyl-tRNA to the A site on the ribosome. They prevent the introduction of new amino acids to the nascent peptide chain.

Uses

Tetracycline has a bacteriostatic activity against aerobic and anaerobic bacterial species. It remains useful in the management of infections by certain intracellular bacterial pathogens such as syphilis, leptospirosis and Lyme disease. It is also active against eukaryotic parasites such as malaria and balantidiasis. Certain infections including anthrax, plague and brucellosis are also susceptible to tetracyclines.

QUESTION BANK

SECTION A

1. Define antibiotics.
2. Draw the structure of penicillin.
3. Explain the types of penicillins available.
4. Draw the structure of tetracycline.

SECTION B

1. What is streptomycin? Mention its structure.
2. Write a note on the structure and functions of penicillin.

SECTION C

1. Describe the structure and functions of streptomycin.
2. What are antibiotics? Give the structure and functions of chloromycetin.
3. Describe the structure, mode of action and uses of tetracycline.

Index

CELL COMPONENTS AND THEIR FUNCTIONS

CELL MEMBRANE

Cell is enveloped by a thin membrane called cell membrane or plasma membrane which allows only selective passage of substances. Membranes are made up of lipids, proteins, and small amounts of carbohydrates. Different membranes have different ratios of protein and lipid. Proteins in the cell membrane are of two types: peripheral proteins and integral membrane proteins. Peripheral proteins exist on the surface of membranes, and they are attached by ionic and polar heads of lipids. Peripheral proteins can be easily removed from the membrane. Integral membrane proteins span across the membrane from one side to another and are attached by hydrophobic bonds. Some of the integral proteins are involved in the transport mechanism. The lipid elements of membrane are composed of phospholipids, glycolipids, and cholesterol. Both polar and non-polar groups are present in the phospholipid molecules. They form a bilayer. The lipid bilayer model was proposed by Davson and Danielli in 1935. According to this model, the membranes are made up of a lipid bilayer holding globular proteins on its either surface by ionic and hydrogen bonds. Structure of the membranes was described as fluid–mosaic model by Singer and Nicholson in 1972. According to this model, the phospholipids are arranged in a bilayer with hydrophobic ends (fatty acid tails) oriented towards the centre of the bilayer while their hydrophilic ends are pointed outside. The cholesterol content of the membrane maintains fluidity. The structure of a cell membrane showing that it is composed mainly of a lipid bilayer of phospholipid molecules is as shown in Figure 1.2.

MEMBRANE TRANSPORT

Lipid-soluble molecules pass through the plasma membrane readily by dissolving in the lipid bilayer. Small molecules pass through membrane pores. The pores are positively charged and allow anions, and neutral molecules pass through more readily than cations. Large polar substances (e.g. glucose and amino acids) are transported through the membrane with the help of carrier molecules.

TRANSPORT MECHANISMS

Diffusion

Diffusion is the movement of a substance from an area of high concentration to an area of low concentration (down a concentration gradient). The concentration gradient is the difference in solute concentration

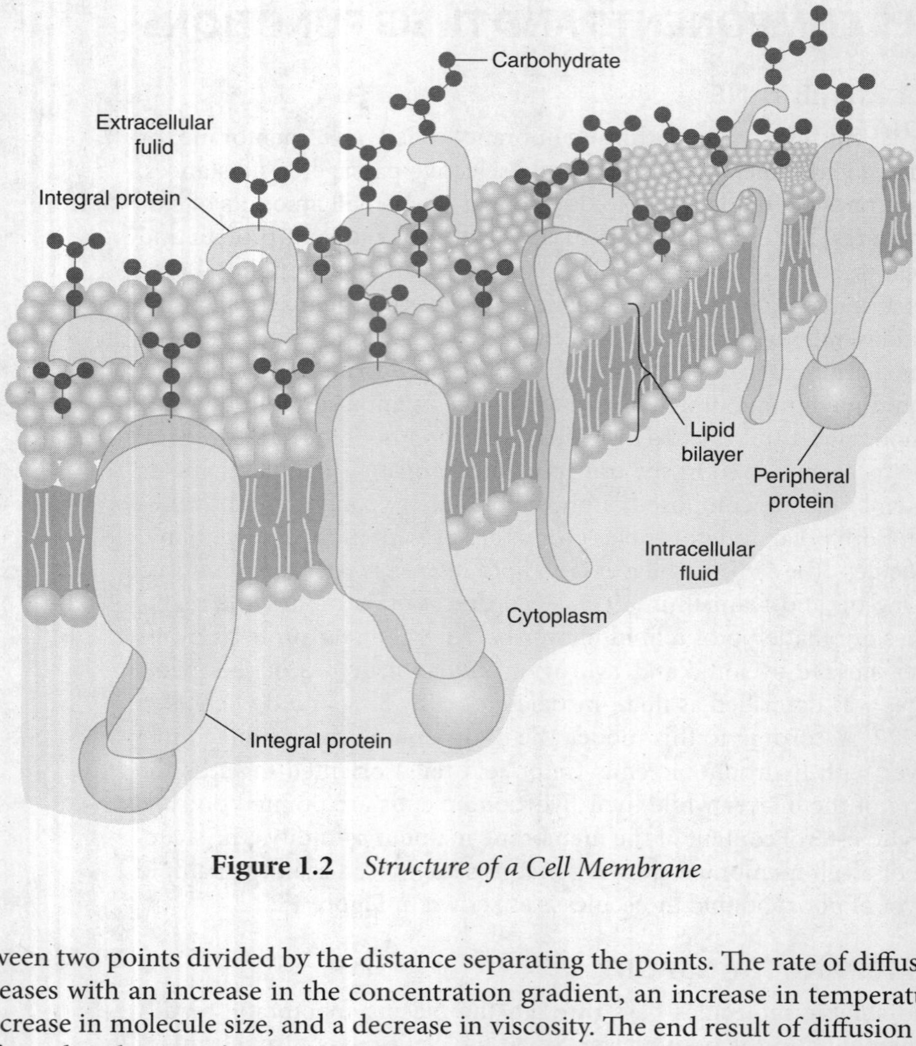

Figure 1.2 *Structure of a Cell Membrane*

between two points divided by the distance separating the points. The rate of diffusion increases with an increase in the concentration gradient, an increase in temperature, a decrease in molecule size, and a decrease in viscosity. The end result of diffusion is a uniform distribution of molecules. Diffusion requires no expenditure of energy. Small, uncharged molecules such as O_2, CO_2, H_2O, and lipid-soluble substances get transported across the membrane through the process of diffusion.

Osmosis

Osmosis is the diffusion of a solvent across a selectively permeable membrane. Osmotic pressure is a measure of the tendency of water (or solvent) to move across the selective permeable membrane. Iso-osmotic solutions have the same concentration of solute particles. Hyperosmotic solutions have higher concentration, and hypo-osmotic solutions have lower concentration of solute particles than a reference solution. When cells are kept in an iso-osmoticc solution, they will neither swell nor shrink. However, when the cells are kept in a hyper-osmotic solution they will shrink, whereas in a hypo-osmotic solution they will swell or may lyse.

MEDIATED TRANSPORT MECHANISMS

Mediated transport is the transport of molecules across the membrane with the help of a carrier. The molecule transported may be large and water soluble. The carriers have the capacity to bind either a single molecule or a group of similar molecules. This process is known as specificity. Similar molecules bind to the same carrier. Transport systems are of two kinds, namely active transport and passive transport. Passive transport does not require energy, whereas active transport requires energy. An example is the movement of Na^+, K^+-ATPase pump, which move in opposite directions across the membrane.

NUCLEUS

Nucleus is the largest cellular organelle. It is spherical in shape, with a diameter of about 5 μM. Nucleus is surrounded by a nuclear membrane. The nuclear membrane is porous in nature and the pores are called nuclear pores. The RNA synthesised in the nucleus enters into the cytoplasm through these nuclear pores. Nucleus consists of thick protoplasm (nucleoplasm), which contains various enzymes, free amino acids, lipids, Ca^{2+}, Mg^{2+} ions, deoxyribonucleic acid (DNA), and ribonucleic acid (RNA). The main component of nucleus is chromatin, which is present throughout the nucleus and is composed of cellular DNA along with proteins—histones. During cell division, chromatin is divided and is shared equally between the daughter cells. The structure of nucleus is as shown in Figure 1.3.

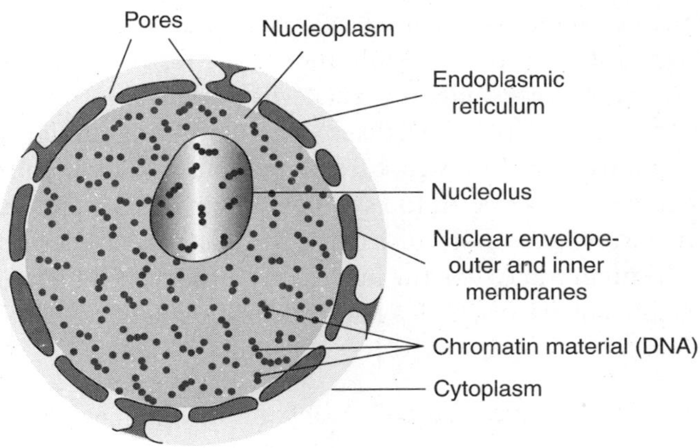

Figure 1.3 *Structure of Nucleus*

Mitochondria

Mitochondria are the largest organelles in cytoplasm. They are spherical, oval, or rod-like bodies, about 0.5 μm – 1.0 μm in diametre and up to 7 μm in length. The structure of mitochondria is shown in Figure 1.4. Mitochondria are considered to be the power houses of the cell because energy is released from oxidation that takes place in these organelles. The energy released is stored in the form of ATP. Mitochondrion is covered by a smooth outer membrane and an inner membrane. The inner core is known as the matrix. The inner membrane is folded to form cristae. Matrix contains enzymes concerned with oxidative phosphorylation and the metabolism of carbohydrate, lipids, and amino acids. The outer membrane controls membrane permeability.

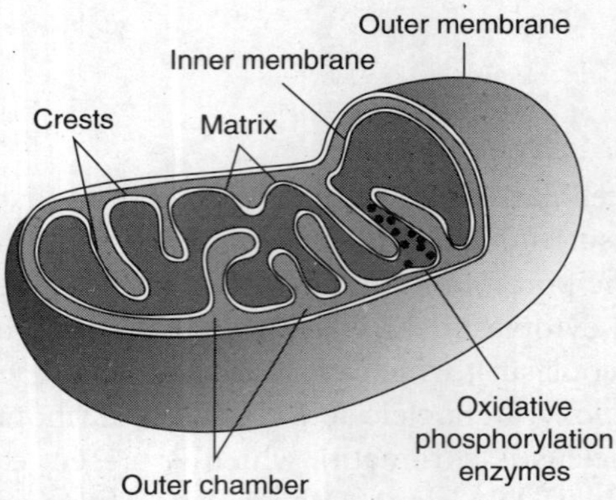

Figure 1.4 *Structure of Mitochondria*

Endoplasmic Reticulum

Endoplasmic reticulum (ER) is a complex system of membranes present in cytoplasm. It is a network of lipoprotein membranes in the form of flattened sacs, tubules, and vesicles, enclosing a space called intracisternal space. Two types of ER are seen: (a) smooth ER without any attached particles and (b) rough (granular) ER with spherical particles called ribosomes. The structure of endoplasmic reticulum is as shown in Figure 1.5. ER is actively involved in the synthesis of proteins, lipids, and polysaccharides and the transport of these molecules to the site of storage or utilisation. It also acts as a precursor of the membranes like the nuclear membrane, cell membrane, and Golgi bodies. Glucose 6–phosphatase acts as a marker enzyme.

Ribosomes

A large portion of the ER is grouped with ribosomes, which are the factories of protein synthesis. Ribosomes are also enriched with lipids, particularly phospholipids. Ribosomes consist of particles of two sizes: one with a sedimentation constant of 50S

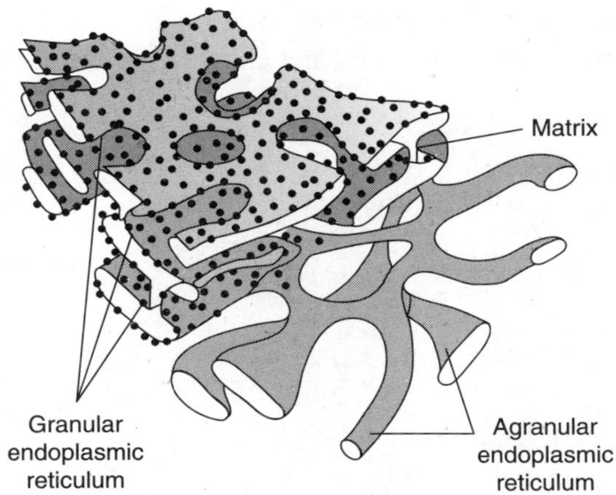

Figure 1.5 *Structure of Endoplasmic Reticulum*

and the other with 30S. The two particles combine to give one with a sedimentation constant of 70S. Ribosomes are the site of final stage of protein synthesis.

Lysosomes

These are subcellular organelles, approximately of the same size as small mitochondria. They are bound by a membrane and appear as bags filled with digestive enzymes such as ribonuclease, glycosidase, acid phosphatase, and aryl sulphates. The structure of lysosome is as shown in Figure 1.6. As long as the membrane of lysosome remains intact, the enzymes within lysosomes are unable to act on substrates in the cytoplasm. Once the membrane is ruptured, release of lysosomal enzymes is followed by dissolution of the cell (autophagy), Therefore, lysosomes are called suicide bags. The main function of lysosomes is intracellular and extracellular digestion. Acid phosphatase is the marker enzyme.

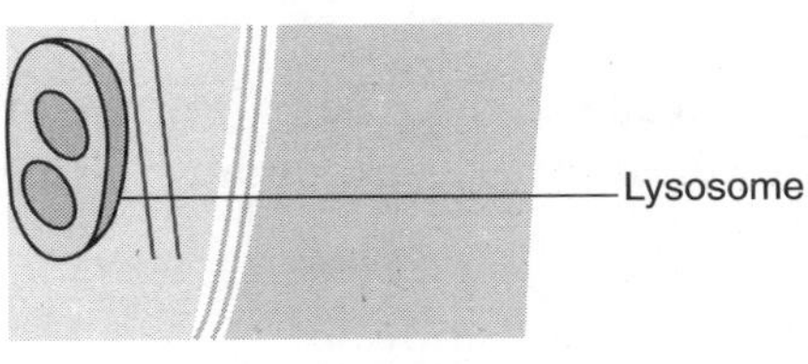

Figure 1.6 *Structure of Lysosome*

Golgi Bodies

Golgi bodies are very similar to ER in structure and function. They are made of a double membrane. The structure of Golgi bodies is as shown in Figure 1.7. Proteins produced by the ribosomes are stored in the form of secretory. Proteins and enzymes are enclosed in the vesicles of Golgi bodies. Golgi bodies are involved in membrane synthesis and transport of proteins. The chief enzyme present in Golgi bodies is alkaline phosphatase.

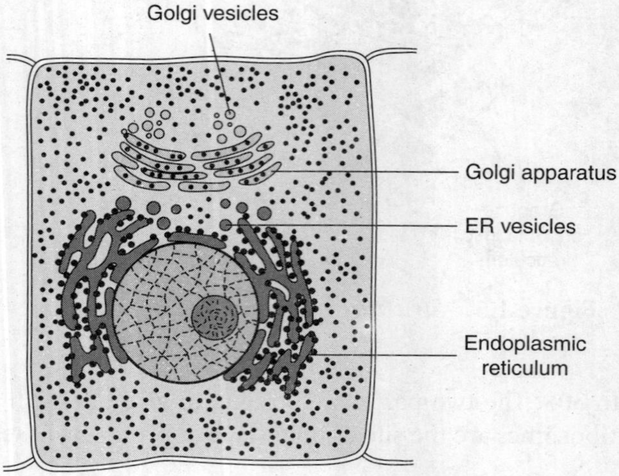

Figure 1.7 *Structure of Golgi Bodies*

Peroxisomes

Peroxisomes are membrane–bound sacs containing enzymes such as peroxidase and catalase. These enzymes catalyse and breakdown the hydrogen peroxide. These enzymes are involved in oxidation and reduction reactions also. Peroxisomes occur largely in hepatic and renal cells. Marker enzyme is catalase.

PROTOPLASM

Protoplasm is the living matter of the cell. The part of the protoplasm present inside the nucleus is nucleoplasm while the part surrounding the nucleus is called cytoplasm or cytosol.

CYTOPLASM

It is an aqueous phase in which all the cytoplasmic organelles are suspended. Many proteins and numerous other solutes are found in cytoplasm. Some important metabolic processes that occur in the cytosol are glycolysis, gluconeogenesis, activation of amino acids, and fatty acid synthesis. The fluidity of cytosol is maintained by the process of hydration of protein molecules and by the presence of microproteins and microfilaments. Lactate dehydrogenase acts as a marker enzyme of cytoplasm.

QUESTION BANK

SECTION A

1. Define lysosomes.
2. What is myosin?
3. What is tubulin?
4. Define cell wall.
5. Explain RER.
6. What are peroxisomes?
7. What is nucleolus?
8. Define plasma membrane.
9. Define active transport.
10. What are microtubules?
11. What is fluid mosaic model?
12. Define extrinsic protein.
13. Define integral protein.
14. Describe Golgi apparatus.

SECTION B

1. Discuss the functions of the Golgi complex.
2. Explain the structure of microfilaments.
3. Write short notes on (a) symport and (b) antiport.
4. Discuss the structure and functions of endoplasmic reticulum.
5. Explain fluid mosaic model.
6. Explain the functions of cytoskeleton.
7. What are the differences between smooth and rough endoplasmic reticulum?
8. Discuss the components of cytoskeleton.
9. Explain the functions of microtubules.
10. Explain the organisation of fluid mosaic model of membrane.

SECTION C

1. Explain the structure of mitochondria and its functions.
2. Write in detail about transport across the membrane.
3. Describe the structure of an eukaryotic cell with a neat diagram.
4. Describe the morphology, structure, genesis and function of nucleus.
5. Discuss the structure, genesis and functions of endoplasmic reticulum.

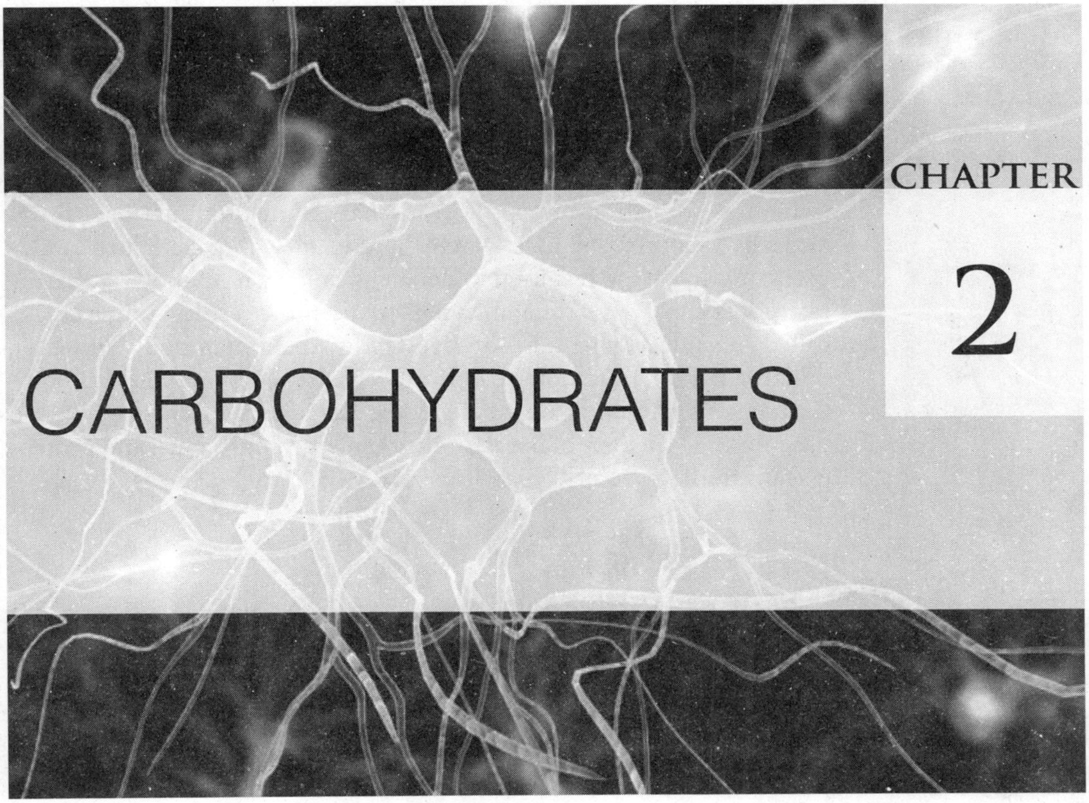

CHAPTER 2

CARBOHYDRATES

Concept Outline

- Introduction
- Definition
- Classification
- Monosaccharides
- Disaccharides
- Oligosaccharides
- Polysaccharides

INTRODUCTION

Carbohydrates are relatively small molecules that are generally called as simple sugars (monosaccharides and disaccharides) as well as macromolecular substances (starch, glycogen, and celluloses). The word sugar emphasises the idea of energy in our minds. From the study of the structure–function relationship of biomolecules, it is apparent that each and every molecule differs structurally and also functionally. Table sugar, which is commonly used as food, is sucrose, whereas the sugar that is mainly present in fruits is fructose. A small change in the structural configuration can lead to changes in the functional properties as well. For example, when there is an α-linkage between glucose molecules, the resulting molecule is called a homopolysaccharide (starch/glycan). When the linkage is β, then the resulting molecule is called a cellulose, which is not digested by the human body. However, only the β-form, and not the α-form, is absorbed by the human body. Carbohydrates are not only considered the source of energy but also used for the synthesis of lipids, amino acids, glycoproteins, and proteoglycans in the system.

DEFINITION

Carbohydrates are defined as polyhydroxy aldehydes or ketones. Carbohydrates are soluble in water and sweet in taste. Carbohydrates are one of the compounds largely present in nature. They perform the functions of storage and energy production and act as one of the membrane constituents.

CLASSIFICATION OF CARBOHYDRATES

Carbohydrates are classified into the following four major groups:

1. Monosaccharides
2. Disaccharides
3. Oligosaccharides
4. Polysaccharides

MONOSACCHARIDES

Monosaccharides are the simplest carbohydrates, which cannot be hydrolysed further. The general molecular formula is $C_nH_{2n}O_n$. They are further subdivided into trioses, tetroses, pentoses, hexoses, and heptoses based on the number of carbon atoms present in them and into aldoses and ketoses based on the aldehyde or ketone group present.

SUBDIVISIONS OF MONOSACCHARIDES

The major subdivisions of monosaccharides are shown in Table 2.1.

Table 2.1 *Subdivisions of Monosaccharides*

Class	Aldose	Ketose
Triose	Glyceraldehyde	Dihydroxyacetone
Tetrose	Erythrose	Erythrulose
Pentose	Ribose	Ribulose
Hexose	Glucose	Fructose
Heptose	Sedoheptose	Sedoheptulose

Monosaccharides are soluble in water and insoluble in organic solvents. They are odourless crystals with a sweet taste. The chemical properties of these sugars are due to the presence of hydroxyl and carbonyl groups. Their sources, biochemical importance, and clinical significance are provided in Table 2.2.

Table 2.2 *Sources, Biochemical Importance, and Clinical Significance of Monosaccharides*

Sugar	Source	Importance	Clinical Significance
D-glucose	It is a chief constituent in starch, glycogen, and cellulose. It is also found in fruits.	Principal source of energy in the body	Present in urine (glycosuria) in patients with diabetes mellitus due to raised blood glucose (hyperglycaemia)
D-fructose	It is found in fruits and honey. It is also the chief source of sucrose and insulin.	Converted to glucose in liver and intestine and is sweeter than glucose	Hereditary fructose intolerance leads to fructose accumulation and hypoglycaemia
D-galactose	It is the chief constituent of lactose (milk sugar).	Converted to glucose in the liver and synthesised in mammary glands to make lactose	Failure to metabolise galactose leads to galactosaemia and cataract
D-mannose	It is found in plant polysaccharides and animal glycoproteins and is obtained by hydrolysis of mannans and gums.	Sugar occurring in glycoproteins	-
D-ribose	It occurs as the main sugar in nucleic acids.	Structural element of nucleic acids and enzymes	-
D-ribulose	It is produced during metabolism.	Intermediates in HMP* shunt	-
D-arabinose	It is present in cherry and gums.	Energy and carbon source in bacteria	-
D-xylose	It is present in wood gums and proteoglycans.	Part of glycoproteins	Found in urine pentosuria

HMP*: Hexose monophosphate

STEREOISOMERS

Stereoisomers are compounds with the same structural formula but different spatial configuration. A carbon atom is said to be asymmetric if it is attached to four different chemical groups. Monosaccharides contain at least one asymmetric carbon. Isomers are formed due to the presence of asymmetric carbon. The number of isomers in a compound depends on the number of asymmetric carbon atoms (n) and is equal to 2^n. For example, glucose has four asymmetric carbon atoms and thus has $2^4 = 16$ isomers.

CONFIGURATION OF D-ALDOSES

The configuration of D-aldoses starting from D-glyceraldehyde is shown in Figure 2.1. This is explained by Kiliani–Fischer synthesis, by increasing the chain length of aldose by one carbon at a time.

EPIMERS

Stereoisomers differing in configuration or arrangement of —OH group on a single asymmetric carbon are known as epimers. For example, mannose and glucose are epimers (Figure 2.2). The position of the —OH group differs only in the C-2 position. Similarly, galactose and glucose are epimers. The position of the —OH group differs only in the C-4 position.

Note: Mannose and galactose are not epimers of each other as they differ in the position of the —OH group at two positions.

ENANTIOMERS

Enantiomers are mirror images of each other. These are otherwise called as D and L sugars; for example, D-fructose and L-fructose are enantiomers. If the OH group to the carbon atom lies on the right-hand side, then it is known as D series, as shown in Figure 2.3.

Most of the sugars in humans are D-sugars, except a few sugars that are of L-forms, such as L-fucose present in glycoproteins and L-iduronic acid found in glycosaminoglycans (GAGs).

OPTICAL ISOMERS (D- AND L-ISOMERS)

Asymmetric carbon atoms confer optical activity to the sugars. When a polarised light passes through a solution showing optical activity, the plane of the light is rotated to the right or the left. If it is rotated towards the right the sugar (compound) is called dextrorotatory (denoted as d or +), and if it is rotated towards the left the sugar (compound) is called laevorotatory (denoted as l or –). If equal amounts of dextrorotatory and laevorotatory isomers are present in the mixture, then there is no optical activity. Such a mixture is known as racemic mixture.

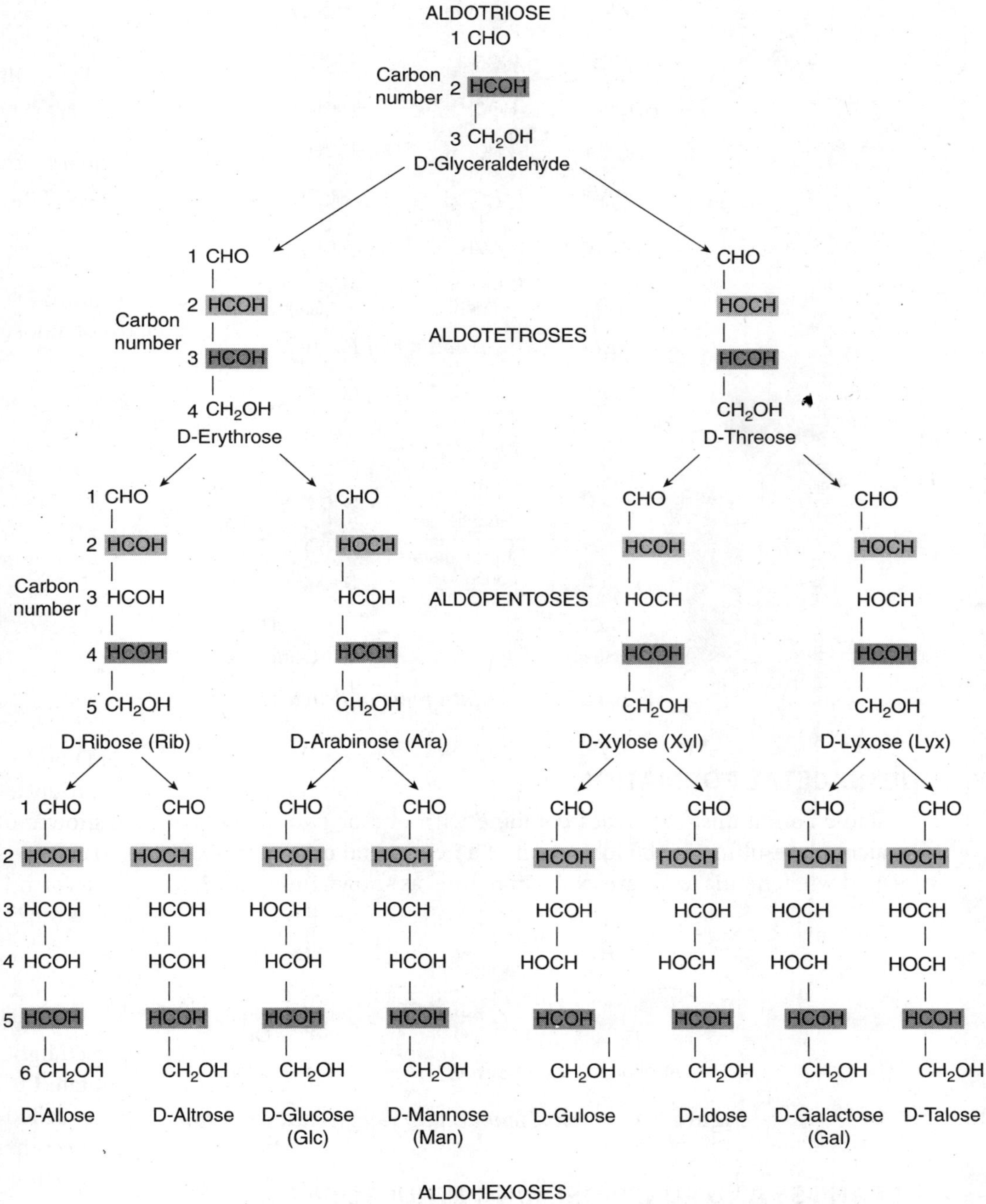

Figure 2.1 *Kiliani–Fischer Synthesis*

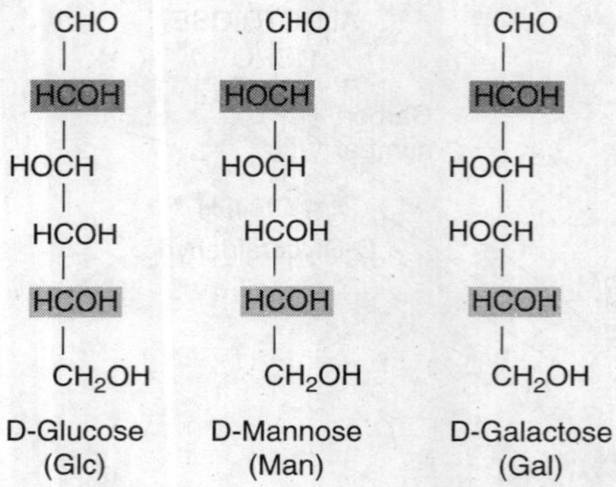

Figure 2.2 *Structures of Epimers*

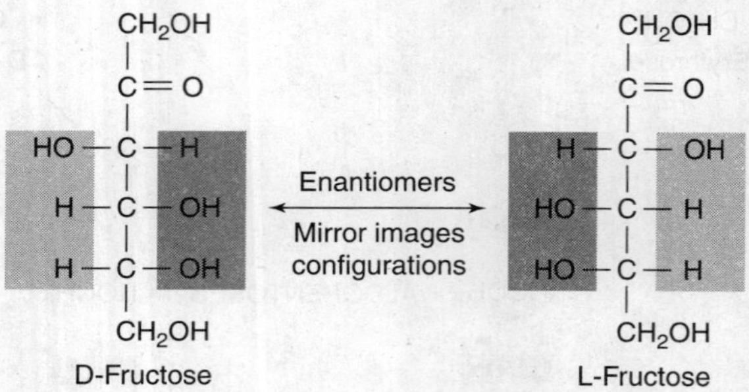

Figure 2.3 *Enantiomers of Fructose*

HEMIACETAL FORMATION

In acidic conditions, aldehydes combine with one or two of the hydroxyl groups of an alcohol, resulting in the formation of a hemiacetal or an acetal. A ring structure is formed when hemiacetals are cyclic in nature, as shown in Figure 2.4.

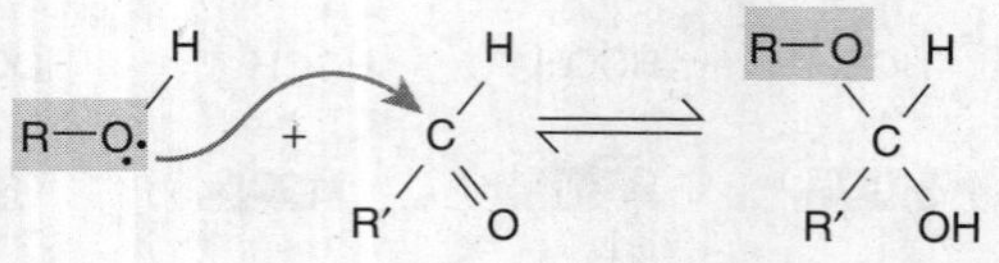

Figure 2.4 *Cyclic Hemiacetals Yield a Ring Structure*

PYRANOSE AND FURANOSE RING STRUCTURES

In 1929, Haworth proposed a six-membered ring structure for sugar called pyranose, because pyran contains a ring of five carbons and one oxygen. Haworth projection formulae are shown in Figure 2.5. Haworth also called sugars containing a five-membered ring as furanose, because furan contains a ring of four carbons and one oxygen. Only 1% of monosaccharides contain five or more carbons in the open chain. In general, monosaccharides in the solution will be in the pyranose and furanose ring forms.

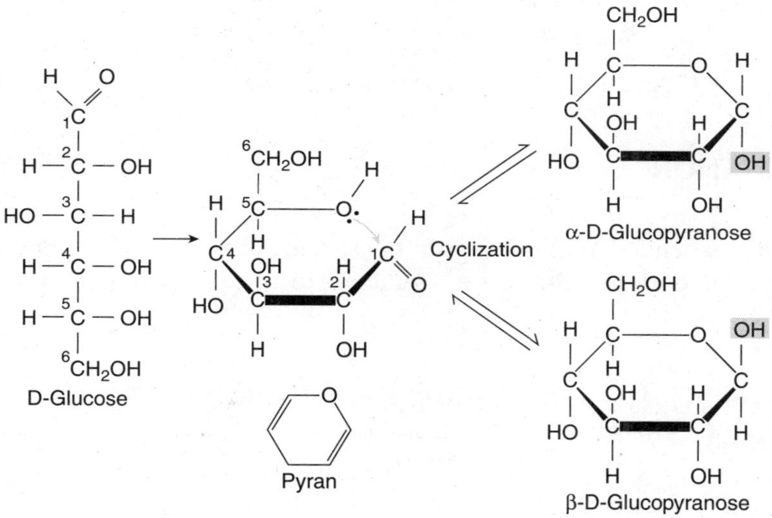

Figure 2.5 *Haworth Projection Formulae*

ANOMERS

Asymmetric carbons (C-1 in the glucose) that are created by cyclisation at the carbon bound to oxygen in hemiacetals formation are anomeric in nature. A ring structure is formed between the carbonyl group and the alcoholic group. This ring formation produces an anomeric carbon at C-1 in aldose and C-2 in ketose sugars.

α- and β-anomers

An α-anomer is formed if in the interacting –CHO group –H is on the left and –OH is on the right; if –OH is on the left and –H is on the right, it is said to be a β-anomer (see Figure 2.6). Enzymes are very specific in their reactions with different configurations. For example, cellulose having a β(1 → 4) linkage cannot be hydrolysed by amylase, which can catalyse only α(1 → 4) linkages. Hemiketal rings are formed by the reaction of the aldehyde and ketone groups with the alcohol group of the same sugar.

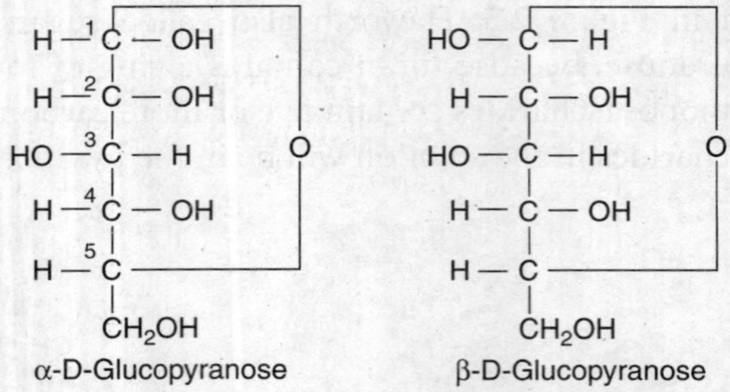

Figure 2.6 *Structure of α- and β-anomers of Glucose*

MUTAROTATION

Definition

Mutarotation is defined as the change in the specific optical rotation representing the inter-conversion of α- and β-forms of D-glucose to an equilibrium mixture.

Example

$$\underset{\alpha\text{-D-glucose}}{[\alpha]_D^{20} = +112.2^\circ} \rightarrow \underset{\text{Mutarational Equimixture}}{[\alpha]_D^{20} = +52.7^\circ} \leftarrow \underset{\beta\text{-D-glucose}}{[\alpha]_D^{20} = +18.7^\circ}$$

Plane Polarised Light

A beam of ordinary light may be regarded as a bundle of electromagnetic waves vibrating in all directions perpendicular to the axis of the beam. When such a beam of light is made to pass through a specially cut crystal of certain minerals or sheet of special plastic polaroid or a crystal of Iceland spar, all vibrations except those in one plane are eliminated. This is called plane polarised light (Figure 2.7).

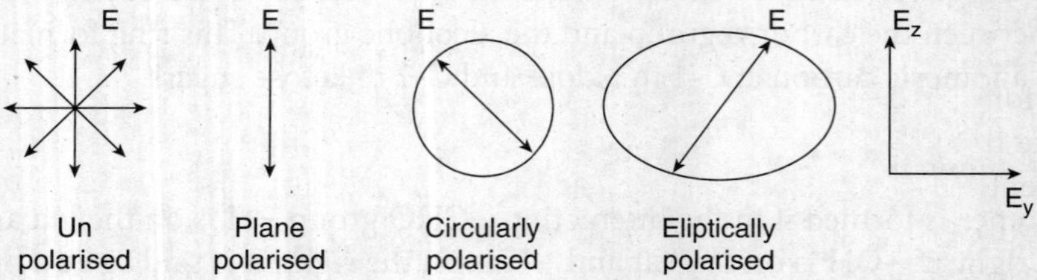

Figure 2.7 *Different Forms of Polarised Light*

When such a beam of plane polarised light is passed through a solution of optical isomer, if the plane polarised light is found to rotate to the left, it is known as laevorotation; if the plane polarised light rotates to an equal number of degrees to the right, it is known as dextrorotation. This phenomenon exhibited by the asymmetric compounds is called optical isomerism.

Expression of optical activity

Optical rotation to the left, that is, laevorotation, is expressed as minus (–) and rotation to the right, that is, dextrorotation, is expressed as plus (+).

Polariscope and Specific Rotation

The actual measurement of the rotation of polarised light involves the use of an instrument called polariscope or polarimeter (Figure 2.8).

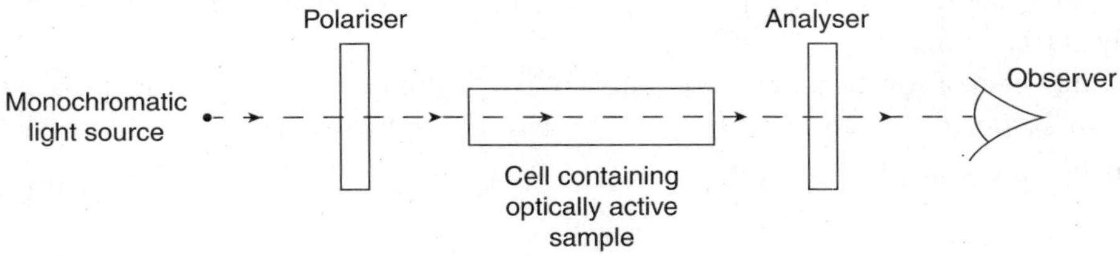

Figure 2.8 *Apparatus for Measurement of Optical Activity*

Because different wavelengths of polarised light are rotated differently, it is necessary to use monochromatic light. Ordinarily, the light from incandescent sodium or mercury is used. Polariscopes often have special filters designed to purify the beam of light as it passes through.

The sodium-D spectral line is composed of two wavelengths D1 (5896 Å) and D2 (5890 Å). The most satisfactory light for polariscope work is the yellow green mercury line (5461 Å), since only one wavelength is involved.

The polarising Nicol prism converts a beam of monochromatic light into plane polarised light. The light passes through a tube, then through the analysing Nichol prism, and finally through the lens. The analysing Nichol prism is mounted in the centre of a large disc with a vernier scale on the edge so that the prism may be rotated about its optical axis and the angle of rotation in degrees and functions may be read from the scale.

Biot formulated practically all the principles of optical rotation. An optically active substance has a definite rotation for a given wavelength. Rotation is proportional to the concentration of the optically active substance and to the thickness of the solution through which the light passes. It also varies with temperature. Some substances show an increase and others decrease with rising temperature. In addition, rotation varies with the nature of the solvent in which the substance is dissolved. Specific rotation can be calculated using the following formula:

$$[\alpha]_D^T = \frac{\alpha_{obs} \times 100}{l \times C}$$

$[\alpha]$ = Specific rotation at temperature T with wavelength sodium lamp D1 = 5896 and D2 = 58900 Mercury lamp

α_{obs} = Observed rotation

I = Length of the polariscope tube in decimetre

C = Concentration of substance in grams/100 ml

Mechanism of Action

Explanation

Mutarotation means changing of optical rotation. Tamet explained that when a solution of freshly prepared crystalline glucose of α-D-form is observed in a polariscope, its initial specific rotation $[\alpha]_D^{20}$ of +121.1° gradually decreases and finally becomes steady at $[\alpha]_D^{20}$ + 52.5°.

If, however, a solution of pure crystals of β-D-glucose is observed in a polariscope, the initial specific rotation of $[\alpha]_D^{20}$ + 19°. For β-D-glucose gradually increases and finally becomes steady at $[\alpha]_D^{20}$ + 52.5°.

$$[\alpha]_D^{20} = +112.2° \longrightarrow [\alpha]_D^{20} = +52.7° \longleftarrow [\alpha]_D^{20} = +18.7°$$

α-D-glucose　　Mutarotational Equimixture　　β-D-Glucose

All reducing sugars (except a few ketoses) undergo mutarotation.

Mechanism of Mutarotation

The phenomenon of mutarotation appears to be due to the changes of α- and β-forms and vice versa via the straight chain aldo to keto form. It is generally accepted that the mechanism of mutarotation involves opening and closing of the hemiacetal ring structure of monosaccharides to form traces of aldehyde and α- to β-ring forms. When equilibrium is attained, the characteristic rotation is reached. All monosaccharides exist in three forms in nature, namely aldehyde form, α-ring form, and β-ring form.

Of the three forms, the aldehyde form is said to extremely unstable and exists only as a transient intermediate. According to Lowry (1925), mutarotation can take place only in the presence of an amphiprotic solvent (i.e. a solvent that can act as both an acid and a base). Water is one such solvent.

BIOCHEMICAL REACTIONS OF MONOSACCHARIDES

Reducing Properties

Sugars that are oxidised at C-1 are known as reducing sugars. Sugars can react with the chemical agents due to the presence of free carbonyl (aldehyde or ketone) groups. In the alkaline medium, they form enediols, which reduce Cu^{2+}, Ag^{+}, $Fe3^{+}$, and Bi^{+3}

and oxidise to the corresponding sugar acids. Such sugars are known as reducing sugars, for example, glucose and fructose. The reducing property of sugars is used for the quantitative and qualitative determination of sugars. Some of the qualitative tests are explained here:

(a) Benedict's test:

Benedict's reagents contain Cu^{2+} (cupric sulphate), alkali (sodium carbonate), and sodium citrate. The role of sodium citrate is to prevent the precipitation of copper hydroxide or carbonate by forming complexes with Cu^{2+} ions. On heating the sugar with Benedict's reagent, the reducing sugar forms enediol in an alkaline medium, which reduces Cu^{2+} to Cu^{+}. This forms a red-coloured precipitate. In Fehling's solution, potassium tartrate is used instead of sodium citrate and KOH is the alkaline medium.

$$\underset{\text{Aldehyde}}{RC(=O)-H} + 2\,Cu^{2+} + 5\,OH^- \longrightarrow \underset{\text{Carboxylate}}{RC(=O)-O^-} + Cu_2O\downarrow + 3H_2O$$

(b) Barfoed's test:

Barfoed's reagents contain a solution of copper acetate in acetic acid. Monosaccharides react with Barfoed's reagent and give a red precipitate. Monosaccharides react faster than disaccharides and this test helps to differentiate between them.

Action of Strong Acids

When sugars react with strong mineral acids, they undergo dehydration and form furfurals. Furfurals condense with phenolic compounds such as α-napthol, thymol, and resorcinol to produce coloured complexes. This property of the sugars helps in the identification of sugars.

(a) Molisch's test:

$$\text{Sugar solution} + H_2SO_4 \xrightarrow{\alpha\text{-napthol}} \text{Red violet ring at the junction of two layers.}$$

(b) Seliwanoff's test:

$$\text{Fructose} \xrightarrow[\text{Resorcinol}]{\text{HCl}} \text{Cherry red colour}$$

Note: This test is used for distinguishing between glucose and fructose. If boiled for a long time, glucose will also give positive results.

(c) Bial's test:

$$\text{Ribose} \xrightarrow[\text{FeCl}_3]{\text{HCl Orcinol}} \text{Green colour (blue/brown on standing)}$$

Note: This test is used for the detection of pentoses.

Oxidation

Aldoses are oxidised to different compounds based upon the reactants. The following are a few examples:

1. D-glucose is oxidised to aldonic acid called gluconic acid. In the same way, mannose and galactose yield mannoic and galactonic acids, respectively, with bromine water.
2. Strong oxidising agents such as concentrated HNO_3 can convert aldehyde as well as primary alcoholic groups to glucosaccharic acid, and galactose produces galactosaccharic and mucic acids.
3. Under specific conditions, sugar is oxidised to uronic acid. Glucose upon oxidisation produces glucuronic acid.

COOH	CHO	COOH
|	|	|
CH_2OH	COOH	COOH
Adonic acid	Uronic acid	Saccharic acid

Glucoses produce glucuronic acid, whereas galactose and mannose yield galacturonic acid and mannuronic acid, respectively.

Reduction

Monosaccharides are reduced to their corresponding alcohols by reducing agents such as sodium amalgam or by hydrogen under high pressure in the presence of a catalyst. For example, aldoses yield the following alcohols:

$$\text{D-glucose} \xrightarrow{H_2} \text{D-sorbitol}$$

$$\text{D-galactose} \xrightarrow{H_2} \text{D-dulucitol}$$

$$\text{D-mannose} \xrightarrow{H_2} \text{D-mannitol}$$

Similarly, ketoses also yield two alcohols due to the formation of a new asymmetric carbon in the C-2 position.

$$\text{Fructose} \xrightarrow{H_2} \text{D-mannitol} + \text{D-sorbitol}$$

The following are the important reactions:

- Reduction of glycerose to glycerol
- Reduction of ribose to ribitol, which forms a part of riboflavin
- Reduction of glucose to ascorbic acid (vitamin C)

Osazone formation

Osazone reaction is possible only when a free carbonyl (aldehyde or ketone) group is present.

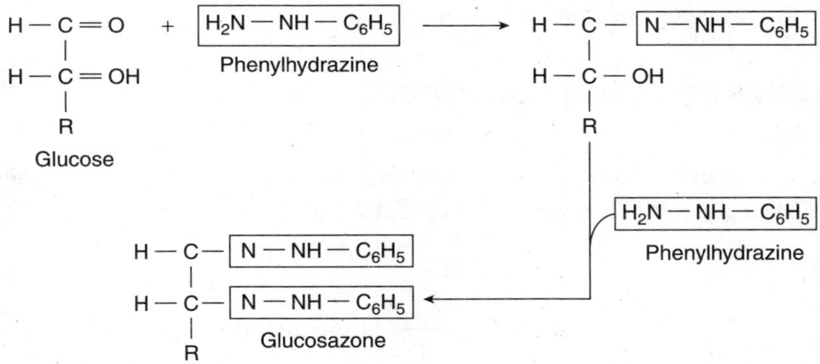

A monosaccharide or reducing disaccharide is first mixed and heated with phenylhydrazine in acetic acid to form sugar-phenylhydrazine, which then reacts with more reagents to form particular osazone-forming characteristic crystals of specific shapes. This test is very important to distinguish and identify the different reducing sugars. Glucose, fructose, and mannose form the same osazones, that is, shape (they form the same needle-shaped crystals). Maltose forms sulphur-shaped crystals and lactose form a ball-shaped crystals.

Formation of glycosides

Glycosides are produced by the interaction of monosaccharide residue with the hydroxyl group of a second compound, which may be a monosaccharide or some other compound. The non-carbohydrate compound is called an aglycone. The glycosidic bond is formed as an acetal link between a hemiacetal and a —OH group.

$$\text{Sugar—H} + \text{R—OH} \longrightarrow \text{Sugar—R (Glycoside)} + H_2O$$

For example, if the sugar is glucose, then it is called glucoside; if it is fructose, it is called fructoside.

1. If the side chain (R) is another sugar, the glycoside becomes a disaccharide.

$$\text{Sugar}_1 - \text{H} + \text{OH} - \text{Sugar}_2 \longrightarrow \text{Sugar}_1 - \text{Sugar}_2 + H_2O$$

2. If the side chain is of sterol group, then it forms the basis for drug formation.

Applied aspects

Glycosides are present in many drugs and spices. They are also the chief constituents of animal tissues. Aglycones are methanol, glycerol, sterol, or phenol. Glycosides play a very important role in medicine because of their action in heart (cardiac glycosides). Examples are the derivatives of digitals and ouabain. Ouabain is a very good inhibitor of Na^+—K^+—ATPase of plasma membranes. Examples of other glycosides are antibiotics such as streptomycin. Phlorizin is a glycoside that inhibits the transport of sugars across the mucosal cells of small intestine.

DERIVATIVES OF MONOSACCHARIDES

Uronic Acids

Uronic acid is formed when glucose is oxidised such that its primary alcohol group is converted to COOH group, without the oxidation of the aldehyde group.

Examples

$$\text{D-glucose} \longrightarrow \text{D-glucuronic acid}$$

$$\text{D-galactose} \longrightarrow \text{D-galacturonic acid}$$

In the body, glucuronic acid is formed from glucose in the uronic acid pathway in the liver.

Applied Aspects

D-glucuronic acid is present as the chief constituent of certain mucopolysaccharides. This is involved in the detoxication mechanism, where bilirubin is linked with D-glucuronic acid.

Deoxy Sugars

Deoxy sugars are the sugars in which the oxygen in the —OH group has been removed leaving the hydrogen. Thus, the —CHOH or —CH_2OH becomes —CH_2 or —CH_3. Deoxy sugars are 2-deoxy-D-ribose found in nucleic acid (DNA) and 6-deoxy-L-galactose found as a constituent of bacterial polysaccharides.

Amino Sugars

Amino sugars are formed when the hydroxyl group is replaced by an —NH_2 group. Examples are glucosamine and galactosamine. Amino sugars occur as N-acetyls in numerous biological important compounds. N-acetyl derivative of D-glucosamine is present in hyaluronic acid, heparin, and some blood group compounds. Galactosamine is present as N-acetylgalactosamine in chondroitin sulphate. Antibiotics such as erythromycin and caromycin also contain amino sugars.

Esters of Sugars

Esterification of carbohydrates with phosphoric acid usually happens in metabolism. For example, glucose-6-phosphate and glucose-1-phosphate are formed during metabolism. ATP donates the phosphate group for esterification.

DISACCHARIDES

Disaccharides are carbohydrates that consist of two molecules of the same or different monosaccharides. During hydrolysis, a disaccharide yields two moles of monosaccharides. The mode of linkage between these monosaccharides is the glycosidic linkage. The general molecular formula is $C_nH_{2n}O_{n-1}$.

Examples

Maltose ⟶ Glucose + glucose (reducing sugar)

Lactose ⟶ Galactose + glucose (reducing sugar)

Sucrose ⟶ Glucose + fructose (non-reducing sugar)

MALTOSE

Occurrence

It is an important component of malt. Maltose does not occur in the human body. It is produced when starch is hydrolysed by α-amylase.

Chemistry

It is composed of two molecules of glucose linked by α(1 → 4) linkage as shown in Figure 2.9. In the structure of maltose, the hydroxyl group on C-1 of the second sugar remains free, so it acts as a reducing disaccharide.

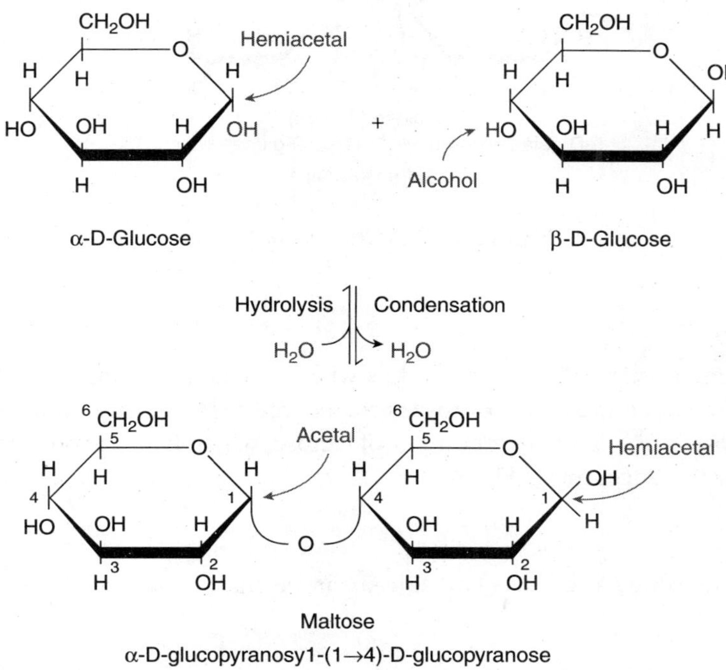

Figure 2.9 *Structure of Maltose*

Identification

It forms characteristic sunflower-shaped crystals with osazone reaction. The anomeric carbon of one glucose is free in maltose, and hence it possesses all the reducing properties of a reducing sugar. Isomaltose is also a reducing disaccharide. It consists of two molecules of glucose units. The type of linkage is $\alpha(1 \rightarrow 4)$ glycosidic linkage. It is formed by the partial hydrolysis of glycogen and starch.

LACTOSE

Occurrence

Lactose is also known as milk sugar and is usually found in milk, occurring as β-lactose. This is produced in mammary glands.

Chemistry

It is composed of two monosaccharides, namely galactose and glucose molecules, linked by $\beta(1 \rightarrow 4)$ glycosidic linkage. The glycosidic linkage is between C-1 of β-galactose and C-4 of α-glucose (Figure 2.10). The anomeric carbon of glucose is free and hence lactose is considered to be a reducing disaccharide.

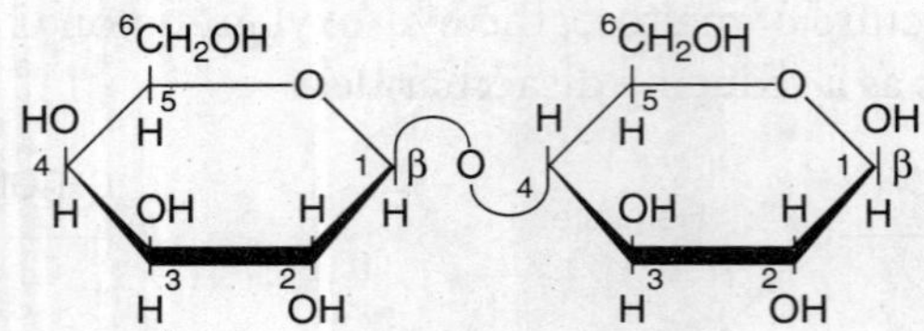

Lactose (β form)
β-D-galactopyranosyl-(1→4)-β-D-glucopyranose
Gal(β1→4)Glc

Figure 2.10 *Structure of Lactose*

Identification

It forms characteristic ball-shaped crystals with osazone reaction. The anomeric carbon of glucose is free and hence lactose is considered to be positive for all the reducing tests. Galactose is one of the components of lactose, which hence produces mucic acid on treating with concentrated HNO_3.

SUCROSE

Sucrose is also named as cane sugar, table sugar, saccharose, or beet sugar.

Occurrence

Sucrose is distributed in plants, particularly in sugarcane and sugar beet.

Chemistry

It is composed of two molecules of monosaccharides, namely glucose and fructose. As both the functional groups (aldehyde and ketone groups) are involved in the formation of glycosidic linkage, there is no free reducing group. Hence, sucrose is known as a non-reducing sugar. The type of linkage in sucrose is α, β(1 → 2) glycosidic between C-1 of α-glucose and C-2 of β-fructose (Figure 2.11).

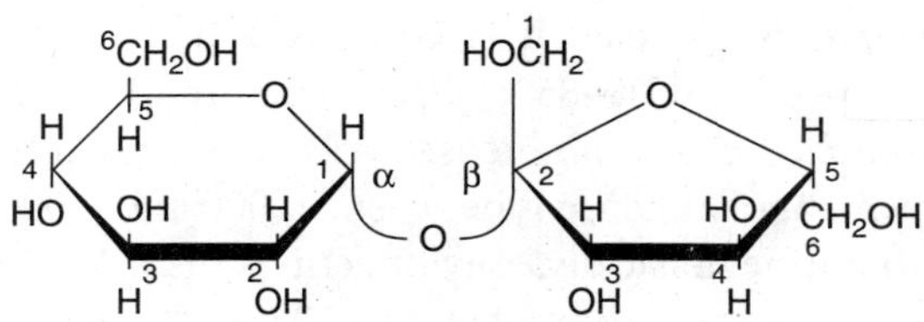

Sucrose
α-D-glucopyranosyl β-D-fructofuranoside
Glc(α1↔2β)Fru

Figure 2.11 *Structure of Sucrose*

Identification

As sucrose is a reducing sugar, it shows negative results to reducing tests. Sucrose is dextrorotatory (+66.5°) and its hydrolytic products are glucose and fructose. The rotation of fructose is laevorotatory (–92°), whereas that of glucose is dextrorotatory (+52.7°). The resulting product of hydrolysis shows laevorotation. This change from dextrorotation to laevorotation is known as inversion. Hence, sucrose is also called as invert sugar. The net result is laevorotatory, resulting in change from +ve to –ve.

$$\underset{+66.5^\circ}{\text{Sucrose}} \longrightarrow \underset{+52.5^\circ}{\beta\text{-D-glucopyranose}} + \underset{-92^\circ}{\beta\text{-D-fructofuranose}}$$

OLIGOSACCHARIDES

Oligosaccharides are carbohydrates that yield two to ten monosaccharide units on hydrolysis. An example is maltotriose. They also serve as parts of blood group antigens.

POLYSACCHARIDES

Polysaccharides produce more than twelve molecules of monosaccharide units on hydrolysis. The molecular formula is $(C_6H_{10}O_5)_n$. Polysaccharides are classified into two types:

1. *Homopolysaccharides:* Polymers with similar monosaccharide units
2. *Heteropolysaccharides:* Polymers of different monosaccharide units or their derivatives, for example, mucopolysaccharides

HOMOPOLYSACCHARIDES

Starch

Occurrence

Starch is a polysaccharide that is largely present in the plant kingdom. The most common sources are cereals, grains, potatoes, carrots, millets, and legumes.

Chemistry

The basic unit of starch is glucose. It is composed of two units, namely amylase (15%–35%) and amylopectin (65%–85%). The structure of amylose looks like a long helically coiled unbranched chain of glucose molecules connected by α(1 → 4) linkages. The molecular weight of amylase is between 10,000 and 1,00,000. Amylose gives a deep blue colour with iodine. Basically, amylopectin is insoluble in water. The molecular weight of amylopectin varies from 50,000 to 10,00,000. Amylopectin is composed of branched chain of glucose and joined by α(1 → 4) linkages (Figure 2.12). Branching occurs through α(1 → 6) linkages between two glucose units of neighbouring chains. The branching point occurs at proper intervals, approximately 25–30 α-D-glucose units. Amylopectin reacts with iodine and gives a violet colour.

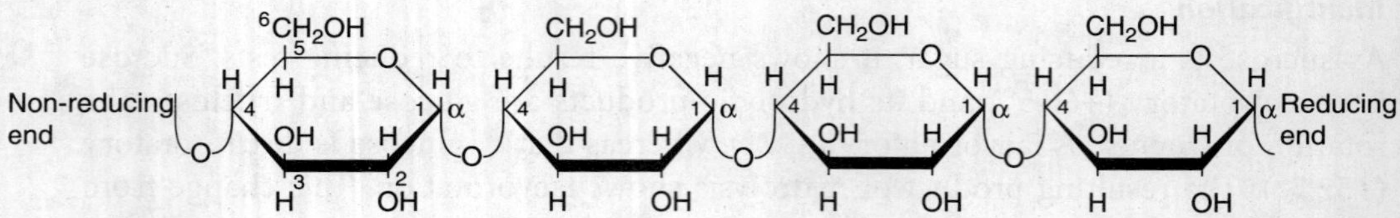

(a) *Amylose*

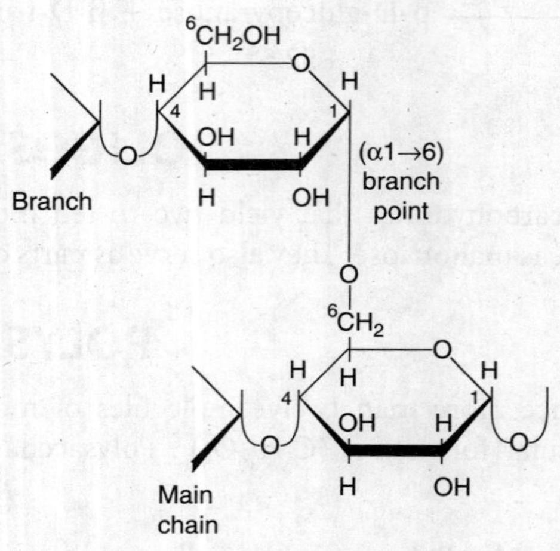

(b) *Amylopectin*

Figure 2.12 *Structural Representation of Amylose and Amylopectin (Starch)*

Glycogen

Occurrence

Glycogen is the storage form of polysaccharide present in animals (muscle and liver). Glycogen in the liver is catabolised to glucose and transferred via blood stream to the tissue for the energy purpose. Glycogen in the muscle acts as a good source of energy.

Chemistry

Glycogen has a branched structure as well. It is a polymer consisting of D-glucose units. The glucose units in the linear structure are joined by $\alpha(1 \rightarrow 4)$ glycosidic linkages and branching occurs via $\alpha(1 \rightarrow 6)$ linkages (Figure 2.13). The branching point occurs at every eight to ten α-D-glucose units. This type of linkage is widely used in determining it quantitatively in the tissues.

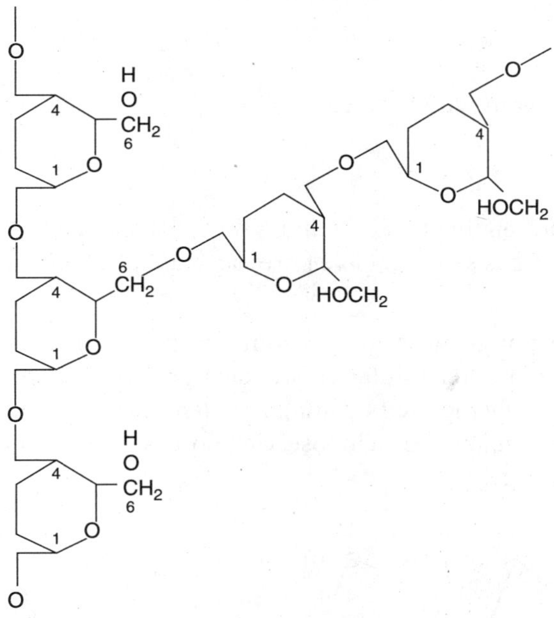

Figure 2.13 *Structural Representation of Glycogen*

Identification

Glycogen shows a deep red colour with iodine.

Dextrins

Dextrins are present in all starch-producing plants. They are also present in honey and are the important constituents of various foods. They are used as adhesives and binders.

Chemistry

Dextrins are produced when starch is partially hydrolysed. This is carried out by acid or enzyme amylase. Dextrins produced from amylose have unbranched chains, whereas those produced from amylopectin have branched chains (refer Figure 2.14). Dextrins are soluble in water and get precipitated by alcohol. They have a sweet taste.

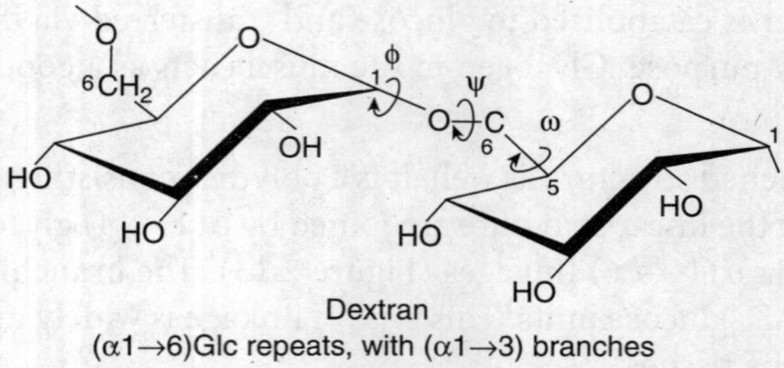

Figure 2.14 *Structure of Dextrin*

Identification

Dextrins give a red colour with iodine.

Cellulose

Occurrence

Cellulose is largely present in plant cells and is the most abundant carbohydrate present in nature. It forms the basis of vegetable fibres such as cotton and jute.

Chemistry

It is composed of repeating units of glucose joined by $\beta(1 \rightarrow 4)$ linkages (see Figure 2.15) Cellulose is insoluble in water, dilute acids, alkalis, and organic solvents.

Cellulose is not easily digested by human system due to the lack of enzyme cellulose, which attacks the linkages in cellulose. Cellulose is as an important part of human diet as it acts as roughage.

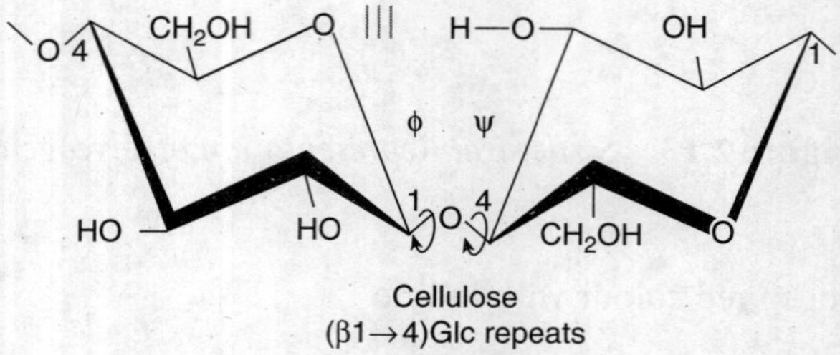

Figure 2.15 *Structure of Cellulose*

Inulin

Occurrence

Inulin is the main constituent of dahlias, onions, and garlics.

Chemistry

Inulin is composed up of D-fructose. It can be catabolised by the enzyme inulinase. It does not show any colour when treated with iodine. Inulin does not have any dietary importance.

Applied aspects

Inulin is used for the determination of glomerular filtration rate (GFR). It is also used to measure body water volume.

Agar

Occurrence

Agar is generally present in sea weeds.

Chemistry

It consists of galactose, which is sulphated as represented in Figure 2.16. It is applied as a laxative in constipation. Purified agar is used as a bacterial culture medium.

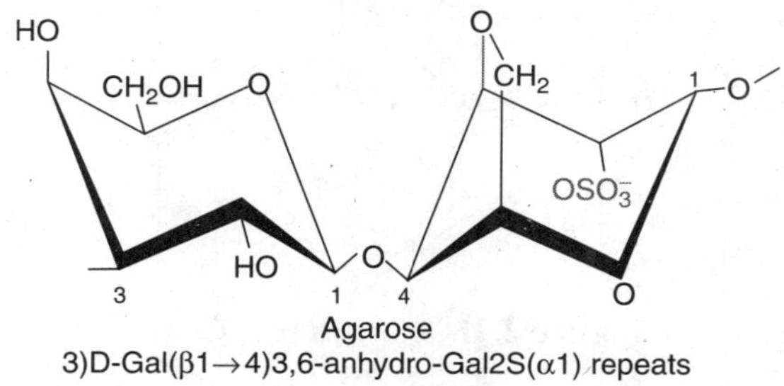

Figure 2.16 *Structure of Agarose*

Dextrans

Dextran is a polysaccharide composed of glucose units. The only difference between dextrans and dextrins is in the structure. Dextrans are made up of a number of D-glucose molecules, having $\alpha(1 \rightarrow 6)$, $\alpha(1 \rightarrow 4)$, and $\alpha(1 \rightarrow 3)$ glycosidic linkages (Figure 2.17).

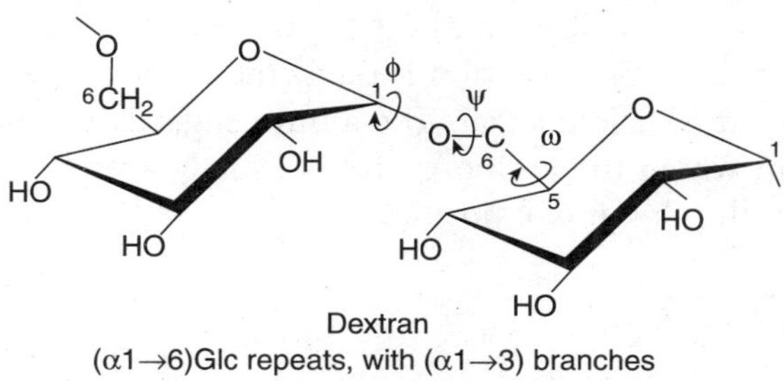

Figure 2.17 *Structure of Dextran*

Applied aspects

Dextran has a molecular weight of approximately 75,000. It is used as a plasma extender when taken intravenously. It increases the blood volume when there is a loss of blood.

Chitin

Occurrence

Chitin is present in the exoskeleton of invertebrates such as crabs, lobsters, and insects.

Chemistry

Chitin is made up of N-acetyl D-glucosamine linked by $\alpha(1 \rightarrow 4)$ glycosidic linkages (Figure 2.18). The structure is very similar to cellulose, except that an acetylated amino group is present in the C-2 position.

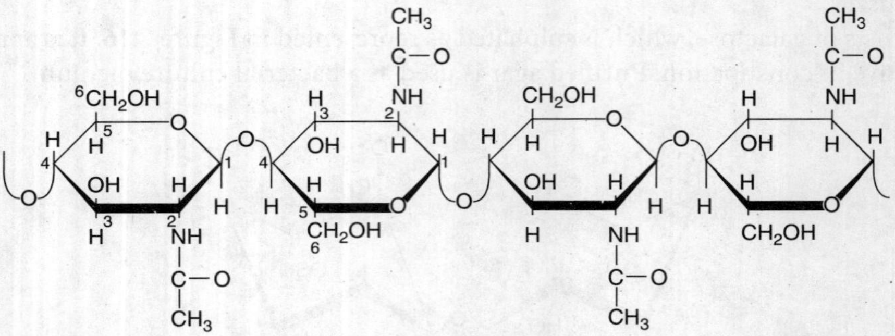

Figure 2.18 *Structure of Chitin*

HETEROPOLYSACCHARIDES

Heteropolysaccharide is a kind of polysaccharide. Heteropolysaccharides containing the repeating units of amino sugars and uronic acid are known as mucopolysaccharides or GAGs. Glycoproteins or mucoproteins are proteins with carbohydrate units made up of a short (2–10 sugar residues) chain. Proteoglycans is a heteropolysaccharide connected to a polypeptide. Glycolipids are lipids containing a carbohydrate group.

Mucopolysaccharides

Mucopolysaccharides are also known as GAGs. Uronic acid is either gluconic acid or its epimer L-iduronic acid; an exception is the keratin sulphate where sulphate galactose is present instead of acidic sugar. The amino sugar is either D-glucosamine or D-galactosamine, in which the amino group is usually acetylated. The amino sugar may be sulphated in the C-4 or C-6 position.

Hyaluronic Acid

Occurrence

Hyaluronic acid is a viscous substance that is present as a ground substance in connective tissues. It is also found in skin, synovial fluid, vitreous humour of the eyes, and umbilical cord. Hyaluronic acid mainly functions as a lubricant and shock absorber. It is the only GAG that is not attached to protein and also not sulphated. It is mostly present in bacteria.

Chemistry

Hyaluronic acid is made up of repeating units of D-glucuronic acid and N-acetyl glucosamine (Figure 2.19). When it is subjected to hydrolysis, it forms D-glucosamine, D-glucuronic acid, and acetic acid. It is also hydrolysed by the enzyme hyaluronidase present in spleen and testicular tissues. The main action of this enzyme is to reduce the viscosity of hyaluronic acid, which as a result rapidly spreads in the tissues. Hence, hyaluronidase is also known as a spreading agent.

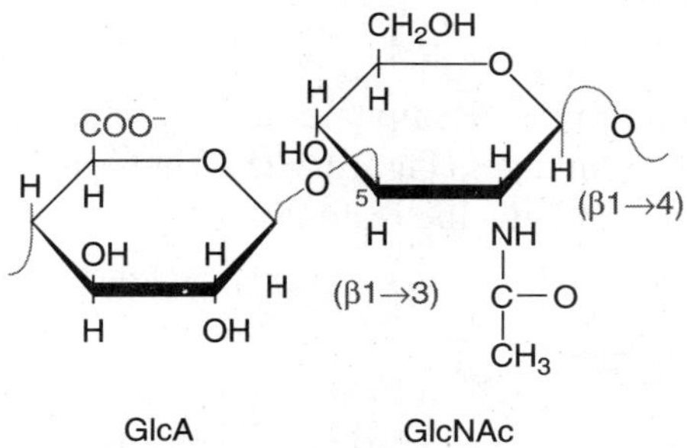

Figure 2.19 *Structure of Hyaluronic Acid*

Chondroitin Sulphate

Occurrence

Chondroitin sulphate is present in cartilage, bone, heart valves, tendons, and cornea.

Chemistry

The structure of chondroitin is very similar to that of hyaluronic acid, as shown in Figure 2.20. The main difference in these structures N-acetylgalactosamine 6-sulphate is present in place of N-acetyl glucosamine.

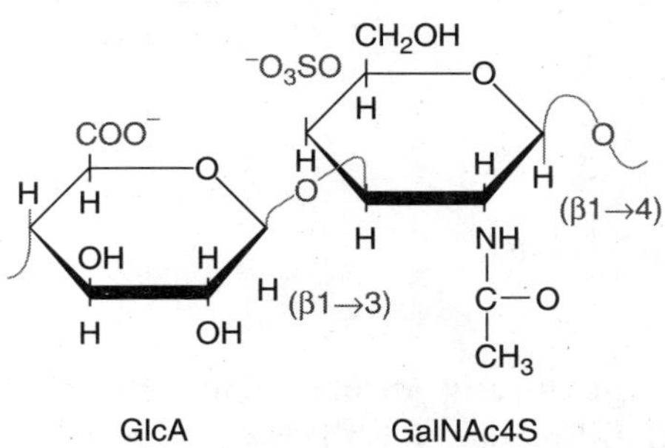

Figure 2.20 *Structural Representation of Chondroitin Sulphate*

Two types of chondroitin sulphate are present, namely A and B. In type A, the sulphate group is present in position 4, whereas in type B it has L-iduronic acid instead of D-glucuronic acid. Chondroitin sulphates are the most abundant GAGs. They are proteoglycans made up of proteins and GAGs. They are present in cartilage, where they bind collagen and hold the fibres tightly.

Heparin

Occurrence

Heparin is present in combination with proteins as proteoglycan. It is one of the abundant anticoagulants present in the liver. It is also present in lungs, spleen, kidneys, and intestinal mucosa.

Chemistry

Heparin consists of glucosamine N-sulphate and sulphate ester of glucuronic acid joined together by $\alpha(1 \rightarrow 4)$ linkages (Figure 2.21). Heparin is unique in structure and is made up of mainly serine and glycine residues.

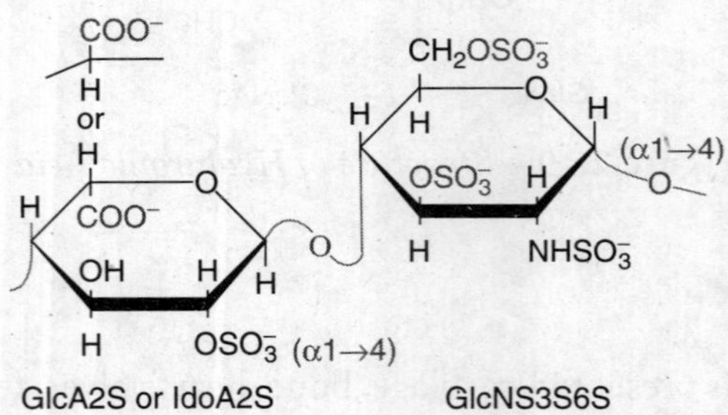

Figure 2.21 *Structural Representation of Heparin*

Applied aspects

Heparin is generally used to prevent blood clotting by inhibiting the conversion of prothrombin to thrombin.

Keratan Sulphate

Occurrence

It is present in cornea and tendons.

Chemistry

This GAG does not contain any uronic acid. The polymer present here is N-acetylglucosamine with β-linkages. It is the most common heterogeneous GAG. The structure of keratan sulphate is shown in Figure 2.22.

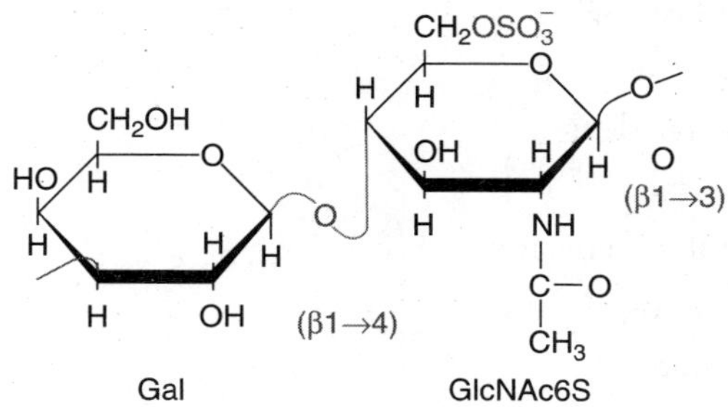

Figure 2.22 *Structural Representation of Keratan Sulphate*

Blood Group Substances

Four types of monosaccharides are found in all types of blood groups. They are galactose, fructose, acetylgalactosamine, and acetylglucosamine. Blood group specificity A, B, and H are associated with the non-reducing end groups of acetylglucosamine, galactose, and fructose, respectively.

Glycosaminoglycans as Ground Substance

Glycosaminoglycans belong to negatively charged heteropolysaccharides combined with a small amount of proteins. GAGs usually absorb large quantities of water and produce a gel-like matrix that forms the basis of a body's ground substances. The viscous and lubricating properties of mucous secretion are caused by the GAGs. Hence, they are also known as mucopolysaccharides. GAGs also help to maintain the water and salt balance of the body. They also contribute to the slippery nature of mucous secretion, synovial fluid, and vitreous humour of eyes. The negative charge makes the GAG chains to spread in solution and repel each other. They are surrounded by water; when brought together, the molecules slip over each other. GAGs can also be compressed and hence the water can be squeezed out; this helps to occupy less volume. They can also be hydrated and hence occupy a larger volume. This is attributed to the resilience of the synovial fluid and vitreous humour of the eyes. GAGs are cleaved by lysosomes. A huge number of hydrolytic enzymes are needed for this purpose.

Applied aspects

Mucopolysaccharidosis is a hereditary disorder due to the accumulation of GAGs in various tissues. Examples are Hunter's syndrome and Hurler's syndrome.

QUESTION BANK

SECTION A

1. Glucose and fructose exhibit similar osazone structures. Justify this statement.
2. Explain inversion of sucrose.

3. What are deoxysugars?
4. Explain Benedict's test.
5. What are disaccharides?
6. Define starch.
7. Explain mucopolysaccharides.
8. Define non-reducing sugars.
9. Define enantiomers.
10. Define heteropolysaccharides.
11. Explain the functions of carbohydrates.
12. What is steroisomerism?
13. Explain asymmetric carbon.

SECTION B

1. Differentiate animal starch from plant starch.
2. Give an account of bacterial cell wall polysaccharides.
3. What is optical isomerism? Describe with examples.
4. Write notes on hte reducing properties of glucose.
5. Describe the structure and properties of starch.
6. Mention five characteristics reactions of an aldose hexose that occurs in our body.
7. Give the structural formula of (a) lactose and (b) fructose.
8. What are optical isomers? Explain with examples.
9. Give an account of the structural configuration of monosaccharides, with special reference to glucose.
10. Discuss the structure and functions of three biochemically important disaccharides.
11. Explain in detail about the pyranose and furanose structures of carbohydrates with an example.

SECTION C

1. Distinguish the chemistry and properties of starch and cellulose.
2. Explain in detail the physical and chemical properties of carbohydrates.
3. How are carbohydrates classified? Discuss the occurrence, structure, and biological importance of cellulose and glycogen.

4. Explain the structure and biological function of any three heteropolysaccharides.
5. Explain the mutarotation of carbohydrates.
6. Describe in brief the structure and important properties of glycogen.
7. Describe in brief the structure and important properties of starch.
8. Describe in brief the structure and important properties of cellulose.
9. Describe the structure and functions of mucopolysaccharides.
10. Define polysaccharides and describe the structure of three homopolysaccharides.

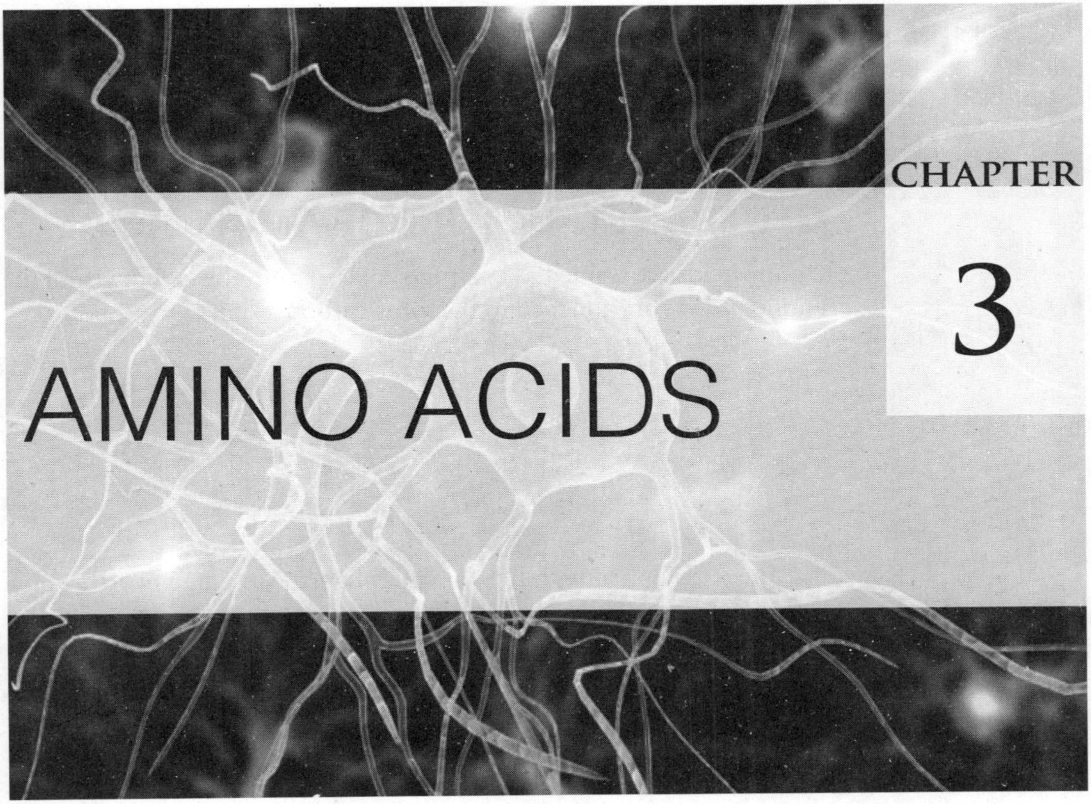

CHAPTER 3

AMINO ACIDS

Concept Outline

- Introduction
- General Structure of Amino Acids
- Classification of Amino Acids
- Peptide
- Classification of Protein
- Protein Configuration
- Denaturation
- Biuret Reaction
- Sequenator
- Quantitative Analysis
- Protein Estimation
- Isolation of Proteins
- Extraction of Proteins from Tissues
- Preparation of Homogenate
- Purification of Protein

INTRODUCTION

Amino acids are a group of organic compounds containing two functional groups – amino and carboxyl. The amino group (NH_2) is the basic, while the carboxyl group (—COOH) is acidic in nature. Amino acids are the simplest unit of protein molecule, and they form the building blocks of protein structure.

GENERAL STRUCTURE OF AMINO ACIDS

Amino acids are termed as α-amino acids, if both the carboxyl and the amino groups are attached to the same carbon atom.

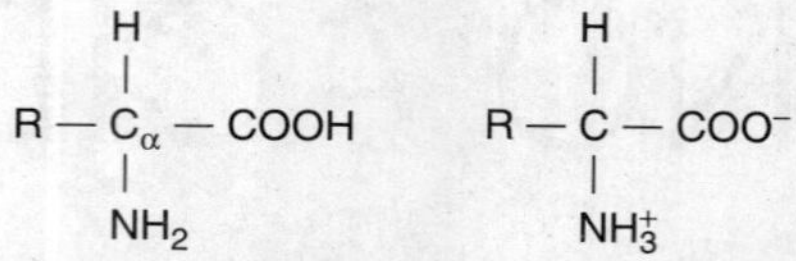

General structure   Exists as ion

The α-carbon atom binds to a side chain represented by R–, which is different for each amino acid found in proteins. Amino acids exist mostly in ionised form in the biological system.

CLASSIFICATION OF AMINO ACIDS

There are different ways to classify the amino acids based on structure, chemical nature, nutritional requirement, metabolic fate, and so on.

CLASSIFICATION OF AMINO ACIDS BASED ON STRUCTURE

A comprehensive classification of amino acids is based on their structure and chemical nature. Each amino acid is assigned a three-letter or one-letter symbol. The twenty amino acids found in protein are divided into seven distinct groups. Example: Glycine.

Glycine has the smallest functional group (hydrogen) of any of the α amino acids

One Letter Code	Three Letter Code	Molecular Wt.(Daltons)	Genetic Code Codons
G	GLY	57.06	GGU, GGC, GGA, GGG

1. On the basis of composition of side chain.
2. On the basis of number of amino and carboxylic groups.
3. On the basis of polarity of the side chain.

4. On the basis of metabolic fate.
5. On the basis of nutritional requirement.

On the Basis of Composition of Side Chain

Fairley and Kilgour, in 1966, grouped twenty amino acids into eight categories.

Simple amino acids: No functional group in their side chain.
Example: Gly, Ala, Val, Leu, and Ileu

Glycine	Gly [G]	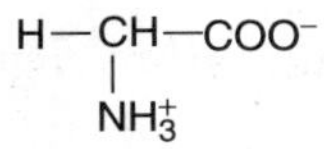
Alanine	Ala [A]	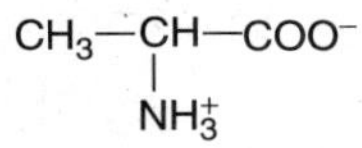
Valine	Val [V]	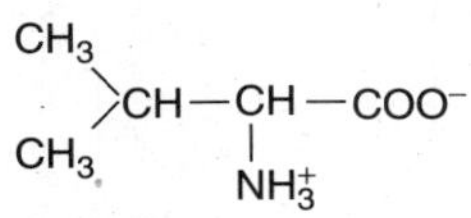
Leucine	Leu [L]	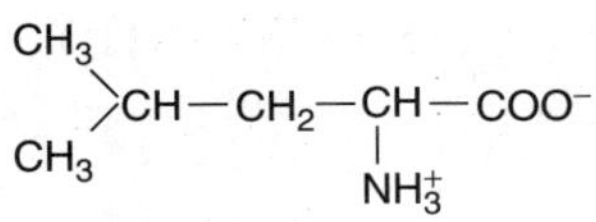
Isoleucine	Ile [I]	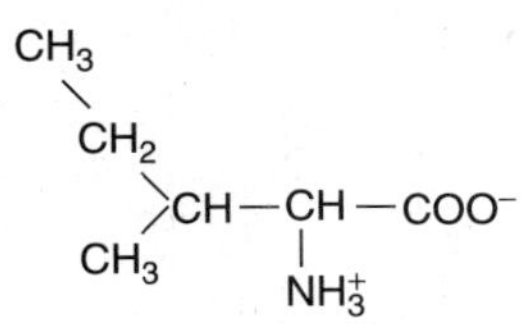

Hydroxy amino acids: Contain hydroxy group.
Example: Ser and Thre

Serine	Ser [S]	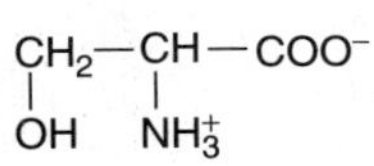
Theronine	Thr [T]	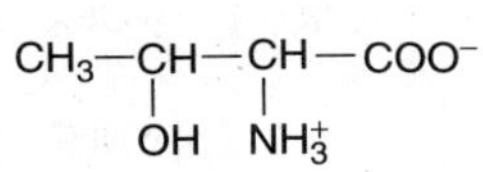

Sulphur amino acids: Contain sulphur group.
Example: Cys and Met

Cysteine	Cys [C]	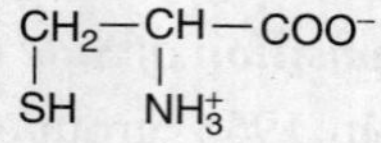
Methionine	Met [M]	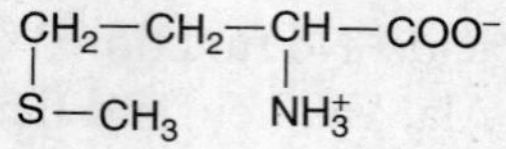

Acidic amino acids: Contain carboxyl group.
Example: Asp and Glu

Aspartic Acid	Asp [D]	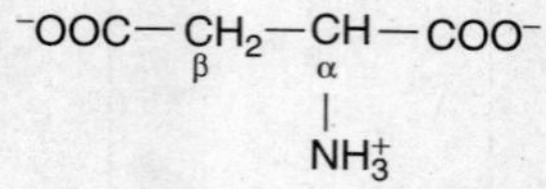
Glutamic Acid	Glu [E]	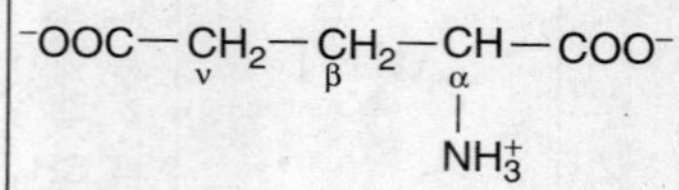

Amino acid amides: Derivatives of acidic amino acid whose one COO^- group transformed into an amide group ($-CO-NH_2^-$).
Example: Asn and Gln

Asparagine	Asn [N]	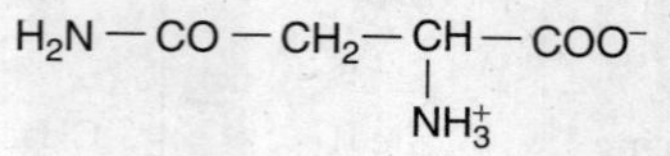
Glutamine	Gln [Q]	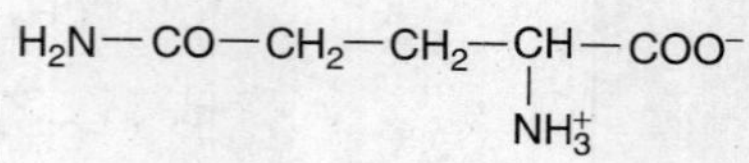

Basic amino acids: Contain amino group.
Example: Lys and Arg

Arginine	Arg [R]	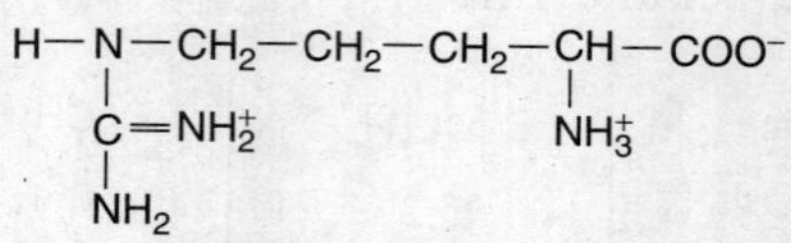
Lysine	Lys [K]	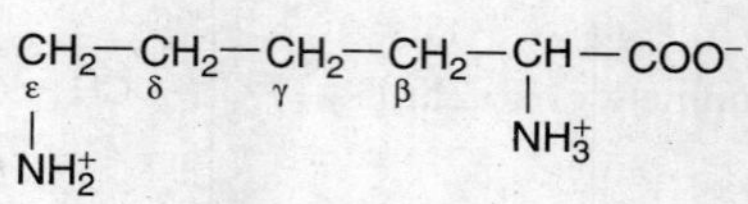

Heterocyclic amino acids: Have a side chain ring, which possesses one atom other than carbon.

Example: Trp, His, and Pro

Histidine	His [H]	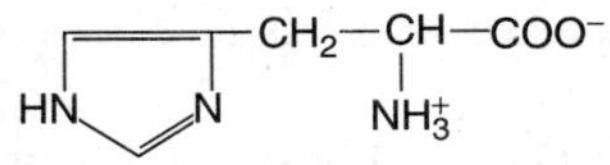
Trytophan	Trp [W]	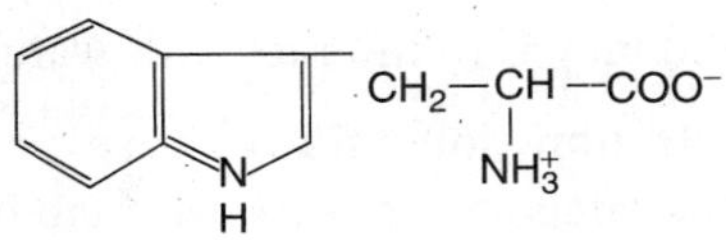
Proline	Pro [P]	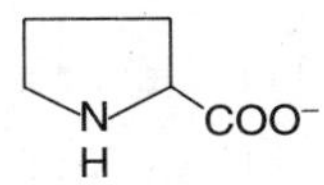

Aromatic amino acids: Have benzene ring

Example: Phe and Tyr

Phenylalanine	Phe [F]	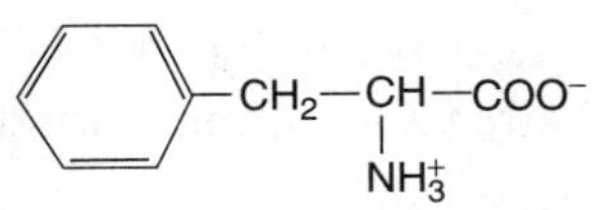
Tyrosine	Tyr [Y]	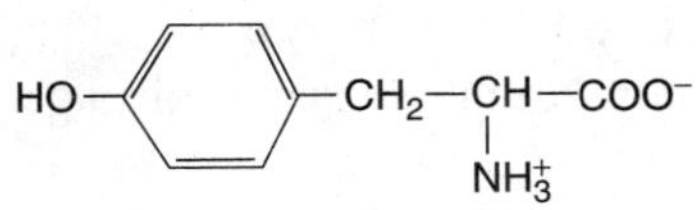

Delimitations of this classification are as follows:

1. *Trp:* This can also be included under aromatic.
2. *His:* This can also be included under basic amino acids.

On the Basis of Number of Amino and Carboxyl Groups

McGilvery and Goldstein-1979-grouped the 20 amino acids into monoamino–monocarboxylic acid, monoamino-dicarboxylic amino acid and diamino monocarboxylic acid.

Monoamino Monocarboxylic Amino Acid

Unsubstitute (Example: Gly, Ala, Val, Leu, Ile, and Leu)
Heterocyclic (Example: Pro)
Aromatic (Example: Phe, Tyr, and Trp)
Thioether (Example: Met)
Hydroxy (Example: Ser and Thr)
Mercapto (Example: Cys)
Carboxamide (Example: Asn and Gln)

Monoamino-Dicarboxylic Amino Acid

Example: Asp and Glu

Diamino Monocarboxylic Amino Acid

Example: Lys, Arg, and His

On the Basis of Polarity of Side Chain

R group: Its tendency is to interact with water at pH 7.

Amino acid with non-polar side chain

R—group in this category is hydrocarbon in nature and thus hydrophobic (water repellent) and lipophilic.

Example: Ala, Val, Leu, Ile, Pro, Phe, Trp, and Met

Amino acid with polar but unchanged R group

R—amino acid is soluble in water because it contains functional group that forms H_2 bond with H_2O.

Example: Gly, Ser, Thr, Cys, Tyr, Gln, and Asn

Amino acid with negative charge

R—contains extra COO^- group with dissociable behaviour.

Example: Asp and Glu

Amino acid with positive charge

The side chain contains extra amino group, which has important basic properties to them.

Example: Lys, Arg, and His

On the basis of Metabolic Fate

Purely ketogenic

Example: Leu, which enters into the metabolic pathway of fat.

Purely glucogenic

Enter into glucogenic pathway.

Example: Glu, Ala, Val, Ser, Thr, Cys, Met, Asn, Gln, Asp, Glu, Arg, His, and Pro.

Both ketogenic and glucogenic

Example: Lys, Ile, Phe, and Tyr are partially ketogenic and partially glucogenic.

On the basis of Nutritional Requirement

Essential or indispensable

Carbon skeletons are not synthesised by human beings, so amino acids are taken in food for normal growth.

Short way to remember essential amino acids
The acronym of the following amino acids is MILLPHTTV.

M–Methionine
I–Isoleucine
L–Leucine
L–Lysine
P–Phenylalanine
H–Histidine
T–Tryptophan
T–Threonine
V–Valine

PARTIALLY ESSENTIAL OR SEMI ESSENTIAL

Histidine and arginine are required for growing children. They are semi or partially essential. Arginine is involved in urea cycle.

NON-ESSENTIAL OR DISPENSABLE

The remaining ten amino acids are non-essential amino acids. The body synthesises all body skeletons, and hence, their absence in food will not adversely affect growth.

NON-STANDARD PROTEIN AMINO ACIDS

In addition to the above twenty amino acids, there are other amino acids, which have limited distribution and high amounts in few proteins.

Examples

Hydroxy proline – 12% collagen
Hydroxy lysine – 1% collagen
N-methyl lysine – Myosin
γ-carboxyglutamate – Blood-clotting protein
α-aminoadipate – Corn protein
α-έ-diaminopimelate – Bacterial protein

NON-PROTEIN AMINO ACIDS

L-ornithine, α-citrulline, β-alanine, and γ-aminobutyrate, are non-protein amino acids.

L-ornithine and *α-citrulline*
Both play an important role in metabolic intermediate of urea cycle.

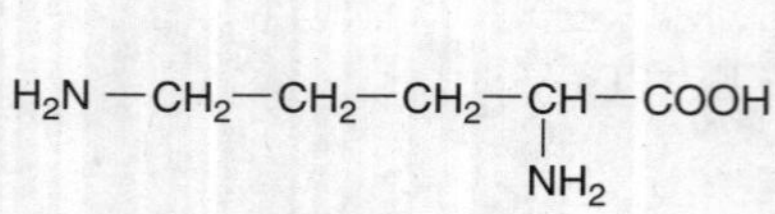

Ornithine

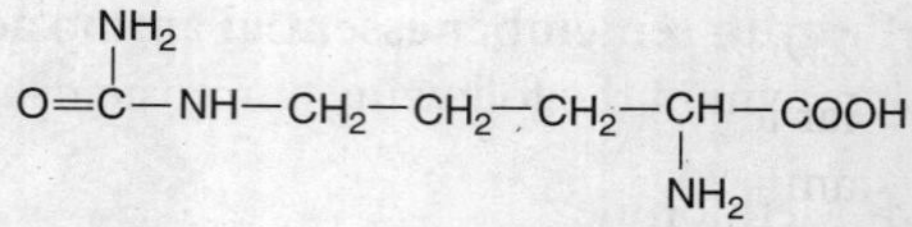

Citrulline

β-alanine – isomer of alanine – vitamin pantothenic acid

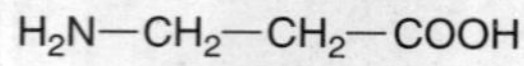

β-alanine

Creatinine – plays an important role in energy storage process.

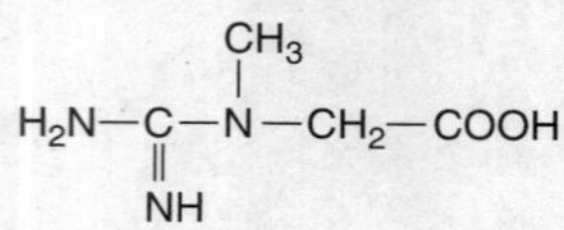

Creatinine

γ-aminobutyrate – free from brain.

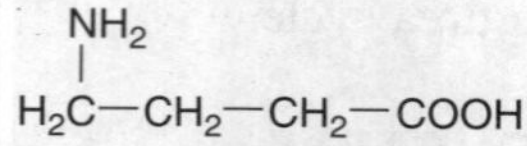

γ-aminobutyrate

PROPERTIES OF AMINO ACIDS

The amino acids differ in their physical chemical properties, which ultimately determine the characteristics of proteins.

Physical Properties

1. Most of the amino acids are soluble in water and insoluble in organic solvents.
2. *Melting points:* Amino acids generally melt at higher temperature, and it is above 200°C.
3. *Taste:* Amino acids may be sweet, tasteless, and bitter.
   Sweet – Gly, Ala, and Val
   Tasteless – Leu
   Bitter – Arg and Ile
   Sodium glutamate is employed as a flavouring agent in food industry to increase taste and flavour.
4. *Optical properties:* All amino acids, except glycine, possess optical isomers due to the presence of asymmetric α-carbon atom.
5. *Ampholytes:* Amino acids contain both acidic (COOH) and basic (NH_2) groups. They can donate a proton or accept a proton. Hence, amino acids are regarded as ampholytes.

6. *Zwitterion* or *dipolar ion:* The name zwitter means hybrid. Zwitterion (or dipolar ion) is a hybrid molecule containing positive and negative ionic groups. The amino acids rarely exist in a neutral form with free carboxylic (—COOH) and free amino acid (—NH_2) groups.

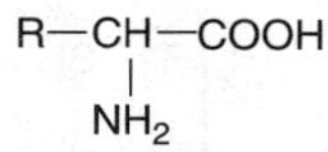

Neutral form

In a strongly acidic pH (< 7 pH) – low pH, the amino acid is positively charged (cation), while in a strongly alkaline pH (>7 pH) – high pH, it is negatively charged (anion). Each amino acid has a characteristic pH (e.g. Leu pH is 6), at which it carries both positive and negative charges and exists as zwitterion.

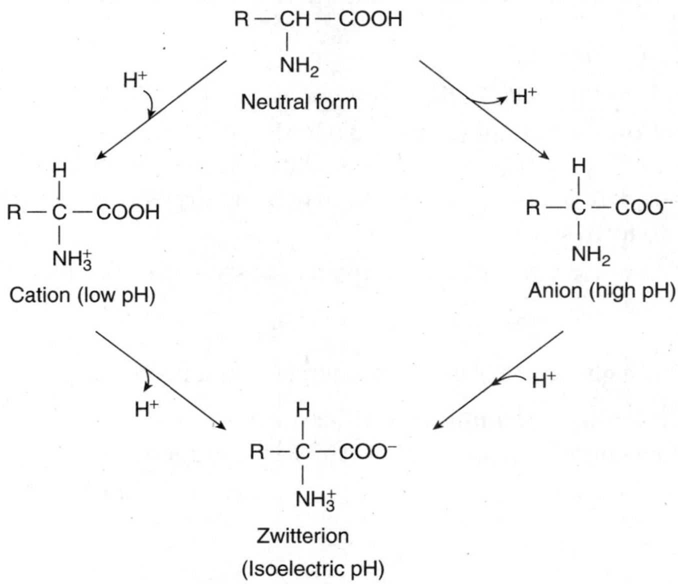

Isoelectric pH

Isoelectric point (pI) is defined as the pH at which a molecule exists as zwitterion or dipolar ion and carries no net charge.

Example: pI for leucine is 6.

Titration of amino acids

The existence of different ionic forms of amino acids can be more easily understood by the titration curves as shown in Figure 3.1. At low pH, leucine exists in a fully protonated form as cation. As titration proceeds with NaOH, leucine loses its proton, and at isoelectric pH (pI), it blames a zwitterion. Further titration results in the formation of anionic form of leucine. Further titration results in the formation of anionic forms of leucine as shown in Figure 3.1.

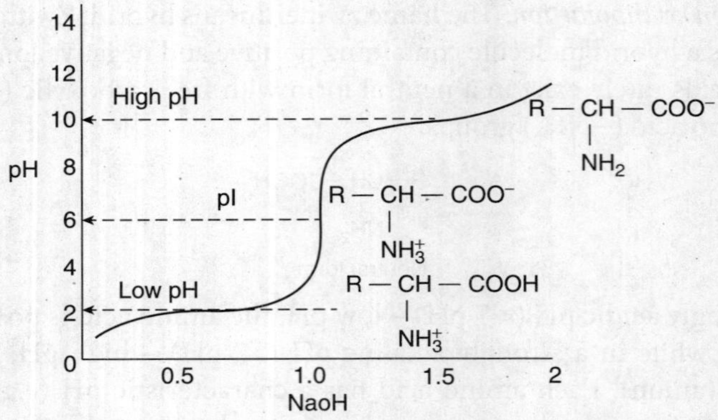

Figure 3.1 *Titration Curve of Amino Acid Leucine*

Chemical Properties

1. *Reactions involving COOH group:*

   Reaction with alkali (salt formation)

   The carboxylic group of amino acids releases H^+ ion with the formation of carboxylate (COO^-) ions. These may be neutralised by the cations like Na^+ and Ca^+ to form salts.

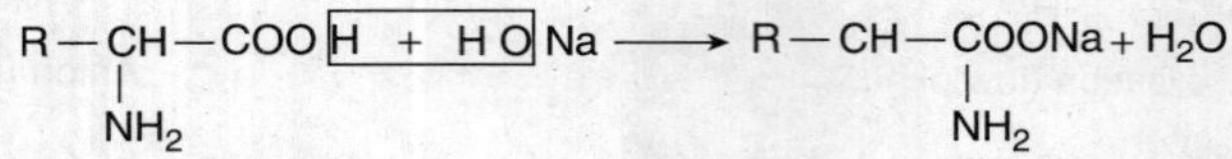

   Sodium glutamate is used commercially as a flavouring agent.

   Reaction with alcohols (esterification)

   With alcohols, corresponding esters are produced.

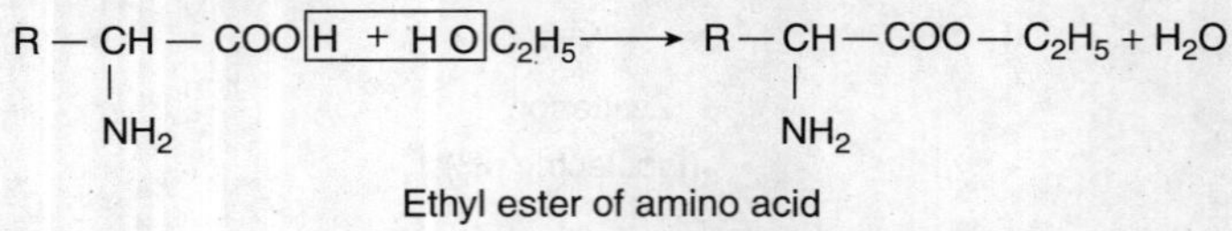

   Significance: Isolation of amino acids in pure form from the protein hydrolysates.

   Reaction with amines

   Amino acid reacts with amines and forms amides.

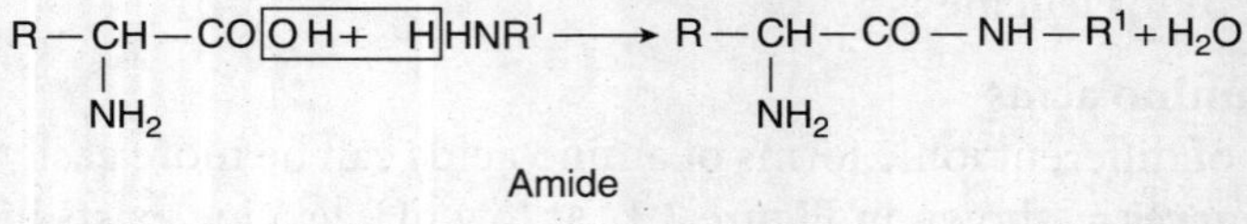

2. *Reactions involving NH_2 group:*

   Reaction with mineral acids (acid salt formation)

   When free amino acids are tested with protein and heated with mineral acids like HCl, acid salt is formed.

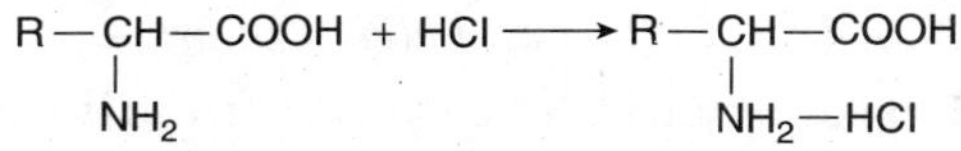

Amino acid hydrochloride

Reaction with formaldehyde

With formaldehyde, the hydroxymethyl derivatives are formed. These derivatives are insoluble in water and resistant to attack by microorganisms.

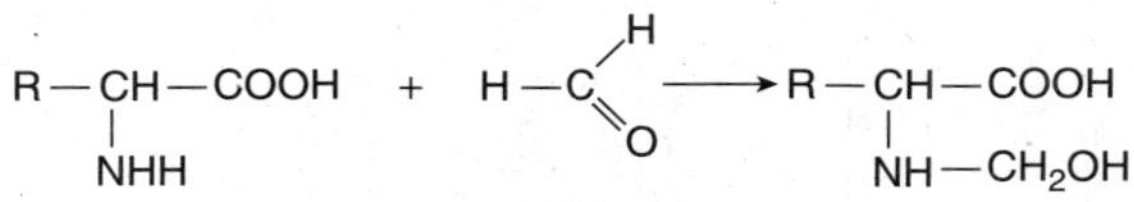

N-Monomethylol derivative

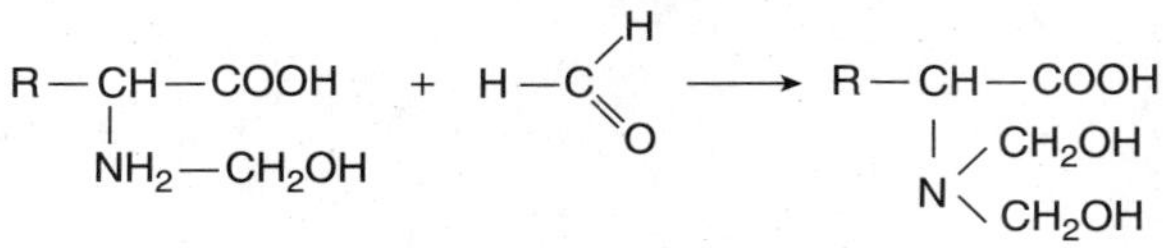

N-Dimethylol derivative

Significance: This reaction is the basis of the Sorensen titration method for determining the purity of individual amino acids.

Reaction with nitrous acid (Van Slyke reaction)

The amino acid reacts with HNO_2 to liberate N_2 gas and produce the corresponding α-hydroxy acids.

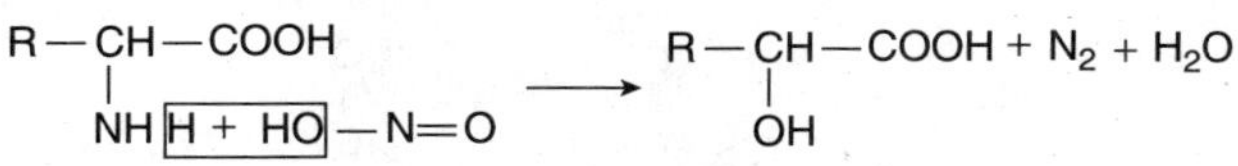

Reaction with acylating agents (acylation)

Acylation is brought about by many acid chlorides and acid anhydrides, when an amino acid in alkaline medium reacts with CH_3CoCl.

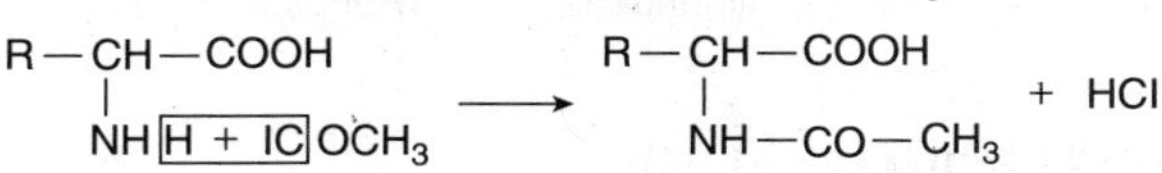

Reaction with FDNB or Sanger's reagent

In mildly alkaline solution, FDNB (1 fluoro-2 4-dinitrobenzene) reacts with α-amino acids to produce yellow-coloured derivative, DNB-amino acid.

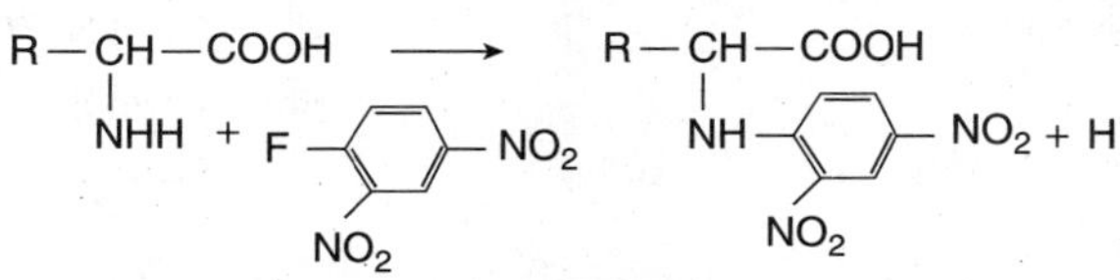

(FDNB) DNB-amino acid (yellow)

Significances: Valuable in elucidation of protein structure. Sanger successfully utilised this method in determining the sequence of amino acids in insulin.

3. *Reaction involving both COOH and NH_2 groups:*

Reaction with triketohydrindene hydrate (ninhydrin)

Ninhydrin is reduced to hydrindantin when treated with amino acids. Amino acid is converted to ammonia, CO_2, and aldehyde. The second molecule of ninhydrin reacts with the former hydrindantin and ammonia to form a purple-coloured complex.

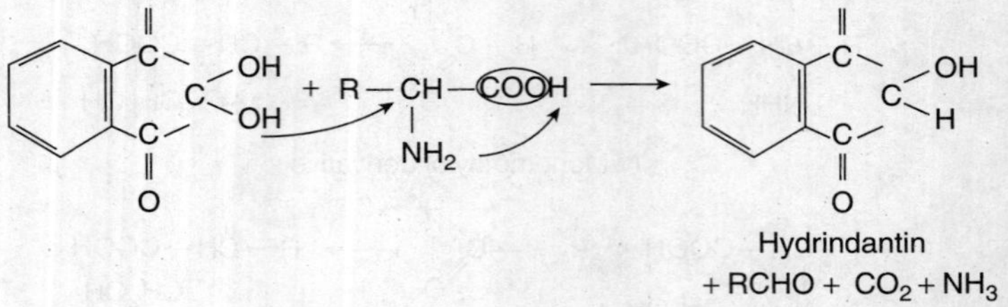

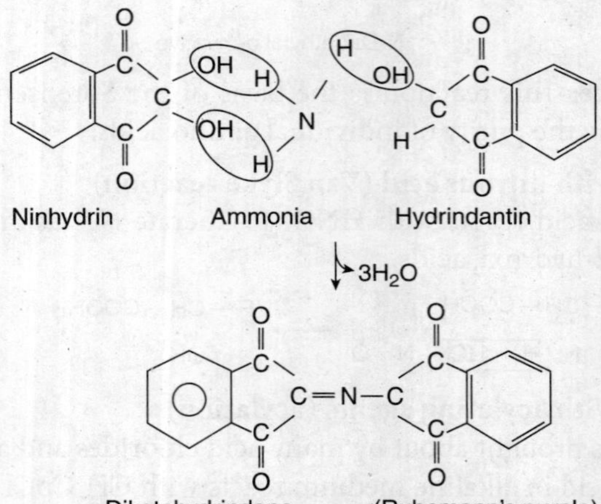

Reaction with phenyl isocyanate

With phenyl isocyanate, hydratonic acid is formed, which, in turn, can be converted to hydrindantin.

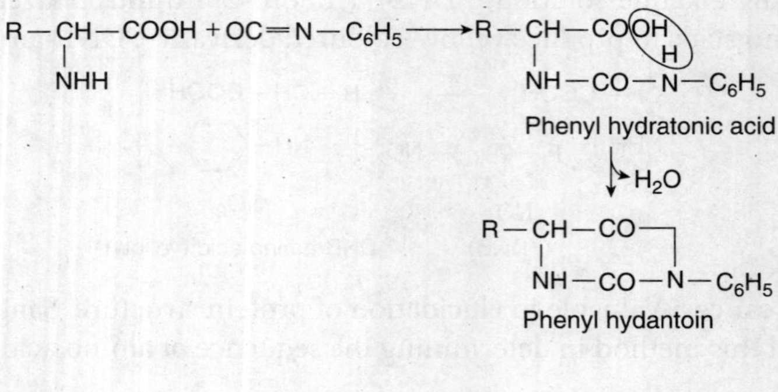

Reaction with phenyl isothiocyanate or Edman reagent

Phenyl isothiocyanate also reacts similarly with amino acids to produce thiohydantoin acid. On treatment with acid in non-hydroxylic solvents, the latter cyclise to thiohydantoin.

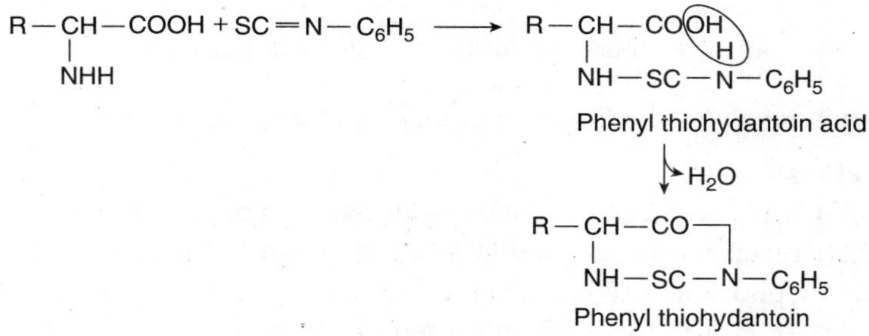

Significance: This reaction is useful in the study of protein structure.

4. *Reaction involving the R or side chain groups or colour reaction:*

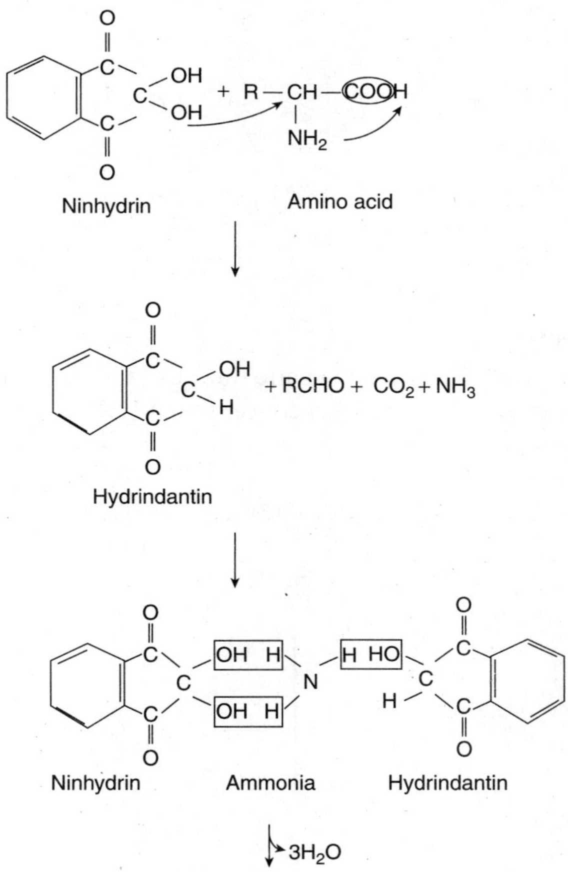

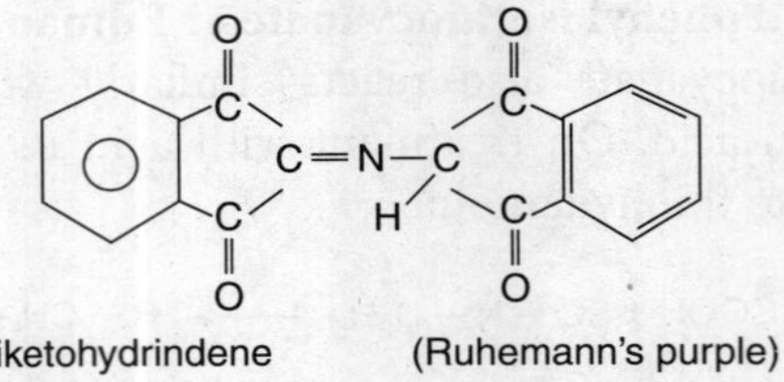

5. *Reaction involving the R or side chain groups or colour reaction:*

Biuret test

This test is a general test for compounds having a peptide bond. Alkaline copper sulphate reacts with compounds containing two or more peptide bonds to give a violet or pink-coloured product, which is due to the formation of coordination complex of cupric ion with unshared electron pairs of peptide nitrogen and oxygen of water.

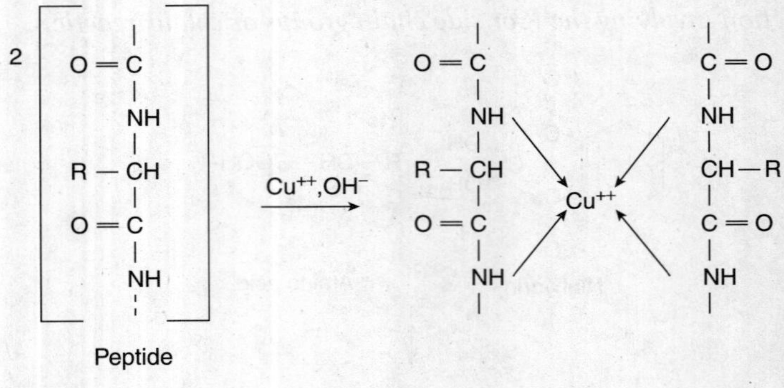

Xanthoproteic test

Amino acid containing an aromatic nucleus (Tyr, Try, and Phe) forms yellow nitro derivative on heating with concentrated HNO_3. The salts of these derivatives are orange in colour.

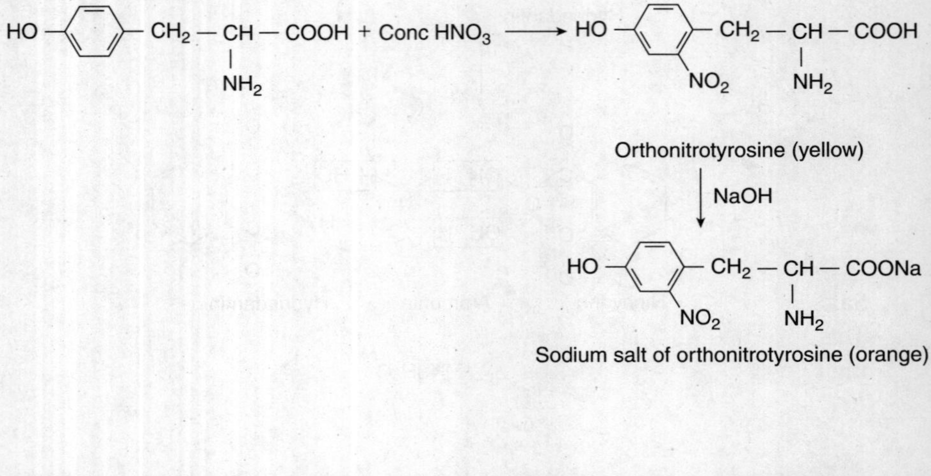

Millon's test

Amino acids or compounds containing hydroxy benzene radical react with Millon's reagent (mercurous and mercuric nitrate containing HNO_3) form a red complex. Thus, this test is specific for tyrosine.

$$Hg + 4HNO_3 \longrightarrow Hg(NO_3)_2 + 2NO_2 + 2H_2O$$

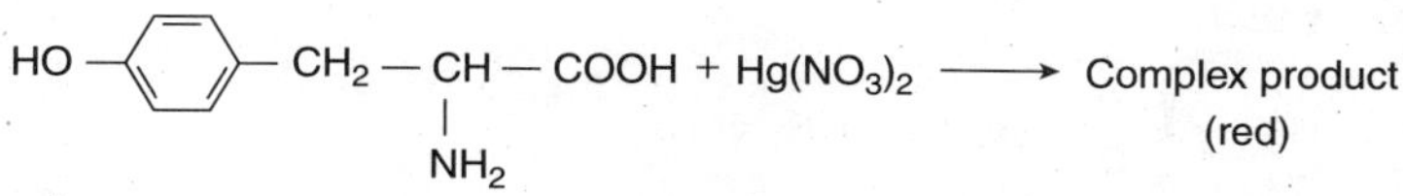

Hopkins-Cole test

Hopkins–Cole reaction is due to the presence of the indole ring of amino acid tryptophan. Tryptophan in the presence of glyoxalic acid and concentrated sulphuric acid gives a purple colour.

Glyoxalic acid can be prepared by reducing oxalic acid with magnesium powder or sodium amalgam or glacial acetic acid, which has been exposed to sunlight. It also contains glyoxylic acid.

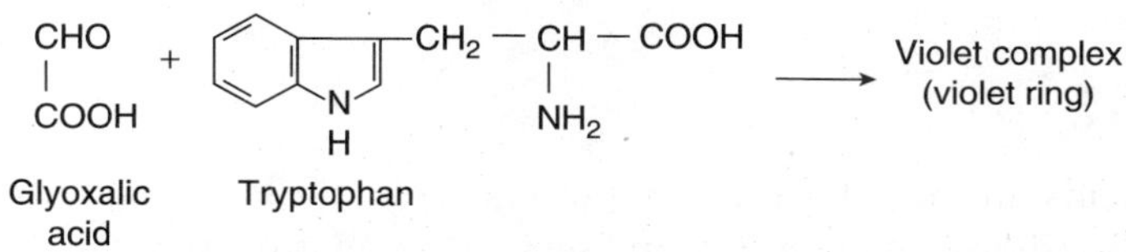

Pauly's Test

Diazotised sulphanilic acid couples with amines, phenols, and imidazole to form highly coloured compounds. The diazonium compound is only formed in the cold condition, and so all solutions must be cooled in ice before diazotisation.

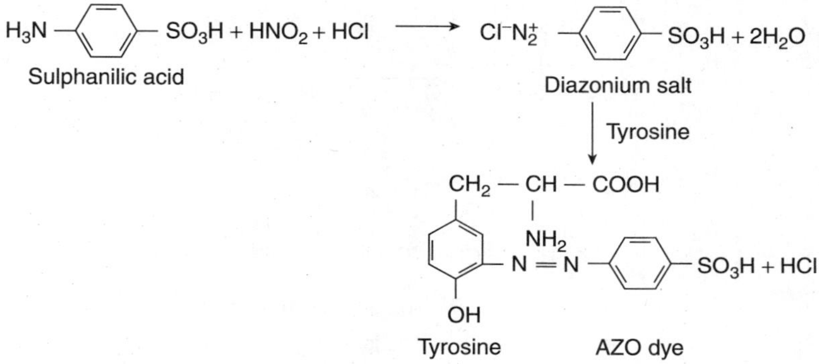

Sakaguchi reaction

This reaction is given by guanidium compounds. The only amino acid containing the guanidine group is arginine.

Arginine + α-napthol + bromine water

↓

Red-coloured product

Sulphur test

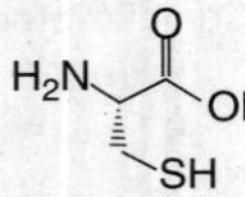

Cysteine

$NaS \rightarrow Na^+ + S^{2-}$

$Pb(CH_3COO^-)_2 + S^{2-} \rightarrow PbS \downarrow + 2CH_3COO^-$

Sodium nitroprusside test

$$\underset{\text{Nitroprusside}}{Na_2(CN)_5Fe(NO)\cdot 2H_2O} + \underset{\text{Cysteine}}{HS-CH_2\underset{\displaystyle NH_3^+}{\underset{|}{CH}}-COO^-} \xrightarrow{OH^-} \text{red complex}$$

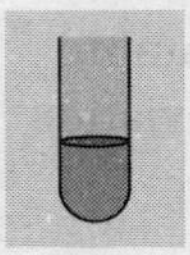

PEPTIDE

The amino acids are held together in protein by a covalent, amide bond or peptide bond. These bonds are rather strong and serve as cementing material between the individual amino acids.

FORMATION OF PEPTIDE BOND

When the amino groups of an amino acid combine with carboxyl group of another amino acid, a peptide bond is formed.

$$^+H_3N-\overset{\displaystyle R_1}{\overset{|}{CH}}-CO\boxed{O^- \;+\; ^+H_3}N-\overset{\displaystyle R_2}{\overset{|}{CH}}-COOH$$

$$\downarrow -H_2O$$

$$^+H_3N-\overset{\displaystyle R}{\overset{|}{CH}}-\underbrace{CO-NH}_{\text{Peptide bond}}-\overset{\displaystyle R_2}{\overset{|}{CH}}-COOH$$

(Dipeptide)

Peptides containing more than ten amino acids are referred as polypeptides.

CHARACTERISTICS OF PEPTIDE BOND

The peptide bond is a rigid planar with partial double bond in character. It exists in transconfiguration. Both —C=O and —NH groups of peptide bonds are polar and are involved in hydrogen bond formation.

WRITING PEPTIDE STRUCTURE

The peptide chains are written with the free amino end (N-terminal residue) at the left and the free carboxyl end (C-terminal residue) at the right. The amino acid sequence is read from N-terminal end to C-terminal end. The protein biosynthesis also starts from the N-terminal amino acid.

SHORTHAND TO READ PEPTIDES

The amino acids in a peptide bond are represented by a three-letter or one-letter abbreviation. This is the chemical shorthand way to write proteins.

NAMING OF PEPTIDE

For naming an amino acid, suffixes *-ine* (glycine), *-an* (tryptophan), and *-ate* (glutamate) are changed to *-yl* with the exception of C-terminal amino acid.

Example: H_3N — Glutamate — Cystine — Glycine — COO^-

Glutathione

One-letter symbol	E — C — G
Three-letter symbol	Glu — Cys — Gly
Peptide name	Glutamyl — Cystine — Glycine

DIMENSIONS OF PEPTIDE BOND

The dimensions of a fully extended polypeptide chain are shown in Figure 3.2. The two adjacent α-carbon atoms are placed at a distance of 0.36 nm.

α-carbon — $N^{(NH_3)}$ = 0.147 nm

α-carbon — $C^{(COO)}$ = 0.153 nm

C—O = 0.123 nm

(Peptide bond) C—N = 0.132 nm

N—H = 0.1 nm

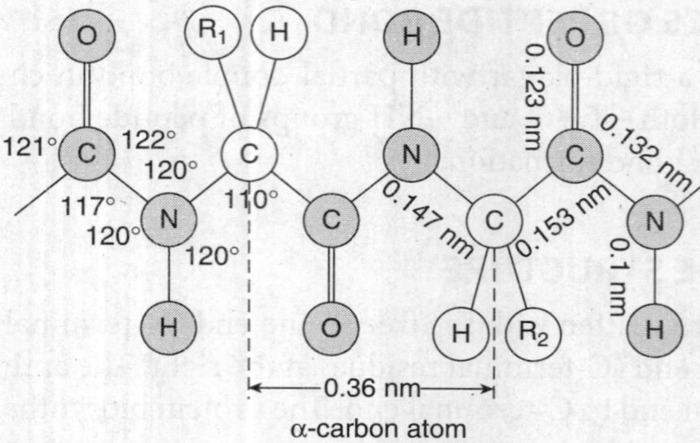

Figure 3.2 *Dimension of a Fully Extended Polypeptide Chain*

BIOLOGICALLY IMPORTANT PEPTIDES

Several peptides occur in the living organisms that display a wide spectrum of biological functions.

Glutathione

It is a tripeptide composed of three amino acids. Chemically, glutathione is γ–glutamyl–cysteine–glycine. It is widely distributed in nature and exists in reduced or oxidised states.

Structure of glutathione is as given below:

$$2G - SH \rightleftharpoons G = S - S - G$$

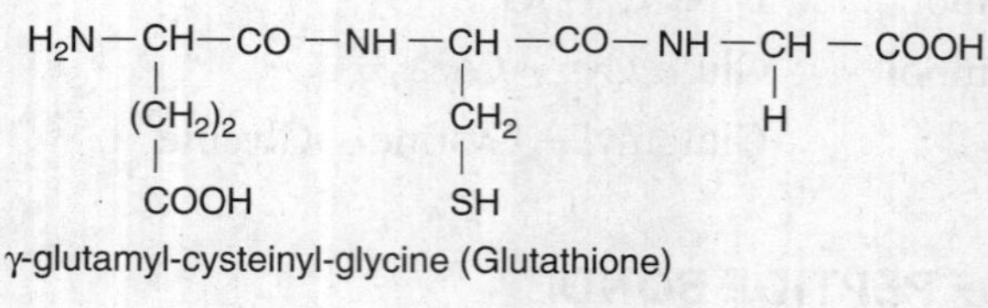

γ-glutamyl-cysteinyl-glycine (Glutathione)

Functions of glutathione

The reversible oxidation–reduction of glutathione is important for many of its biological functions. It maintains a ratio of about 10011 of GSH to G–S–S–G. Glutathione (reduced) performs specialised functions in erythrocytes. It maintains RBC membrane structure in integrity. It protects haemoglobin from getting oxidised by agents such as H_2O_2. It involves in the transport of amino acids in the intestine and kidney tubules. Toxic amounts of peroxide and free radicals produced in the cells are scavenged by glutathione peroxidase.

$$2GSH + H_2O_2 \xrightarrow{\text{Peroxidase}} G-S-S-G + 2H_2O$$

It is involved in detoxification (removing toxic substances). It acts as coenzyme (PGE_2 synthase and glyoxalase). It involves in the formation of correct disulphide bonds in several proteins.

Thyrotropin Releasing Hormone (TRH)

It is a tripeptide secreted by hypothalamus.

$$-\text{OOc}-\text{Glu}-\text{His}-\text{Pro}-\text{NH}_2$$

Functions: It stimulates pituitary gland to release thyrotropic hormone.

Oxytocin

It's a hormone secreted by posterior pituitary gland and contains nine amino acids (nanopeptide).

S—S (between Cys 1 and Cys 6)

Cys — Tyr — Ile — Gln — Asn — Cys — Pro — Leu — Gly
1 2 3 4 5 6 7 8 9

Functions: It stimulates the contraction of smooth muscles, especially during childbirth.

Vasopressin (Antidiuretic Hormone (ADH)):

ADH is also a nanopeptide produced by posterior pituitary gland.

S—S (between Cys 1 and Cys 6)

Cys — Tyr — Phe — Gln — Asn — Cys — Pro — Arg — Gly
1 2 3 4 5 6 7 8 9

Functions: It stimulates the kidney to retain water and thus increases blood pressure.

Angiotensin

Angiotensin 1 is a decapeptide (ten amino acids), which is converted to angiotensin 2 (eight amino acids). The latter has more hypertensive effect and also stimulates the release of aldosterone from adrenal gland. Methionine enkephalin is a pentapeptide found in brain.

Functions: It has opiate-like function and inhibits the sense of pain. Bradykinin and kallidin are mono and decapeptides.

Peptide Antibiotics

Antibiotics such as gramicidin, bacitracin, tyrocidine, and actinomycin are peptides in nature. Gastrointestinal hormones like gastrin, secretin, and so on are gastrointestinal peptides which serve as hormones. Substance P is a decapeptide which acts as neurotransmitter.

CLASSIFICATION OF PROTEIN

The proteins are all remarkably similar in structure as they contain amino acids. Proteins are classified based on the following:

1. Shape
2. Composition and solubility
3. Function
4. Nutritional status

CLASSIFICATION OF PROTEIN BASED ON THE SHAPE OF A PROTEIN MOLECULE

Globular or corpuscular proteins:

They have an axial ratio (length: width) of less than ten (usually 3–4) and hence possess a relatively spherical or orbit shape. These are usually soluble in water or in aqueous media, containing acids, bases, salts, or alcohol, and diffuse readily. Nearly all enzymes are globular proteins.

Example: Protein hormones, blood transport proteins, antibodies, and nutrient storage proteins.

CLASSIFICATION OF PROTEIN BASED ON COMPOSITION AND SOLUBILITY

Based on the composition and solubility, proteins are broadly classified into three major groups.

1. *Simple proteins:* These proteins are composed of only amino acid residues. These are further classified into two classes.
   (a) Globular proteins: These are spherical or oval in shape, soluble in water or other solvents and digestible.
      (i) Albumins: These are soluble in water and dilute salt solutions and are coagulated by heat; examples are serum albumin, ovalbumin (egg) and lactalbumin (milk).
      (ii) Globulin: These are soluble in neutral and dilute salt solutions; examples are serum globulin and vitelline (egg yolk).
      (iii) Glutelins: These are soluble in dilute acids and alkalis and are mostly found in plants; examples are glutelin (wheat) and oryzenin (rice).
      (iv) Prolamines: These are soluble in 70 per cent alcohol; examples are gliadin (wheat) and zein (maize).
      (v) Histones: These are strongly basic proteins and are soluble in dilute ammonium hydroxide; examples are thymus histones and histones of codfish sperm.

(vi) Globins: These are generally considered along with histones. However, globins are basic proteins and are not precipitated by ammonium hydroxide.

(vii) Protamines: These are strongly basic and resemble histones but are smaller in size and soluble in ammonium hydroxide. Protamines are also found in association with nucleic acids. Examples are sperm proteins.

(b) Fibrous proteins: These are fibre-like in shape, insoluble in water and resistant to digestion. Albuminoids or scleroproteins constitute the most predominant group of fibrous proteins.

(i) Collagens: These are connective tissue proteins lacking tryptophan. On boiling with water or dilute acids, collagens yield gelatin, which is soluble and digestible.

(ii) Elastins: These proteins are found in elastic tissues such as tendons and arteries.

(iii) Keratins: These are present in exoskeletal structures such as hair, nails and horns. Human hair keratin contains as much as 14 per cent of cysteine.

2. *Conjugated proteins:* Besides the amino acids, these proteins contain a non-protein moiety known as prosthetic group or conjugating group.

(a) Nucleoproteins: Nucleic acid (DNA or RNA) is the prosthetic group. Examples are nucleohistones and nucleoprotamines.

(b) Glycoproteins: The prosthetic group is carbohydrate, which has less than four per cent protein. The term mucoprotein is used if the carbohydrate content is more than four per cent. Examples are mucin (saliva) and ovomucoid (egg white).

(c) Lipoproteins: Proteins found in combination with lipids are the prosthetic group. Examples are serum lipoproteins and membrane lipoproteins.

(d) Phosphoproteins: Phosphoric acid is the prosthetic group. Examples are casein (milk) and vitelline (egg yolk).

(e) Chromoproteins: The prosthetic group is coloured in nature. Examples are haemoglobins and cytochromes.

(f) Metalloproteins: These proteins contain metal ions such as iron, cobalt, zinc, copper and magnesium. Examples are ceruloplasmin (Cu) and carbonic anhydrase (Zn).

3. *Derived proteins:* These proteins are the denatured or degraded products of simple and conjugated proteins. Derived proteins are of two types. The primary derived are the denatured, coagulated or first hydrolysed products of proteins. The secondary derived are the degraded (due to breakdown of peptide bonds) products of proteins.

(a) Primary derived proteins:

(i) Coagulated proteins: These are the denatured proteins produced by agents such as heat, acids, and alkalis. Examples are cooked proteins and coagulated albumin (egg white).

(ii) Proteans: These are the earliest products of protein hydrolysis by enzymes, dilute acids, alkalis, and so on, which are insoluble in water. Examples are fibrins formed from fibrinogen.

(iii) Metaproteins: These are the second-stage products of protein hydrolysis obtained by treatment with slightly stronger acids and alkalis. Examples are acid and alkali metaproteins.

(b) Secondary derived proteins: These are the progressive hydrolytic products of protein hydrolysis. These include proteoses, peptones, polypeptides and peptides.

CLASSIFICATION OF PROTEIN BASED ON FUNCTION

Based on the function they perform, proteins are classified into the following groups:

1. Structural proteins—keratin of hair and nails and collagen of bone
2. Enzymes or catalytic proteins—hexokinase and pepsin
3. Transport proteins—haemoglobin and serum albumin
4. Hormonal proteins—insulin and growth hormone
5. Contractile proteins—actin and myosin
6. Storage proteins—ovalbumin and glutelin
7. Genetic proteins—nucleoproteins
8. Defence proteins—snake venoms and immunoglobulins
9. Receptor proteins—hormones and viruses

CLASSIFICATION OF PROTEIN BASED ON NUTRITIONAL STATUS

The nutritive value of proteins is determined by the composition of essential amino acids. From the nutritional point of view, proteins are classified into three categories.

1. *Complete proteins:* These proteins have all the ten essential amino acids in the proportion required by the human body to promote good growth. Examples are egg albumin and milk casein.
2. *Partially incomplete proteins:* These proteins, which lack one or more essential amino acids, help in moderate growth. Examples are wheat and rice proteins (Lys and Thr)
3. *Incomplete proteins:* These proteins are called incomplete proteins as they do not have any of the essential amino acids. Hence, they are not involved in growth. Examples are gelatin (lacks Trp) and Zein (lacks Lys).

PROTEIN CONFIGURATION

Linderstrom-Lang first defined the four basic structural levels of organisation, which are referred to as primary, secondary, tertiary, and quaternary. Three of these structural levels (primary, secondary, and tertiary) can exist in molecules composed of a single polypeptide chain, whereas the fourth (quaternary) involves interactions of polypeptide within a multi-chained protein molecule.

The basic structure of a protein consists of one or more linear chains of amino acid units. This linear structure forms a helical shape to produce the secondary structure. The helical structure folds in several patterns to produce the three-dimensional structure or tertiary structure. Finally, certain proteins are made up of subunits of similar or dissimilar types of the polypeptide chains. These subunits interact with each other to give rise to the so-called quaternary structure of protein as shown in Figure 3.3.

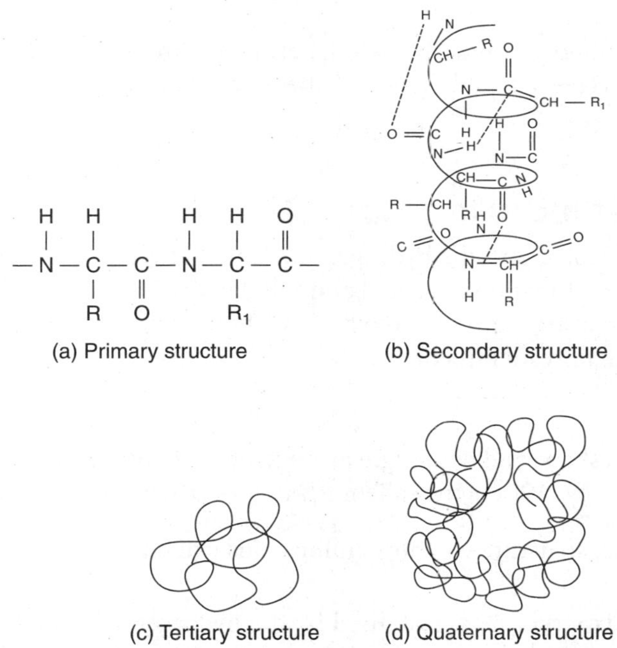

Figure 3.3 *Structure of Primary, Secondary, Tertiary and Quaternary Structure of Protein*

PRIMARY STRUCTURE – AMINO ACID SEQUENCE

The primary structure of protein refers to the number sequence of amino acids and the constituent units of polypeptide chain. The main mode of linkage of amino acids in proteins is the peptide bonds which link the α-carboxyl group of amino acid residue to the α-amino group of the other.

Linus Pauling and Robert Corey in the late 1930s demonstrated the α-carbon of adjacent amino acids, which are separated by three covalent bonds arranged as $C_\alpha - C - C - C_\alpha$ This is as shown in Figure 3.4.

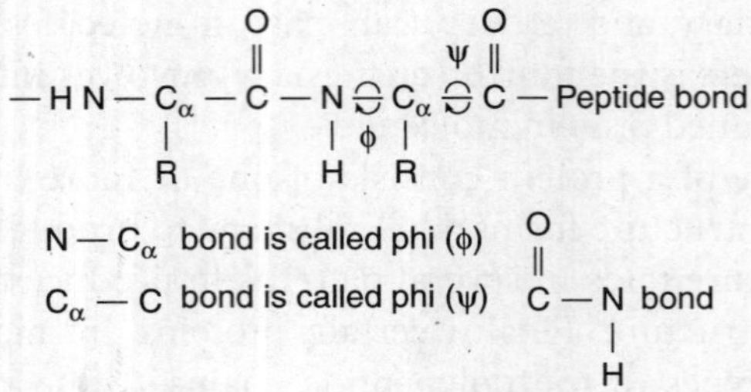

Figure 3.4 *Structure of Bond in Peptide*

Rotation is permitted around the bond between nitrogen and α-carbon atom of the main chain (N$-C_\alpha$) and between the α-carbon atoms and carbonyl atoms ($C_\alpha-$C).

SECONDARY STRUCTURE

The protein gets folded, and the hydrogen bonding between the neighbouring amino acids results in the formation of a rigid and tubular structure called helix. This constitutes the secondary structure of proteins. Two types of secondary structure, α-helix and β-sheet, are mainly identified.

α-helix

α-helix is spiral in shape and has a rigid arrangement of polypeptide chain as proposed by Pauling and Corey (1981). The salient features of α-helix are as follows:

1. The α-helix structure contains amino acids extending outwards from the central axis.
2. The α-helix structure is stabilised by the hydrogen bonding formed between the H atom attached to peptide N and O atom attached to peptide C. They are individually weak but together they are strong to stabilise the helix.
3. All amino acids, except the first and last ones, are involved in hydrogen bonding.
4. A single α-helix turn contains 3.6 amino acids and the distance covered by this is 0.54 nm. The space between each amino acid is 0.15 nm.
5. Low energy is utilised for the α-helix formation.
6. The stability is more for the α-helix than for the left-handed helix.
7. Proline is the only amino acid that disrupts the α-helix.

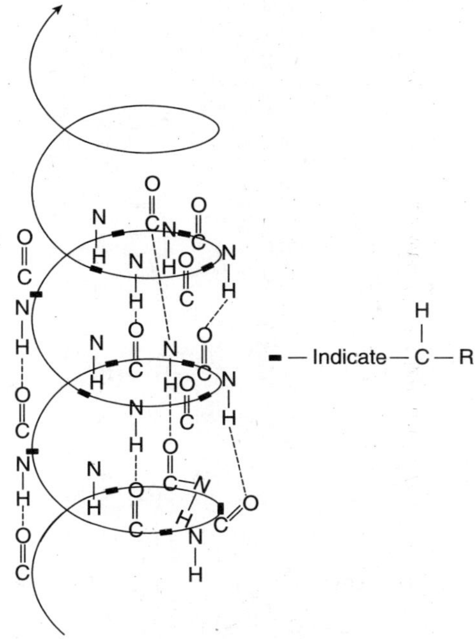

Figure 3.5 *Structure of α-helix*

β-Pleated Sheets

β-pleated sheets form when two or more polypeptide chain segments line up side by side. Each individual segment is referred to as β-strand. Rather than being coiled, each β-strand is fully extended. β-pleated sheets are stabilised by hydrogen bonds that are formed between the polypeptide backbones N—H and carbonyl group of adjacent chain.

There are two β-pleated sheets:

1. Parallel
2. Anti-parallel

Parallel β-Pleated Sheets

The polypeptide chains are arranged in the same direction if the N-terminal ends of all the participating polypeptide chains lie on the same edge of the sheet with all C-terminal ends on the opposite edge. In other words, the hydrogen-bonded neighbouring polypeptides are aligned in the same direction of the N— to —C terminals. Parallel β-pleated sheet structure is given below.

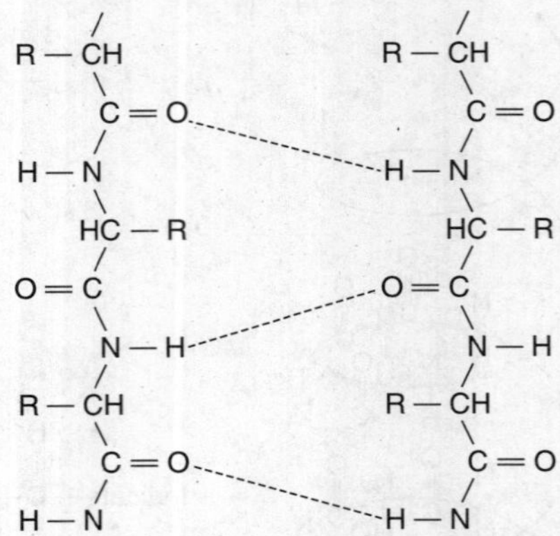

Anti-parallel β-Pleated Sheets

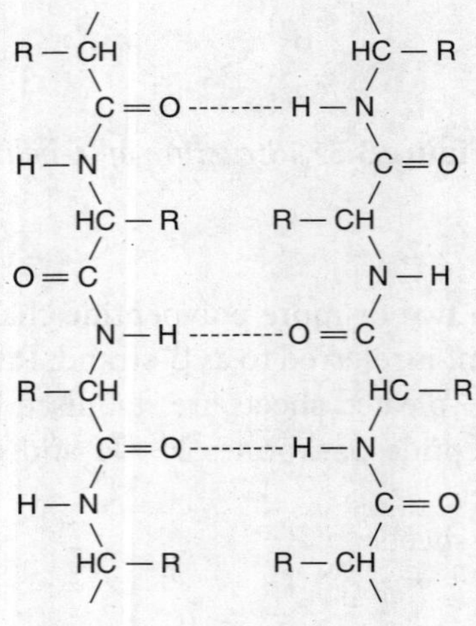

The chains run in opposite direction; that is, the hydrogen-bonded neighbouring polypeptides are aligned in opposite direction. Collinear hydrogen bonds are formed in anti-parallel β-sheets, which make them more stable than the parallel β-sheets.

Silk fibroin is one example of a protein that has anti-parallel pleated sheet structure. Anti-parallel β-pleated sheet structure is as given below.

Mixed parallel and anti-parallel β-sheets are observed in many proteins. These proteins contain the combined α-helix and β-pleated sheet structure. This is called a super secondary structure. In β α β units, α-helix is connected to two β-pleated sheets.

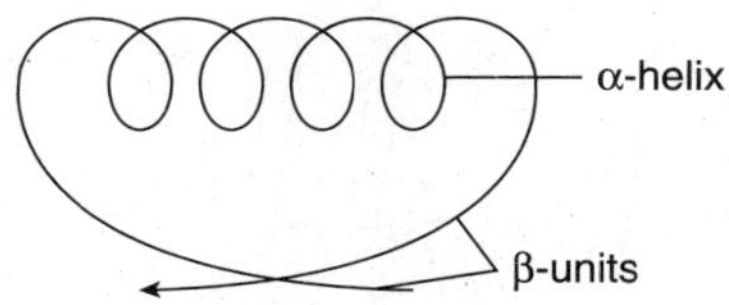

In the β-meander structure, two anti-parallel β-sheets are joined by polar amino acids and glycine to produce a polypeptide chain called β-turns, as shown in Figure 3.6.

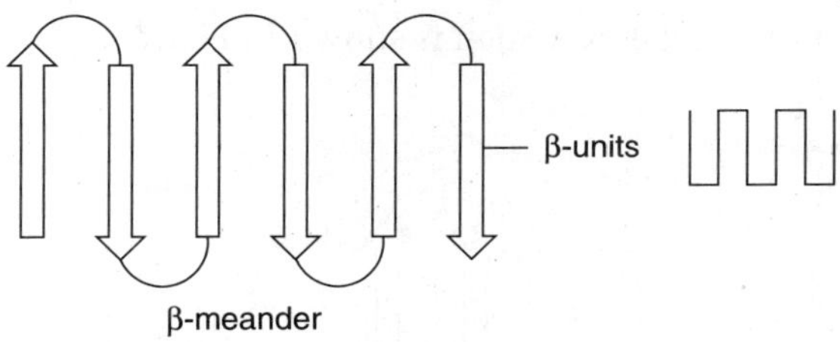

Figure 3.6 β-*Meander Pattern*

In α-α units, two structure α-helices are separated by a loop or non-helical segment as shown in Figure 3.7.

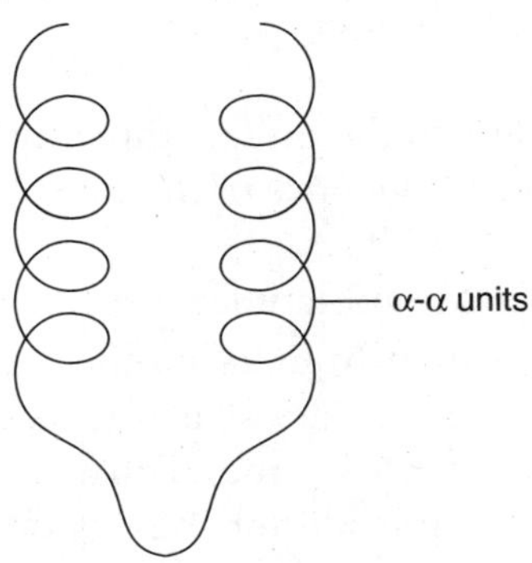

Figure 3.7 α-α *Units Pattern*

β-barrel arrangement is formed when various β-sheet configurations fold back on themselves. β-barrel arrangement is formed when various β-sheet configurations fold back on themselves as shown in Figure 3.8.

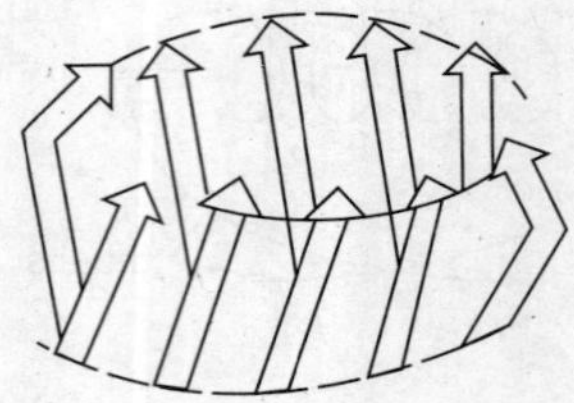

Figure 3.8 β*-Barrel Arrangement*

A Greek pottery pattern design contains an anti-parallel β-sheet folded back on itself and resembles the Greek key, which is shown in Figure 3.9

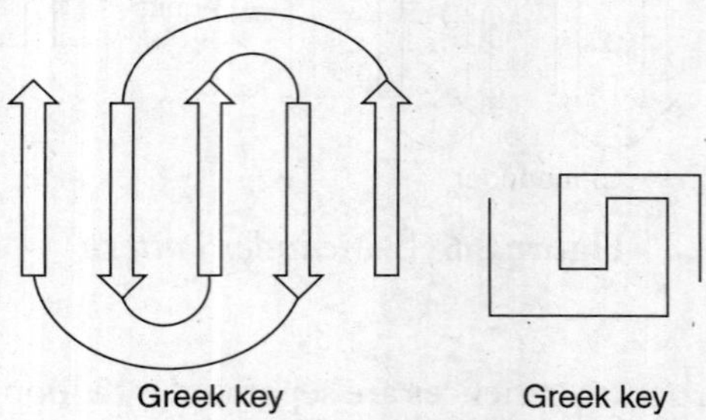

Figure 3.9 *Greek Key Pattern*

TERTIARY STRUCTURE

The term *tertiary structure* refers to the unique three-dimensional confirmations that globular protein assumes as a consequence of the interaction between the side chains in their primary structure.

Tertiary structure has several important features.

The amino acids that are distant from each other in the primary structure come closer during the formation of the tertiary structure. As a result of this property, the polypeptide chain folds in a compact manner. During this process, most water molecules are excluded from the protein's interior, making interactions between both polar and non-polar groups.

Large globular proteins (i.e. those with more than two hundred amino acid residues) often contain several compact units called domains. Domains are typically structurally independent segments that have specific functions (e.g.: binding an ion on small molecule).

The following three domains are found in several proteins:

The EF hand, which consists of a helix–loop–helix configuration, binds specifically to Ca^{2+} as shown in Figure 3.10.

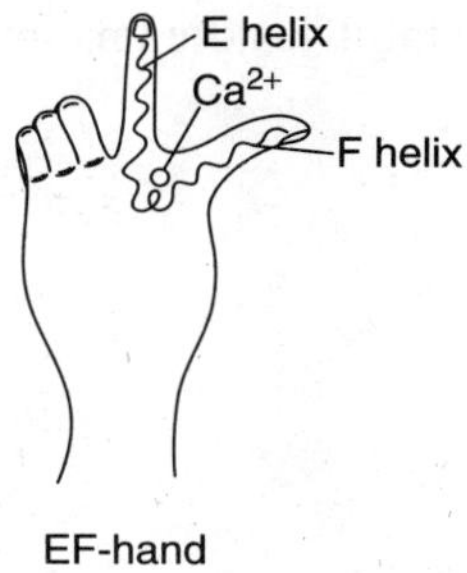

Figure 3.10 *Structure of Helix-loop-helix Pattern*

The zinc finger motif is commonly found in DNA-binding proteins as shown in Figure 3.11. Zinc finger promotes protein DNA interaction.

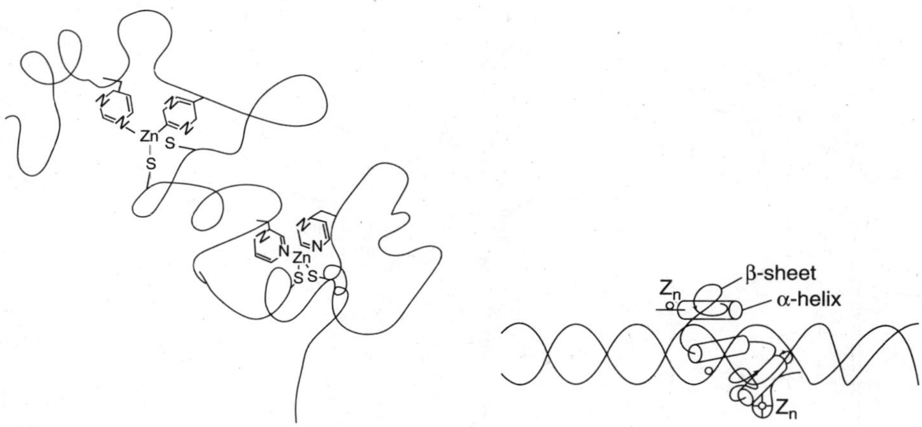

Figure 3.11 *Structure of Zinc Finger Motif*

Leucine zipper – DNA-binding domain as shown in Figure 3.12.

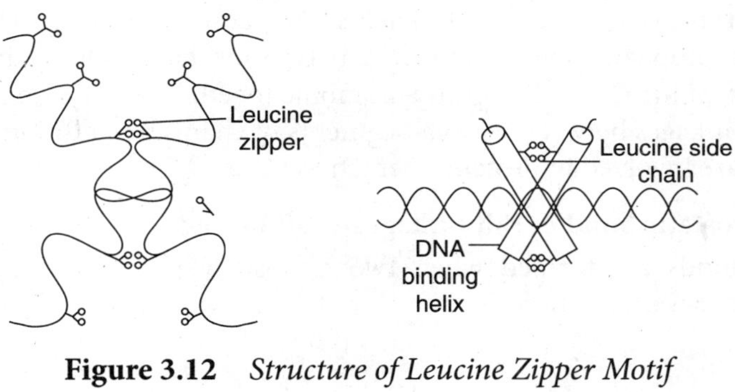

Figure 3.12 *Structure of Leucine Zipper Motif*

The following type of interaction stabilises the tertiary structure:

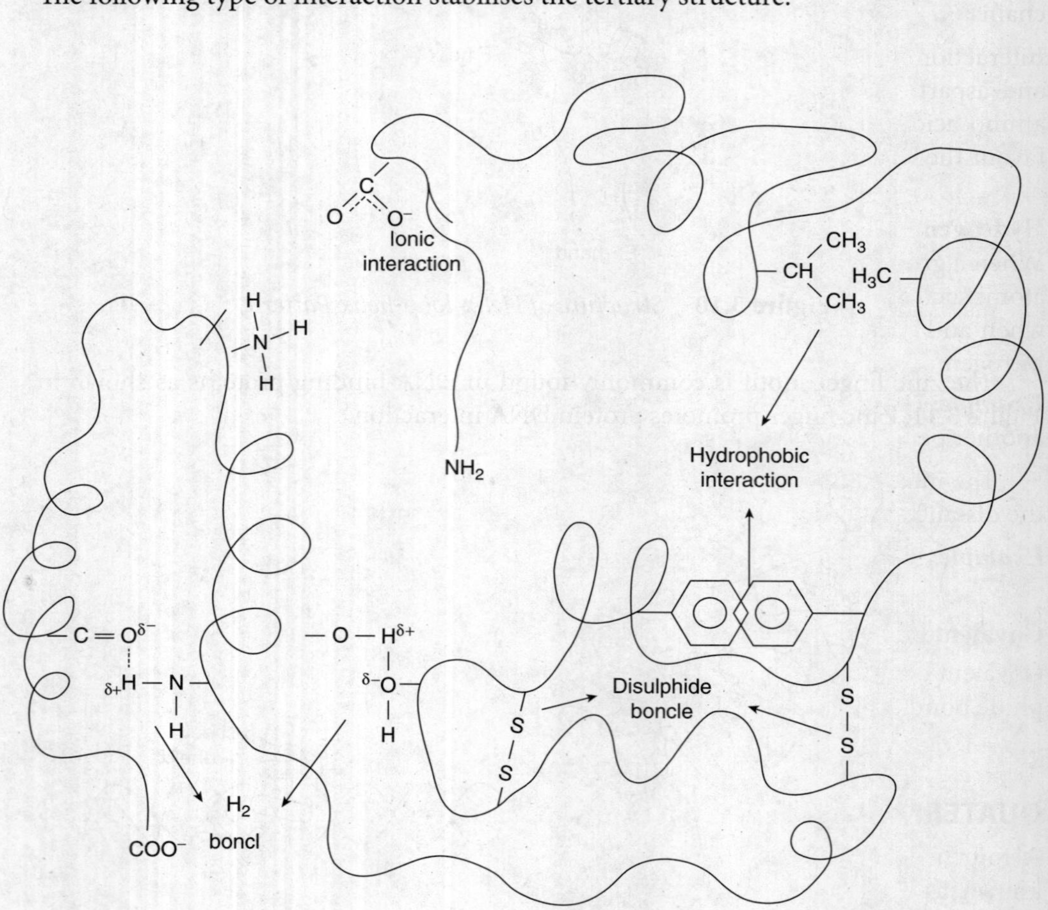

Hydrophobic interactions: Non-polar or hydrophobic bond

As polypeptide folds, hydrophobic R—groups are brought into close proximity as they are excluded from water. Many amino acids like valine, alanine, leucine, isoleucine, methionine, tryptophan, phenylalanine, and tyrosine have side chain or R groups which are hydrophilic. These R— groups combine together with the removal of hydrogen and form linkages between various segments of chains. In addition to this, hydrophobic bonds are involved in protein interaction.

Ionic or electrostatic bond or salt linkage or salt bridge

Electrostatic bonds are formed when two oppositely charged groups come close together and attract each other.

Example: Divalent cations like Mg^{2+} may form electrostatic bonds with two acidic side chains.

Interaction between acidic and basic groups of amino acid: The R—group of glutamic and aspartic acid contributes negatively charged carboxylate groups, and the basic amino acids (arginine, histidine, and lysine) contain positively charged groups and favour the ionic bonds in between these amino acids.

Hydrogen bonds: (CO—NH)

When a group containing a hydrogen atom is covalently bonded to an electronegative atom, such as O_2 or nitrogen favours hydrogen bonding. A hydrogen bond is formed when an H_2 atom shares electrons with two neighbouring atoms, that is, oxygen and nitrogen.

Example: Hydrogen atom of one peptide shares its electron with the carbonyl atom of another peptide.

The interaction occurs between C—O and proton of an NH— or OH group with the distance of about 2.8Å.

Example: Silk fibrion

Covalent bond – Disulphide bond (—S—S—)

Covalent bond is formed between two cysteine amino acid residues; as a result, disulphide bonds are formed. An example is cystine.

QUATERNARY STRUCTURE

A fourth degree of complexity in protein structure has been recognised, which is known as quaternary structure. High molecular weight proteins are composed of several polypeptide chains. Each polypeptide components is called subunit. Subunits in a protein complex may be identical or quite different. The identical peptides in a polypeptide are referred to as oligomers. Mostly oligomers contain two or four subunits referred to as dimers and tetramers. Synthesis of individual subunits is more efficient than the lengthy polypeptide because in supra molecular structures such as collagen, replacement of worn-out or damaged components can be managed more efficiently.

Quaternary structure is stabilised by non-covalent interactions such as hydrophobic interactions, electrostatic interactions, and hydrogen bonds.

During protein folding, hydrophilic interactions are very important because the structure of complementary interfacing surfaces is the same as those observed in the interior globular protein domains.

Example of quaternary structure is globular protein – haemoglobin. Haemoglobin represented as Hb—is a roughly spherical molecule found in RBC where its primary function is to transport O_2 from lungs to every tissue in the body. Its tetramer protein contains four polypeptide chains. The four chains are held together by non-covalent interactions. Each chain contains a haem group and a single O_2 binding site as shown in Figure 3.13.

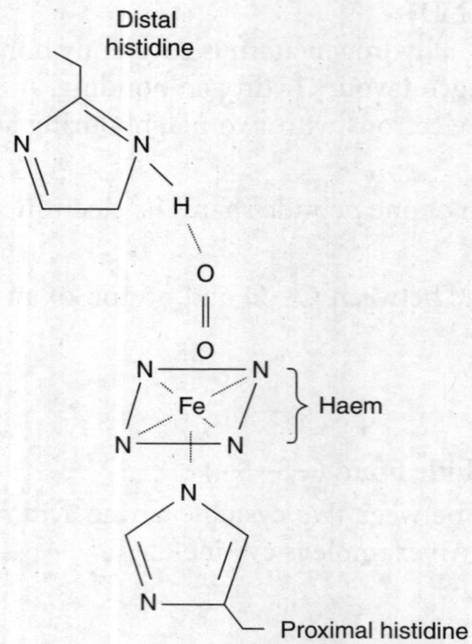

Figure 3.13 O_2 *Binding Site of Haem Created by a Folded Globin Chain*

Haemoglobin A[1], the principle Hb in adults, consists of two alpha (α) chains and two beta (β) chains. Adults also have delta (δ) chains in place of two beta (β) chains of Hb A[2] in $\alpha_2\delta_2$.

The first Hb appearing during embryonic development has two subunits $\zeta_2\varepsilon_2$; the two zeta (ζ) chains are analogous to the α-chains and the two epsilon (ε) chains are analogous to the β-chains.

When after about six weeks, ζ chain production increases, the tetramer $\alpha_2\ \varepsilon_2$ appears; that is, ζ is replaced by α. Third embryonic haemoglobin is $\zeta_2\ \gamma_2$, where ε is replaced by γ. These last two Hb represent transition phases, leading to the appearance of fetal haemoglobin (Hb F), whose tetrameric composition is $\alpha_2\ \gamma_2$. The α and ζ chains contain 141 residues each. The β, γ, and δ chains contain 146 residues each have homologous amino acid sequence as shown in Figure 3.14. The γ and δ chains differ from the β chain at 39 and 10 amino acid residues.

Human Hb protein consists of four polypeptide chains of two types; two α-chains is two β-chains. The polypeptide portion is collectively called globin.

The α-chain has valine at the N-terminal and arginine at the C-terminal, whereas in the β-chain, valine in situated at the N-terminal and histidine at the C-terminal.

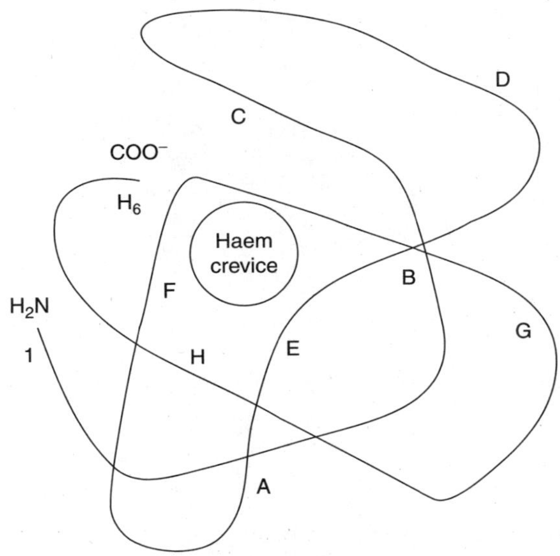

Figure 3.14 *Structure of Single β-Chain of Haemoglobin*

The α-chains are in contact with the β-chains. Apart from this contact, there are also interactions between the two α-chains and the two β-chains. The haem prosthetic group fits into the crevice near the exterior of the globin chain. O_2 binds to the haem groups.

The α-chain contains 141 residues and the β-chain contains 146 residues. So, totally there are 574 ($141 \times 2 + 146 \times 2 = 574$) amino acid residues. Individual α-chains and β-chains have a characteristic tertiary structure, in which the chain is folded.

The bond between the α-chains and the β-chains are ionic bonds and hydrogen bonds. The two pairs of chain are joined by ionic, hydrogen, and hydrophobic forces. Thus four polypeptides are joined together into tetrahedral structure and form a quaternary structure, as shown in Figure 3.15.

The haem group is occupied in a pocket lined with hydrophobic R—groups. It is fixed to its polypeptide chain through a coordination bond of the iron group to the R—group of histidine residue.

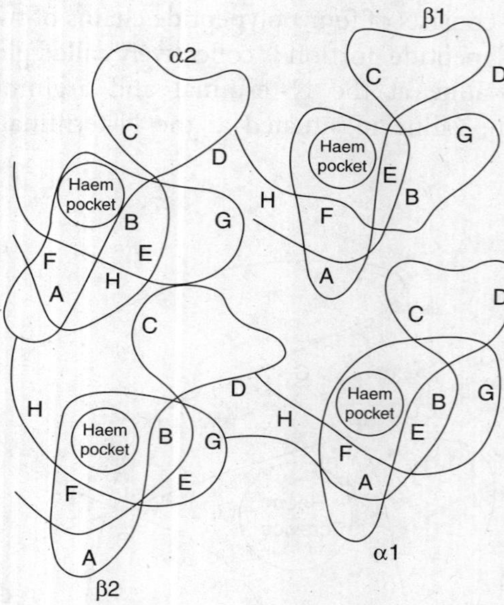

Figure 3.15 *Quaternary Structure of Haemoglobin*

Haemoglobin

Protein contains four subunits designated to α and β. Each subunit contains a haem group that binds reversibly with O_2.

The nine highly conserved amino acid residues in haemoglobin are given in Table 3.1.

Table 3.1 *Highly Conserved Amino Acid Residues in Haemoglobin*

Position	Amino acid	Role
F_8	Histidine	Proximal haem-linked histidine
E_7	Histidine	Distal haem-linked histidine
CD_1	Phenylalanine	Haem contact
F_4	Leucine	Haem contact
B_6	Glycine	Allows close approach between B and E helices
C_2	Proline	Helix formation
HC_2	Tyrosine	Cross links the F and H helices
C_4	Threonine	Uncertain
H_{10}	Lysine	Uncertain

Fibrous protein

Fibrous proteins contain secondary structures such as α-helices or β-pleated sheets. As a result of the rod-like or sheet-like shapes, fibrous proteins have structural than dynamic roles.

α-keratin

In fibrous proteins, bundles of helical polypeptide are commonly twisted together into larger bundles. The structural unit of the α-keratins—a class of proteins found in the hair, wool, skin, horns, and finger nails is an α-helical polypeptide. Each peptide has three domains: an amino terminal head domain, a central rod-like α-helical domain, and a carboxyl terminal tail.

There are two families of α-keratin polypeptides: type 1 and type 2. One peptide from each family associates to form a coiled coil dimer. Two anti-parallel rows of theses dimers form a supercoiled structure called protofilament. Disulphide bridges and hydrogen bonds are the basic interactions between protofilaments. Four protofilaments are packed together to form hundreds of filament. All the filaments collectively are called as macrofilaments. The hair cell, also called a fibre, is made up of several microfibrils, as shown in Figure 3.16.

The α-helical rod-like domains of a type 1 polypeptide and type 2 polypeptide form a coiled coil. Two staggered anti-parallel rows of these dimers form a supercoiled protofilament. Hundreds of filaments, each containing four protofilaments, form a microfibril.

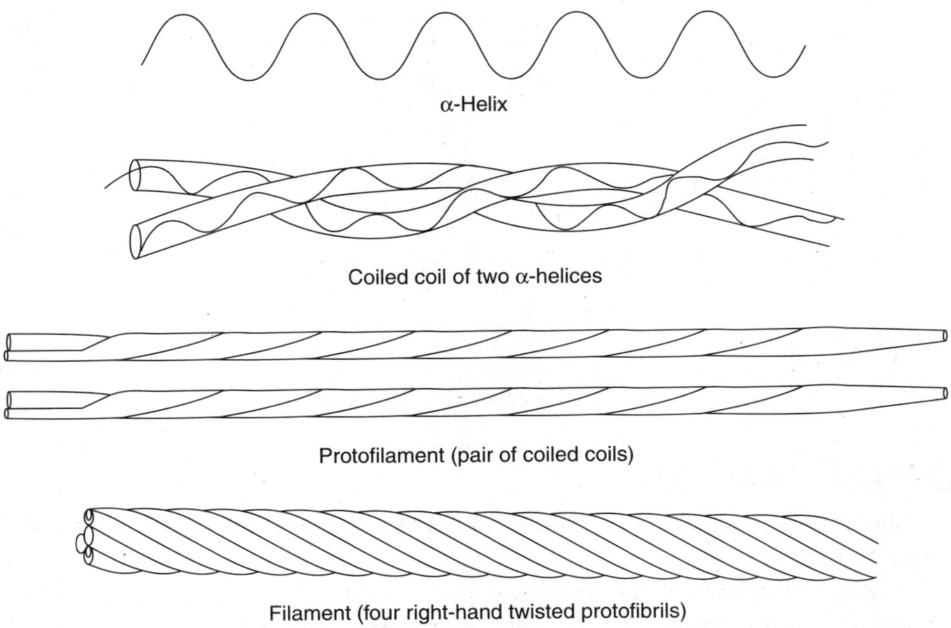

Figure 3.16 *Diagrammatic Representation of α-keratin Structure*

Collagen

Collagen is the most abundant protein in the vertebrates. It is present in skin, bones, blood vessels, and corneas. Collagen is composed of three left-handed polypeptide helices as shown in Figure 3.17 are twisted around each other to form right-handed super helix. The amino acid composition of collagen is different. Glycine constitutes one-third of amino acid residues. Proline and 4-hydroxy proline accounts for 30% of amino acid residues. A small amount of 3-hydroxy proline and 5-hydroxy lysine also contributes to the structure of collagen. Triplets of sequence Gly—*x*—*y* is repeated in large numbers in a collagen structure, where *x* and *y* are mostly proline and hydroxy proline.

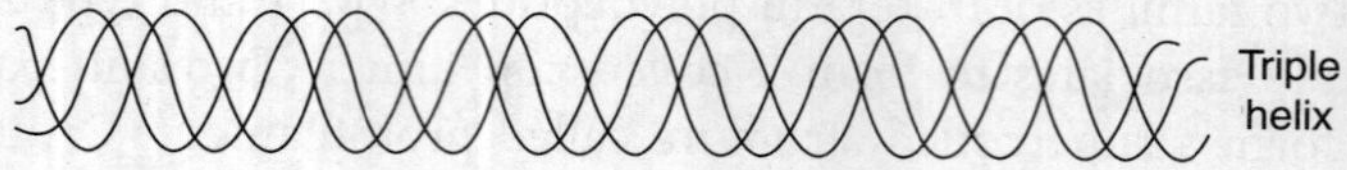

Figure 3.17 *Diagrammatic Representation of Triple Helix Structure of Collagen*

Silk fibroin

Several insects and spiders produce silk substance that contains a fibrous protein, fibroin. In fibroin, which is considered to be a β-keratin, the polypeptide chains are arranged in anti-parallel β-pleated sheet confirmations because the pleated sheets are loosely bonded to each other (van der Waals force) as shown in Figure 3.18. They slide over each other easily. This arrangement gives silk fibers their flexibility.

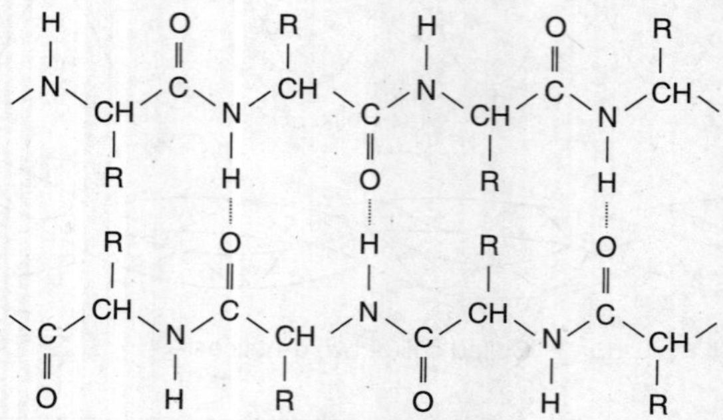

Silk fibroin

Figure 3.18 *Structure of Silk Fibroin*

DENATURATION

'The phenomenon of disorganisation of native protein structure is known as denaturation.'

Denaturation results in the loss of secondary, tertiary, and quaternary structures of proteins. This involves a change in physical, chemical, and biological properties of protein molecules.

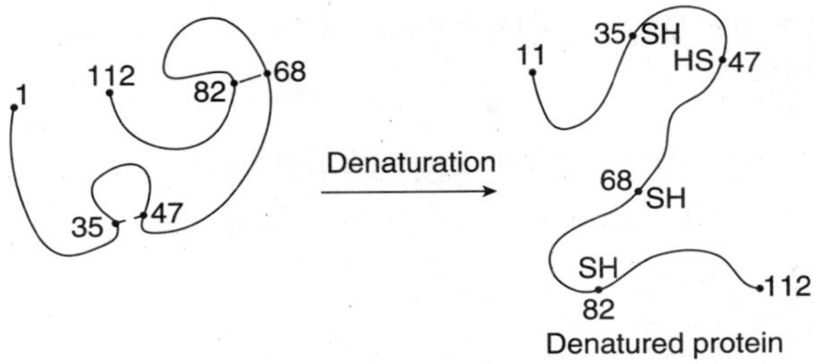

AGENTS OF DENATURATION

Physical agents: Heat, violent shaking, X-rays, UV radiation, and high pressure (5,000–10,000 atm).
Chemical agents: Acids, alkalis, organic solvents (ether and alcohol), aromatic anions (salicylates), salts of heavy metals (Pb and Hg), urea, and salicylates.

CHARACTERISTICS OF DENATURATION

The native helical structure of protein is lost.

The primary structure of a protein with peptide linkages remains intact; that is, peptide bonds are not hydrolysed.

Protein loses its biological activity.

Denaturated protein becomes insoluble in the solvent in which it was originally soluble.

The viscosity of denaturated protein is increased.

Denaturation is associated with increase in ionizable and sulfhydryl groups of protein. This is due to loss of hydrogen and disulphide bonds.

Denaturated protein is more easily digested, that is, due to increased exposure of peptide bonds to the enzyme.

Cooking causes protein denaturation and, therefore, more easily digested.

Denaturation is usually irreversible.

Example: Omelet can be prepared from an egg, which is irreversible.

Careful denaturation is sometimes reversible, which is known as renaturation.

Example: Hb undergoes denaturation in the presence of salicylates by removing salicylates; Hb is renaturated.

Denaturated protein is not crystallised.

PROPERTIES OF PROTEINS

Physical properties

Colour and taste: Proteins are colourless and usually tasteless. These are homogenous and crystalline.

Shape and size: Proteins range in shape from simple crystalloid spherical structure to long fibrillar structure.

1. *Globular proteins:* Pepsin, albumin, globulin, and insulin.
2. *Fibrillar proteins:* Thread like or ellipsoidal in shape.

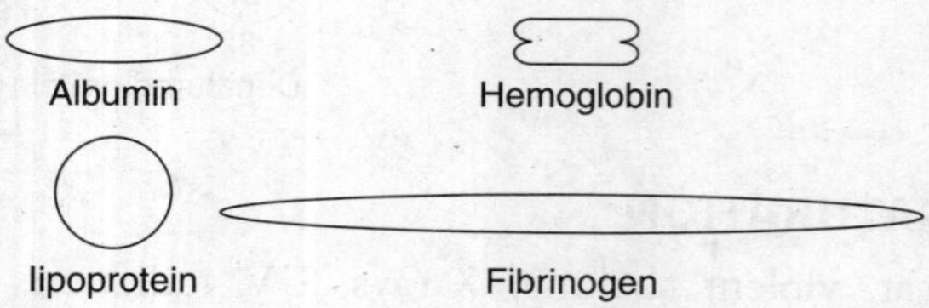

Molecular weight

Proteins vary in their molecular weight, which, in turn, is dependent on the number of amino acid residues. Each amino acid on an average contributes to a molecular weight of about 110. Majority of protein polypeptides may be composed of 40 to 40,000 amino acids with a molecular weight ranging from 4,000 to 4, 40,000.

Example: Insulin – 5,700
Myoglobin – 17,000
Haemoglobin – 64,450

Colloidal nature

Proteins exhibit colloidal properties. They show considerable light-scattering effect in solution and hence result in visible turbidity (Tyndall effect).

Amphoteric nature

Proteins are amphoteric in nature; that is, they have both acidic and basic properties. In the electric field, their migration depends upon the net charge carried by the molecule.

The net charge is influenced by the pH value. Each protein has a specific pI value at which it will not move, because at this point the net charge of the protein is always zero. However, the total charge of the protein at this point is always maximum. Thus, proteins exist as dipolar ions, internal salts, or zwitterions at pI and are present in solution as

$$(+H_3N)_m - R - (COO^-)_n$$

Ion-binding capacity

As proteins are amphoteric in nature, they can form salts with both cations and anions. Many ions form insoluble salts with proteins and therefore they are used as precipitating agents for protein.

Example: Anions of some acids like phosphotungstic, trichloroacetic acid, picric acid, and so on form insoluble acids with proteins. Heavy metals are used for precipitating proteins on the alkaline side of pI, the proteins behaving as anions. Ions of Hg, Cu, Ag, Zn, and so on are frequently used for this purpose.

Solubility

The solubility of proteins usually depends on the pH. It is lowest at the isoelectric point and increases with increasing alkalinity or acidity. The reason is that as protein molecules exist as either cations or anions, repulsive forces between the ions are high and hence the proteins possess excess charges of the same sign. Thus, the proteins are more soluble in these conditions than in the isoelectric state.

Salting in effect

The addition of small quantities of neutral salts increases the solubility of proteins. This process is called salting. Due to the diminished protein, protein interaction is at low salt concentration.

Salting-out effect

The process of protein precipitation by the addition of neutral salts such as ammonium sulphate or sodium sulphate is known as salting out. This phenomenon is explained on the basis of dehydration of protein molecule by salts. This causes increased protein, protein interaction, resulting in molecular aggregation and precipitation.

Optical activity

All proteins rotate the plane-polarised light to the left; that is, these are laevorotatory.
Example: Specific rotation $[\alpha]_D$ for ovalbumin is near –30° over the pH range 3.5 × 11. However at lower or higher pH values the rotation becomes more negative eg at pH 13, the $[\alpha]_D$ is about –60°.

Colour reactions of proteins

Proteins give several colour reactions and are often useful in identifying the nature of amino acids.

BIURET REACTION

Biuret is a compound formed by heating urea at 180°C.

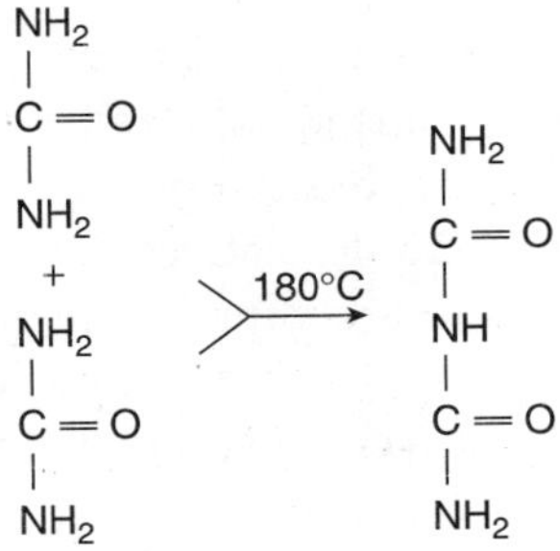

Urea Biuret

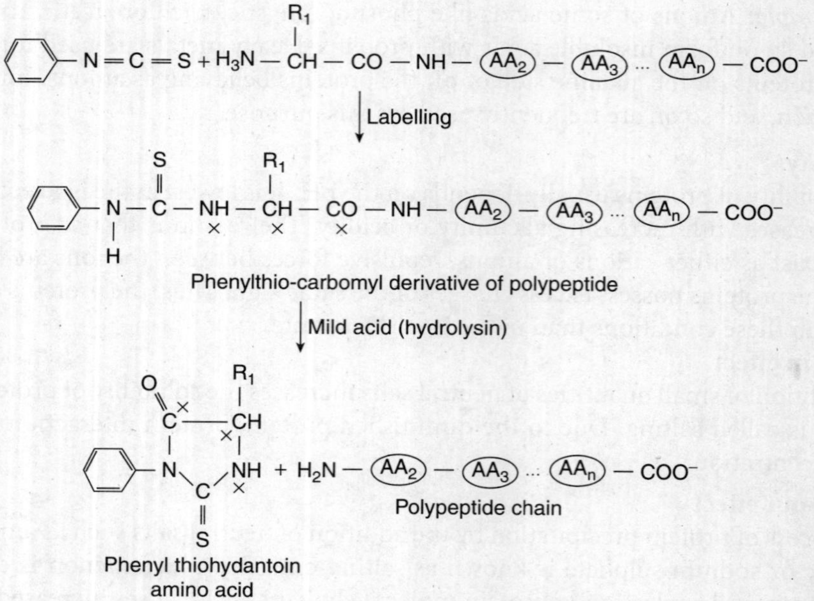

SEQUENATOR

It is an automatic machine to determine the amino acid sequence in a polypeptide chain. The individual amino acids obtained by hydrolysis using Edman's–Sanger's method can be separated by cation-exchange chromatography. In this method, amino acids are poured to a column that contains a resin to which a negatively charged group is bound. The amino acids are sequentially released from the charged group in the column by eluting with solutions of increasing pH. As the pH increases, the amino acids release hydrogen ion from the α-carboxyl groups and then from the side chains of the α-amino groups.

QUANTITATIVE ANALYSIS

The amino acids that are separated from the column should be quantitated by heating them with ninhydrin, which forms purple-coloured compounds.

The amount of each amino acid is determined spectrophotometrically by measuring the absorbance of the amino acid derivatives, which is recorded continuously, on a chart recorder or fed directly into a computer for quantitative analysis. Each peak corresponds to an individual amino acid. The analysis described above is usually performed by using an amino acid analyser, an automated machine, which is described in Figure 3.19

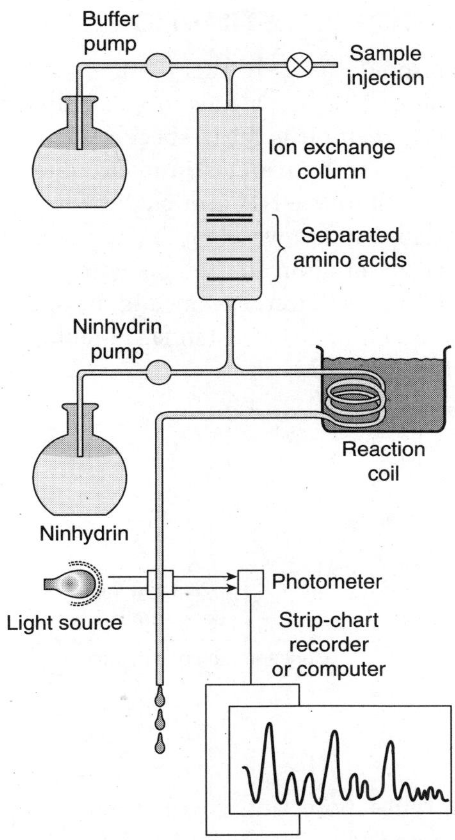

Figure 3.19 *Diagrammatic Representation of Amino Acid Analyser*

PROTEIN ESTIMATION

During protein purification, the concentration of the protein in the solution should be determined. Colorimetric methods are generally used to determine the protein concentration in the purified sample. The amount of this coloured product is then measured spectrometrically, and the amount of colour related to the amount of protein is present by appropriate collaboration.

Various methods to estimate protein are as follows:

1. UV absorption method
2. Lowry's method
3. Bradford's method
4. Biuret method
5. Kjeldhal's method

BIURET METHOD OF PROTEIN ESTIMATION

Biuret is the method commonly used. In this method, the CO—NH groups of proteins form a purple complex with copper ions in an alkaline medium. As all proteins contain peptide bonds, this method is fairly specific. Biuret reagent is a mixture of alkaline copper sulphate and sodium potassium tartarate. The cupric ions form a CO— ordination complex with four —NH groups present in peptide bonds, giving an absorption maximum at 540–560 nm wavelength. Standard protein is also simultaneously prepared bovine serum albumin in the concentration 5 mg/ml and run along with protein sample including the Biuret reagent, and the optical density of each sample and standards are read at 520 nm. Draw a standard graph, and from this, the protein (sample) concentration is calculated.

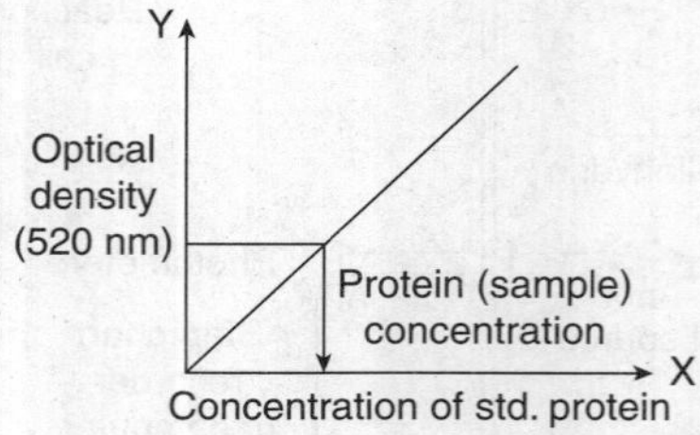

LOWRY'S METHOD

This method is about ten times more sensitive than Biuret method. The reagent Folin-Ciocalteu is quite complex and contains phosphomolybdic acid and tungstate. The proteins which contain tyrosine and tryptophan react with these and produce a dark blue colour. A standard protein (BSA) was prepared and used for the assay. This method is based on both the Biuret reaction, where peptide bonds of proteins react with Cu^{2+} under alkaline condition producing Cu^{+}, which reacts with Folin's reagent and reduces phosphomolybdate tungstate to heteropoly-molybdenum blue by the copper catalysed oxidation of aromatic amino acid. A graph is drawn using standard values, and from this, unknown (protein sample) is calculated. This shows how much of protein was present in the isolated protein sample.

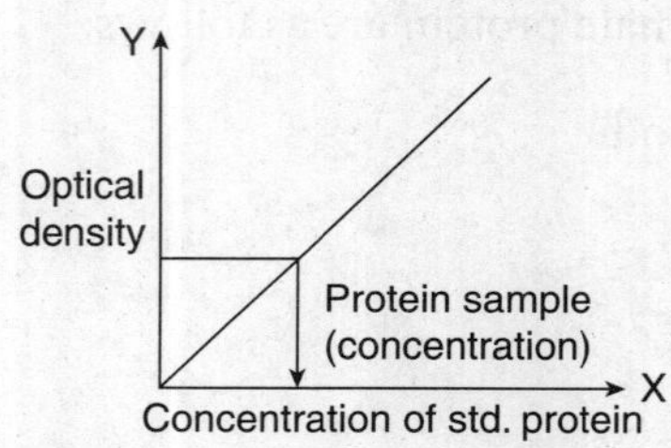

ISOLATION OF PROTEINS

Isolation of proteins without alteration in their biological function is very important.

EXTRACTION OF PROTEINS FROM TISSUES

In biological systems, proteins exist in a variety of forms; some are in free soluble form; some are bound to nucleic acids; and some are bound to lipid-rich membranes. Each type has to be extracted differently.

PREPARATION OF HOMOGENATE

Prepare a homogenate of a biological tissue, liver. Dissect out the liver from an albino rat, dry it, folding between folds of filter paper sheets, and drop it in a beaker kept in ice. Using a pair of scissors, mince it into small pieces. Add about 10 ml of cold distilled water to the mince, transfer to a homogeniser, and grind well. The operation should be carried out in cold, and frothing should be avoided. The homogenate so obtained is filtered through two to three layers of muslin to remove any lumps that may be present. The filtrate is diluted to 20 ml and stored in cold. This homogenate is kept for use in all experiments. Determine the protein content of the various fractions (O–E) by any suitable method, preferably by the modified Biuret method.

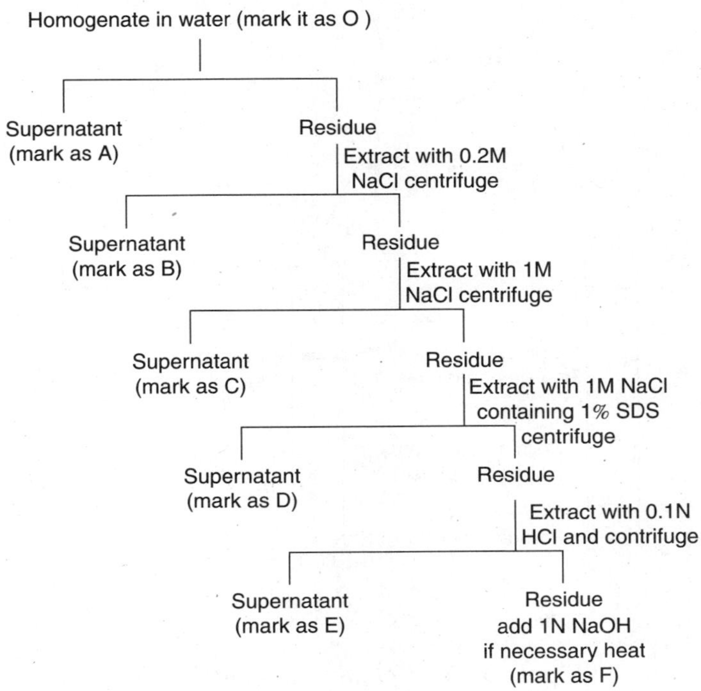

PURIFICATION OF PROTEIN

A pure preparation of a protein is essential, before studying its properties, amino acid composition and amino acid sequencing. The source of a protein is generally tissue or microbial cells. The first step in purifications is to open the cells and to release the proteins into a solution known as a crude extract. Next, differential centrifugation is carried out to collect the subcellular fractions. The subcellular fractions are now available for purifying one or more of the proteins present in them. Based on the size and charge, the proteins are separated and this process is known as fractionation. The solubility of protein is generally lowered at high salt concentrations—an effect called 'salting-out'. The addition of salt in right amounts can selectively precipitate some proteins, while others remain in solution. Ammonium sulphate $(NH_4)_2\ SO_4$ is often used for this purpose because of its high solubility in water.

The most appropriate method for fractionating the proteins makes use of column chromatography as shown in Figure 3.20. Here, the proteins are separated based on their charge, size, binding affinity, and other properties. A stationary phase with porous solid material is packed in the column, and a mobile phase (buffer solution) is percolated through the column.Proteins in the eluents are placed on the top of the column

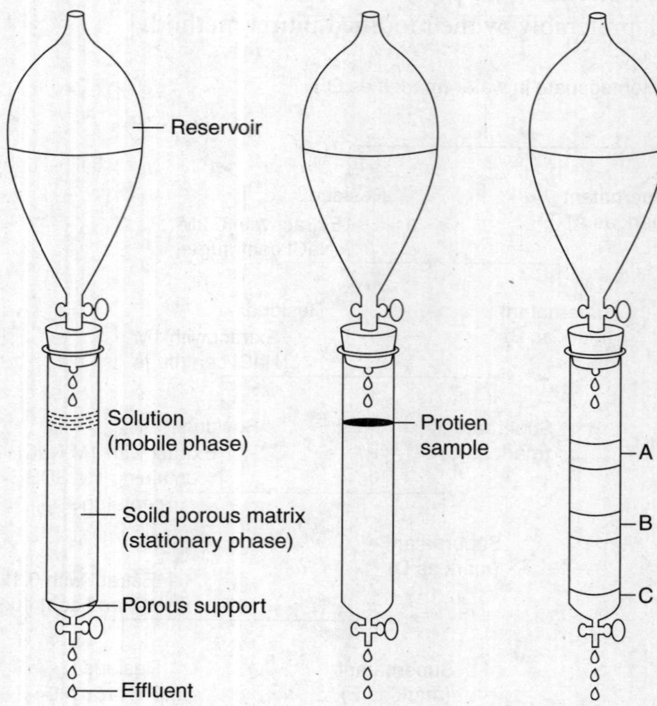

Figure 3.20 *Column Chromatography*

and then percolated through the solid matrix. Proteins in the solutions migrate quickly or slowly based on their size and charge.

For example, in cation exchange chromatography, the solid matrix contains negatively charged groups. Proteins with net positive charge in the mobile phase migrate through matrix more slowly than those with net negative charge; this is because the migration of the former is retarded by the stationary phase. Two bands can be located when two types of proteins are present in the sample. As the length of the column increases, the resolution of the proteins will improve.

QUESTION BANK

SECTION A

1. Haemoglobin exhibits quaternary structure. Justify this statement.
2. Give an account of the features of peptide bonds.
3. Write an account of the structure of collagen.
4. How are proteins classified based on their solubility?
5. What is peptide bond? Comment on its nature.
6. Mention the forces stabilising the secondary, tertiary, and quaternary structures of proteins.
7. Write the structure of two aromatic amino acids.
8. What is a Zwitter ion?
9. What is a peptide bond?
10. List any two forces stabilising the protein structure.
11. Give the structure of methionine and histidine.
12. What are essential amino acids?
13. Explain the use of Sanger's reagent in protein chemistry.

SECTION B

1. Explain the secondary structure of protein.
2. Write briefly on the biological importance of glutathione and oxytocin.
3. Explain the structural characteristics of insulin.
4. Write an account of the chemical properties of protein.
5. What is meant by denaturation and renaturation of protein?
6. Explain the importance of Edman reaction.
7. Describe the primary structure of proteins.

SECTION C

1. How would you identify and quantify the N-terminal residue of protein?
2. Give a detailed account of forces involved in the stabilisation of protein structure.
3. Describe the different levels of protein structure.
4. Explain in detail the determination of amino acid sequence.
5. Discuss the chemical reactions of amino acids involving amino and carboxyl groups.
6. How is the amino acid sequence of a polypeptide determined?

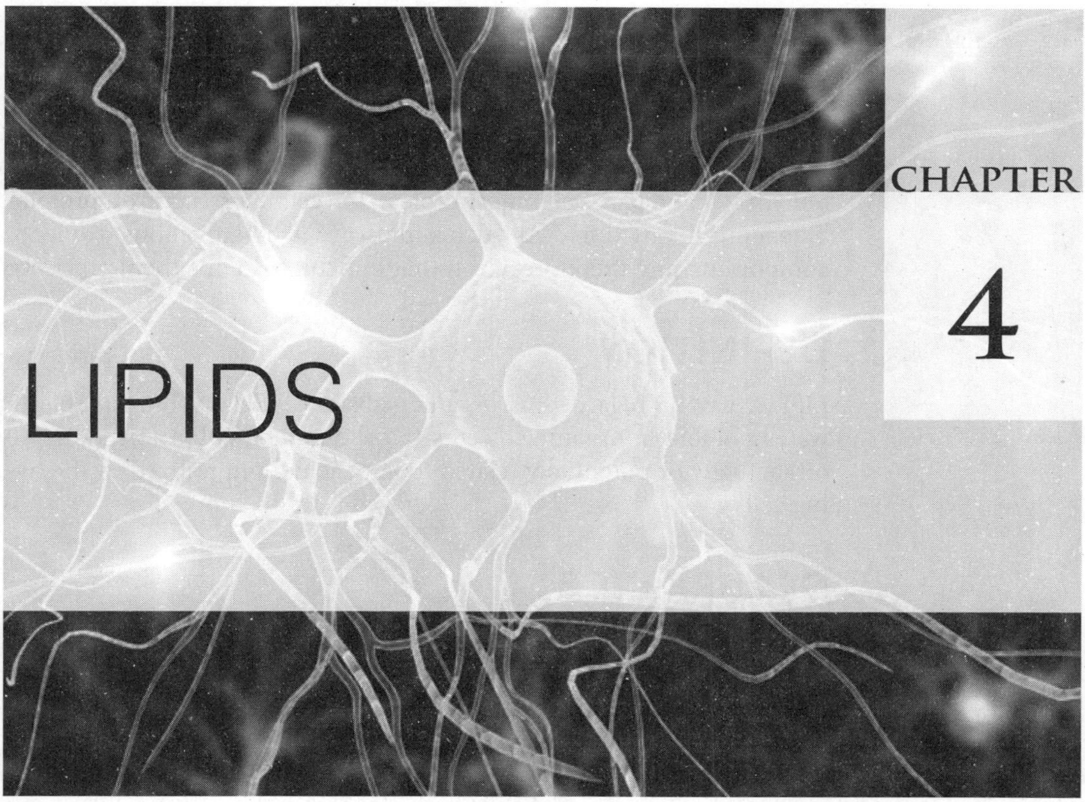

CHAPTER 4
LIPIDS

Concept Outline

- ❑ Introduction
- ❑ Definition
- ❑ Occurrence
- ❑ Biological Significance
- ❑ Classifications of Lipids
- ❑ Simple Lipids
- ❑ Fatty Acids
- ❑ Glycerol
- ❑ Compound Lipids
- ❑ Choline
- ❑ Derived Lipids
- ❑ Sterols
- ❑ Bile Acid
- ❑ Bile Salt
- ❑ C_{21} Steroids—Adrenal Cortical Hormones
- ❑ Mineralocorticoids
- ❑ Isoprenoids
- ❑ Biomedical Importance of Lipids
- ❑ Biomedical Importance of Emulsification
- ❑ Biomediacl Importance of EFA
- ❑ BIomedical Importance of Sterols
- ❑ Function of Prostaglandins
- ❑ Functions of Leukotrienes
- ❑ Lipoproteins
- ❑ Prostaglandins

INTRODUCTION

Lipids (Greek: *lipo*—fat) are efficient energy storage forms next to carbohydrates and play a significant role in the cellular structure and other biochemical functions. As such, lipids are a heterogenous group of compounds, and therefore, it is rather difficult to define them precisely.

DEFINITION

Lipids may be regarded as organic substances relatively insoluble in water, soluble in organic solvents (alcohol, ether, chloroform, and so on) actually or potentially related to fatty acids, and utilized by the living cells.

OCCURRENCE

Fats are widely distributed in plants and animals tissues, from which, they can be extracted by fat solvents like alcohol and ether. In plants, they are particularly present in nuts, seeds, and certain leaves. In animals, cholesterol, phospholipids, and glycolipids are abundantly present in the nervous system. Fat is also deposited in subcutaneous tissues, mesenteric tissues, and fatty tissues around kidney and yellow bone marrow.

FOOD SOURCES RICH IN FAT

Animal sources: Milk, egg, meat, liver, and fish oils.
Plant sources: Nuts, seeds, and oils.

BIOLOGICAL SIGNIFICANCE

1. Lipids are one of the main types of foodstuffs. They act as a major energy fuel to the body and give more heat and energy than carbohydrates.
2. Certain normal breakdown products of fatty acids in the body such as acetic acid and bile acids form important building blocks of biologically active and complex materials like cholesterol, sex hormones, and steroids.
3. Another function of the dietary lipid is to supply essential dietary constituent known as the essential fatty acids (EFA). They resemble vitamins, and they cannot be synthesised by the body and, therefore, should be supplied in the diet in small

quantities for optimal growth. Examples of EFA are linoleic acid, linolenic acid, and arachidonic acid.

4. Fat deposited underneath the skin exerts insulating effect to the body. It protects the body from excessive cold and heat. Obese people can withstand cold and heat better than lean people. The mesenteric fat acts as padding and protects the internal organs.
5. Vitamins A, D, E, and K, which are fat-soluble vitamins, are dietary essentials.

CLASSIFICATIONS OF LIPIDS

The classification proposed by Bloor is the most accepted classification for lipids.

- Simple lipids
- Derived lipids
- Compound lipids
- Substances associated with lipids

Lipid classification is elaborated in Figure 4.1.

SIMPLE LIPIDS

Simple lipids are those which contain esters of fatty acids and alcohol. They are further classified into fats and waxes. The alcohol in fats is glycerol, and the alcohol in waxes is anything other than glycerol. This alcohol has a higher molecular weight.

Neutral lipid, fat, triacylglycerol, or triglycerides are esters of glycerol with fatty acids. Triacylglycerols are the most abundant group of lipids that primarily function as fuel reserves of animals. The fat reserve of normal humans is sufficient to meet the body caloric requirements for two to three months. Fat primarily occurs in adipose tissue: Adipocytes of adipose tissue predominantly found in the subcutaneous layer and in the abdominal cavity are specialised for storage of triacylglycerols.

STRUCTURE OF ACYLGLYCEROLS

Triglycerides or neutral lipids are composed of three molecules of fatty acid, esterified to glycerol. The fatty acids may be of the same type; the resulting glyceride is called simple glycerides. If the three fatty acids in a given glyceride are different, then the resulting glyceride is called a mixed glyceride. The structure of triglycerides is given in Figure 4.2.

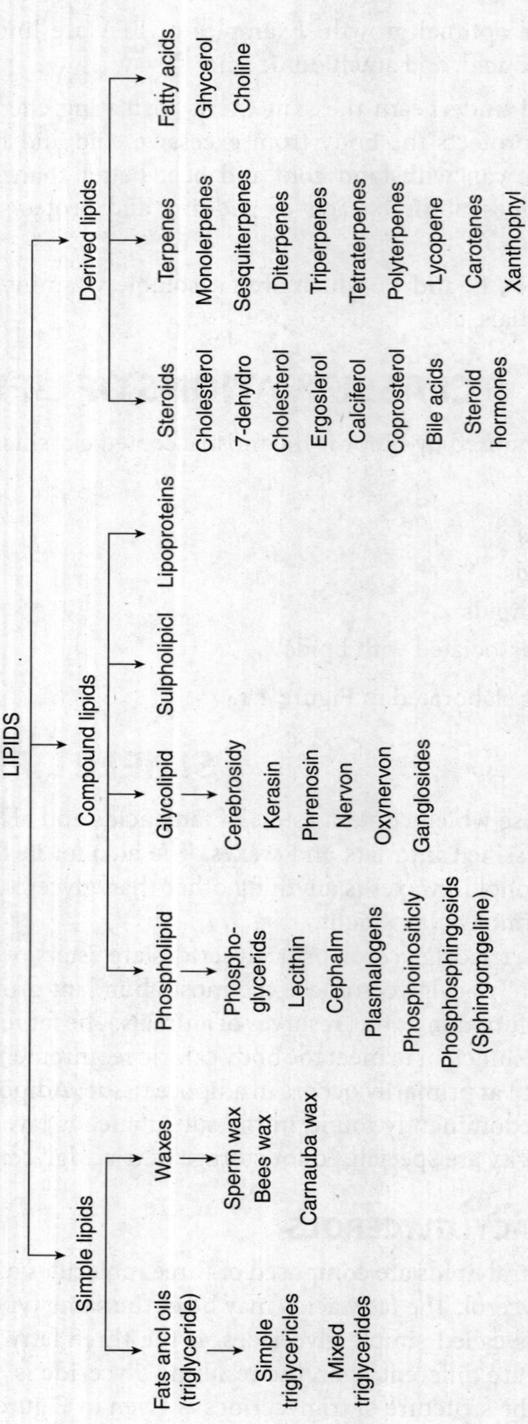

Figure 4.1 *Classification of Lipids*

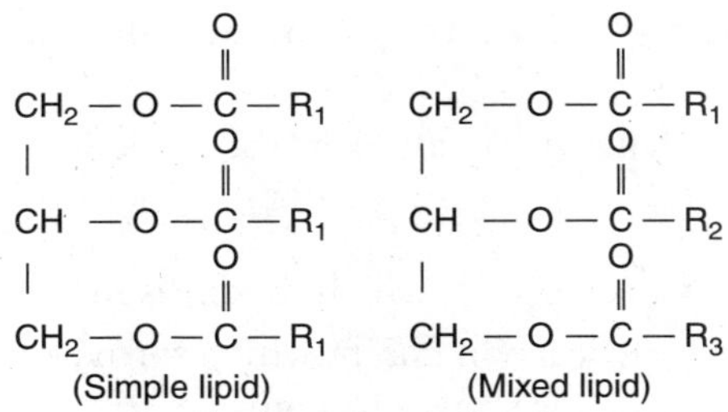

Figure 4.2 *Structure of Triglycerides*

PROPERTIES OF FATS

Physical Properties

1. They are insoluble in water but are readily soluble in fat solvents like alcohol, ether, chloroform, and so on.
2. They are greasy to touch and leave an oily impression on paper.
3. Pure glycerides are tasteless, odourless, and colourless.
4. Fats used in food have specific flavour and colour. The flavour is attributed to the presence of certain foreign substances absorbed by the fat from its natural environment during the processing of the fat; for example, the flavour of butter is due to the presence of bacterial flora, which is carefully controlled to impart special flavour to butter.
5. The quantity of saturated and unsaturated fatty acids present in the fat determines its hardness. For example, fat containing saturated fatty acids will be solid in nature at room temperature, whereas fat containing unsaturated fatty acids, for example oils, will be liquid in nature at room temperature.
6. Fats possess definite melting points. The melting point of fat is usually higher than the temperature at which it solidifies. The melting points of natural fats are not sharp because they are mixtures of several fats.

   Example: Melting point of beef fat is 49.5°C.
   Solidification temperature of beef fat is 36°C.

7. Fats have lesser specific gravity than water and, therefore, float in water. The specific gravity of solid fat is less than liquid fat.

   Specific gravity of solid fat is 0.86.
   Specific gravity of liquid fat is 0.91.

8. *Spreading of fat:* When liquid fat is placed on water, it spreads uniformly over the surface of water. The effect of this property is to lower the surface tension

and help the transport of fat. The phenomenon of spreading is explained as follows:

Two groups in the structure of fats exert their role in the mechanism of spreading. These are the following:

- A hydrophilic carboxyl group, which dissolves in water and is referred as C_1.
- A hydrophobic hydrocarbon chain, which forms a layer on the surface and is referred as C_2 which is shown in Figure 4.3.

Thus, CH_3COOH, a fatty acid, contains the hydrocarbon chain (CH_3), which is referred as C_2 and carboxyl group (COOH), represented as C_1. The two groups are arranged above and below the surface of water.

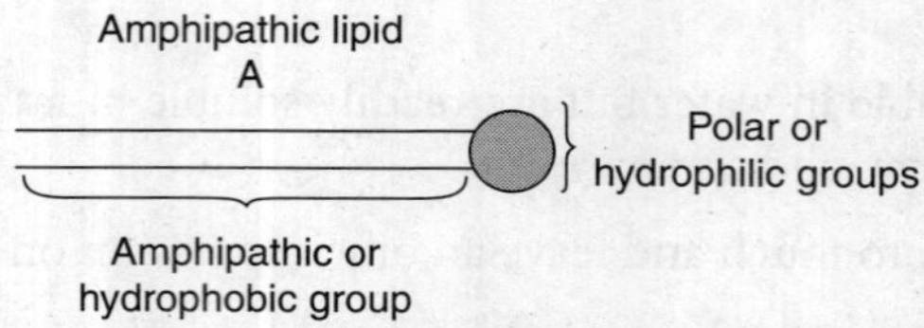

Figure 4.3 *Diagrammatic Representation of Fat Structure*

9. *Emulsification:* Fats are insoluble in water; they can be broken into minute droplets and dispersed in water. This is called emulsification as shown in Figure 4.4. This greatly increases the surface area of fats, and this is an essential requisite for digestion of fats in the intestines. Emulsification is brought about by the following:
   - *Mechanical action:* This can be facilitated by mixing mechanically or passing superheated steam.
   - *Chemical action:* This is accomplished by the action of bile salts, which lower the surface tension.

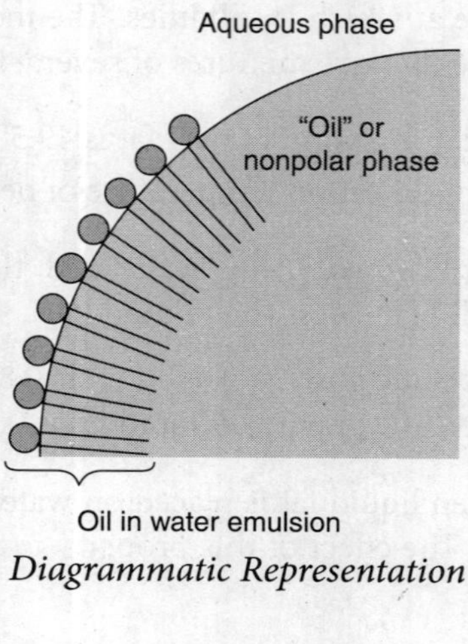

Figure 4.4 *Diagrammatic Representation of Emulsion*

Chemical Properties

1. *Hydrolysis*

Fats, being esters of glycerol and fatty acids, are readily hydrolysed to glycerol and fatty acids by acids, alkalis, superheated steam, and enzymes known as lipases, which act on the ester linkages. Two kinds of hydrolysis are known as enzyme hydrolysis and alkali hydrolysis.

Enzyme hydrolysis is affected by lipase of pancreatic juice. This is important in the digestion of fats. The process of alkali hydrolysis is called saponification, which means 'soap making'. The alkali salt resulting from saponification is soap.

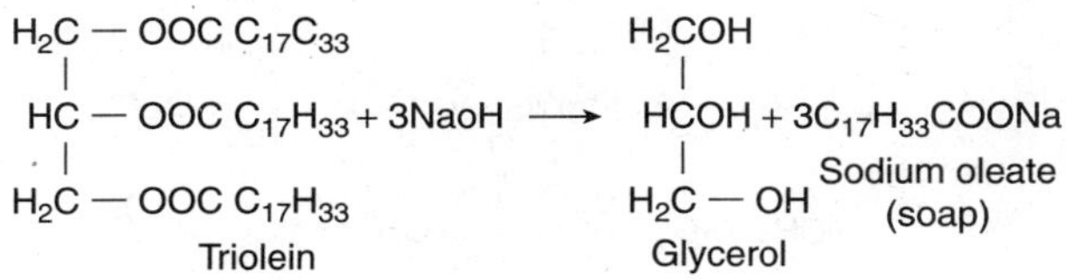

Soaps

The common fats used for making soaps contain predominantly palmitic, stearic, and oleic acids. The soaps we use for washing consist largely of the sodium or potassium soaps of these fatty acids. The potassium soaps are soft and soluble, and the sodium soaps are hard and less soluble in water. When soluble soap is used in hard water, which contains calcium and magnesium ions, the efficiency of producing lather is diminished. This is because the fatty acids in the soap are precipitated by the calcium and magnesium salts of the hard water. The precipitation continues until all the calcium and magnesium ions combine with the fatty acids of the soap.

Soaps are good cleansing agent because they are effective emulsifying agents. Dirt is held to the surface by greasy substances. Soaps, being good emulsifying agents, lower the surface tension also. The greasy materials are emulsified by the soaps, and dirt is carried away with them in washing.

Reactions of soap in dilute solution: A strong solution of pure soap in water is clear whereas, when this is diluted with more water, the solution becomes cloudy and opalescent. This is because on dilution, soaps undergo partial hydrolysis with liberation of insoluble fatty acids, which make the solution cloudy.

$$RCOONa + H_2O \longrightarrow (RCOOH + NaOH)$$

The liberation of NaOH in this way accounts for the alkalinity of ordinary washing soap, which, if used on the skin, has a deleterious effect. This disadvantage is overcome in the preparation of toilet soaps by the addition of excess of fatty acids or by using large proportion of sodium oleate, which is not so readily hydrolysed in water.

2. *Autoxidation*

   Unsaturated fatty acids in fats react with ozone and oxygen to undergo a reaction called autoxidation, which is the presence of appropriate amount of moisture, light, warmth, and catalytic substances. Ozonides, peroxides, aldehyde, and ketones are formed.

   - With ozone, an unstable ozonides is formed, which later cleaves by water to give rise to two aldehyde groups.

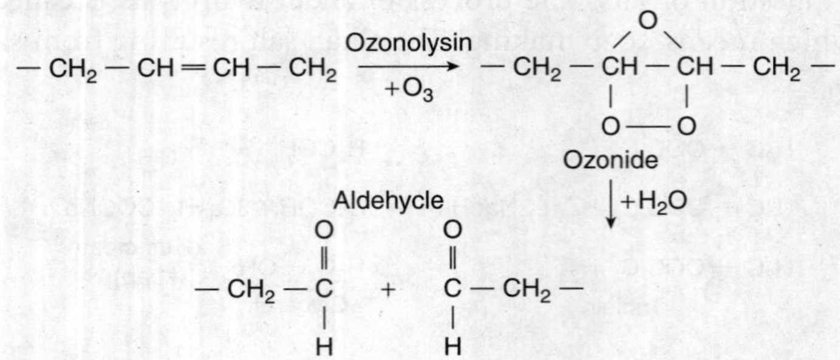

   - With $KMnO_4$, under mild condition, the glycols are formed at the sites of double bonds.

$$CH_3(CH_2)_7CH{=}CH(CH_2)_7COOH \text{ (Oleic acid)} \xrightarrow{KMnO_4} CH_3(CH_2)_7CH(OH){-}CH(OH){-}(CH_2)_7COOH$$

9, 10 Dihydroxy stearic acid

   - Under vigorous conditions, the same reagent cleaves the molecule at the double bond and oxidises the terminal portions to the carboxyl group.

$$CH_3(CH_2)_7CH{=}CH(CH_2)_7COOH \xrightarrow{KMnO_4} \underset{\text{Pelargonic acid}}{CH_3(CH_2)_7COOH} + \underset{\text{Azelaic acid}}{HOOC(CH_2)_7COOH}$$

Rancidity

Rancidity is the term used to represent the deterioration of fats and oils, resulting in an unpleasant taste. Fats containing unsaturated fatty acids are more susceptible to rancidity. The chemical changes which occur during rancidification are known as rancidity, and this takes place in two stages and are accordingly referred as the following:

- Hydrolytic rancidity
- Oxidative rancidity

Hydrolytic rancidity

Involves partial hydrolysis of the glycerides to mono- and diglycerides while glycerol and free fatty acids are also liberated. The hydrolysis is hastened by the presence of moisture, warmth, and lipases present in fats. As a result of the liberation of volatile butyric acid, hydrolytic rancidity produces a disagreeable odour and taste in fats that contain a high amount of volatile fatty acids, for example, butter. Butter undergoes rancidification more easily in summer.

Oxidative rancidity

Oils that are rich in unsaturated fatty acids frequently undergo oxidation at ordinary temperature. The oxidation proceeds slowly and forms short chain fatty acids (C_4 to C_{10}) and aldehydes, which give a rancid taste and odour to the fats. This process is also known as oxidative rancidity because it undergoes autoxidation.

Autoxidation takes place by free radicals in which the α-methylene group is first attacked. A hydrogen atom is removed and this leads to a chain of reactions leading to oxidation.

Oxidative rancidity takes place more frequently in animal fats than in vegetable fats. This is because natural antioxidants such as tocopherols, phenols, and napthols present in the vegetable oils will check autoxidation.

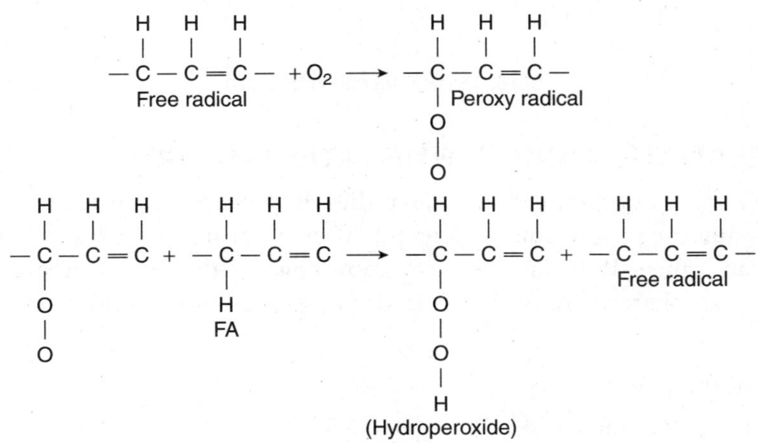

Prevention of rancidity

Rancidification may be prevented by the following:

- Proper storage of the fats in air-light non-metallic receptacles kept in cool place, away from light, moisture, and warmth. Storing in non-metallic vessels is necessary because metals like copper and lead catalyse rancidity.
- Addition of antioxidants like Vitamin C and E.

Kries test is a delicate test used to detect oxidative rancidity. In this test, the fat is treated with ether, phloroglucinol, and hydrochloric acid. Positive test is indicated by the development of a red colour. This is due to the presence of epihydrin aldehyde, which is one of the oxidation products.

Hydrogenation

Unsaturated fatty acids either free or combined in lipids react with gaseous hydrogen to yield the saturated fatty acids.

$$\underset{\text{Oleic acid}}{CH_3(CH_2)_7CH=CH(CH_2)_7COOH} + H_2$$

$$\downarrow$$

$$\underset{\text{Stearic acid}}{CH_3(CH_2)_7CH_2-CH_2-(CH_2)_7COOH}$$

Halogenation

Unsaturated fatty acids and esters can take up halogens like Br_2 and I_2 at their double bonds at room temperature in acetic acid or methanol solution. This reaction is the basis of the iodine number determination.

$$\underset{\text{Linoleic acid}}{CH_3(CH_2)_4CH=CH\,CH_2\,CH=CH(CH_2)_7COOH} + 2I_2$$

$$\downarrow$$

$$CH_3(CH_2)_4\underset{\displaystyle I}{\underset{|}{CH}}-\underset{\displaystyle I}{\underset{|}{CH}}-CH_2-\underset{\displaystyle I}{\underset{|}{CH}}-\underset{\displaystyle I}{\underset{|}{CH}}-(CH_2)_7COOH$$

9, 10, 12, 13 Tetraiodostearic acid

CHARACTERISATION AND IDENTIFICATIONS OF FATS

Fats and oils form essential constituent of diet. It is necessary to identify a pure fat or to determine the proportion of different types of fat mixed in edible oils and fats like butter and ghee. With an adequate knowledge of the characteristic properties of fats, it is possible to identify the fat under investigation by applying the following methods:

1. Physical methods
2. Chemical methods

Physical Methods

- *Specific gravity determination:* Different oils show different specific gravities. The variation from the normal value indicates a mixture of oils.
- *Refractive index determination:* Fats have definite angles of refractive index. Variation from normal value indicates a mixture of fats.
- *Solidification point or setting point:* Solidification point is more specific than melting point. It is the temperature at which the fat melts and sets to solid. Each fat has a unique solidification point.

Chemical Methods

Saponification number

Definition: Saponification number is defined as the milligrams of KOH required to hydrolyse (saponify) one gram of fat or oil.

Example:		
	Human fat:	195 – 200
	Butter:	230 – 240
	Coconut oil:	250 – 260

Significance: The saponification number thus provides information on the average chain length of fatty acids in the fat. It varies inversely with the chain length of the fatty acids. The shorter the average chain length of the fatty acids, the higher is the saponification number.

Acid number

Definition: Acid Number is defined as the number of milligrams of KOH required to neutralise the free fatty acid present in 1 gram of fat.

Example:		
	Butter:	0.5 – 0.6
	Coconut oil:	2.5 – 6

Significance: In normal circumstances, refined oils should be free from any free fatty acids. Oils, on decomposition due to chemical or bacterial contamination, yield free fatty acids. Therefore, oils with increased acid number are unsafe for human consumption.

Iodine number

Definition: Iodine number is defined as the number of grams of iodine absorbed by 1 gram of fat.

Example:		
	Human fat:	65 – 69
	Butter:	25 – 40
	Coconut oil:	7 – 10

Significance Iodine number measures the degree of unsaturation of fatty acids in the fat. The iodine number for solid fats is lesser than that for oils. It should be noted that the iodine number does not provide information about the number of double bonds in the fatty acids.

Polenske number

Definition: Polenske number is defined as the number of millilitres of 0.1 N KOH required to neutralise the insoluble fatty acids (those which are not volatile with steam distillation) obtained from 5 gm of fat.

Example: The Polenske number is usually negligible or very low in pure ghee.

Reichert–Meissl number

Definition: Reichert–Meissl number is defined as the number of 0.1 N KOH required to completely neutralise the soluble volatile fatty acids distilled from 5 gm of fat.

Example: Butter: 25 – 30

Significance: RM number is useful in testing the purity of butter since it contains a good concentration of volatile fatty acids. Butter has RM number in the range of 25–30 while it is less than 1 for most other edible oils. Thus any adulteration of butter can be easily tested by this RM value.

Acetyl number

Definition: Acetyl number is defined as the number of milligram of KOH required to completely neutralise the acetic acid obtained by saponification of 1 gm of fat after it has been acetylated. (The treatment of fat or fatty acid mixture with acetic anhydride results in acetylating of all alcoholic OH groups.)

Example: Butter: 17 – 35
Coconut Oil: 2

Significance: The acetyl number is, thus, a measure of the number of OH groups in the fat. For example, castor oil has a high acetyl number (146) because of high content of a hydroxyl acid.

WAXES

Waxes are esters of fatty acids with alcohol other than glycerol. One such alcohol is cholesterol. They are insoluble in water but soluble in fat solvents. They are not easily hydrolysed like fats or digested by lipases. Therefore, they are of no value nutritionally.

Examples of waxes are the following

1. *Bees wax:* This is secreted by honeybees to from honeycomb. This chief ingredient is myricyl palmitate.
2. *Chinese wax:* This is secretion of an insect.
3. *Carnauba wax:* This is found on the leaves of Carnauba palm of Brazil. They are of value in the manufacture of shoe polish, floor waxes, varnishes, and candles.
4. *Lanoline:* This is also known as wool wax. It is obtained from wool and is used in making ointments. The common wax in human body is found in the blood plasma and suprarenal and sebaceous glands. They are all esters of cholesterol, that is, cholesterol palmitate.

Significance: The importance of waxes in plants and animals is that they serve as protective agents. The waxy coating on the surface of plants prevents excessive loss of moisture. The waxy coating over fruits like apple and citrus fruits prevents them from drying and from being infected with fungi and bacteria. In birds, the presence of wax on the surface of feathers and hairs help to keep them soft and pliable. The waxes prevent the aquatic animals from becoming wet.

FATTY ACIDS

Fatty acid is an important constituent of fat. It is a monocarboxylic acid with a hydrocarbon chain. The length of the chain may vary from 2 to 34 carbon atoms. It is obtained by the hydrolysis of fats by acids, alkalis, or enzymes. Almost all the fatty acids found in nature have straight chain and even number of carbon atoms. The chain may be saturated or unsaturated. The commonest fatty acids that occur in nature are palmitic, stearic, and oleic acids. Among these fatty acids, oleic acid is unsaturated.

Fatty acids are carboxylic acids with hydrocarbon side chains. They are the simplest form of lipids.

Occurrence: Fatty acids occur mainly in esterified form as the major constituents of various lipids. They are also present as free fatty acids. Fatty acids of animal origin are much simpler in structure in contrast to those of plant origin, which often contains groups such as epoxy, keto, hydroxy, and cyclopentane rings.

TYPES OF FATTY ACIDS

Fatty acids are divided into three types:

1. Saturated fatty acids
2. Unsaturated fatty acids
3. Cyclic fatty acids

Even and odd carbon fatty acids

Most of the fatty acids that occur in natural lipids are of even carbons (14C–20C). Among the odd chain fatty acids, propionic acid (3C) and valeric acid (5C) are well known.

Difference between saturated and unsaturated fatty acids

Saturated fatty acids do not contain double bonds, while unsaturated fatty acids contain one or more double bonds, whereas saturated fatty acids do not contain double bonds. Saturated and unsaturated fatty acids are equally present in the natural lipids. If fatty acids contain one double bond, they are known as monounsaturated and if they contain more than one double bond, they are known as polyunsaturated fatty acids (PUFA).

Saturated Fatty Acids

These are fatty acids which do not contain double bonds. They do not exhibit addition reactions, and their iodine values are nil. Generally, the presence of saturated fatty acid makes the fat solid. They are built up on acetic acid, which is the first member of the series. General formula for saturated fatty acids is (CnH_2n H COOH). Refer Table 4.1. They contain even number of carbon atoms.

Table 4.1 *Saturated Fatty Acids*

Common Name	Systemic Name	Abbreviation	Structure
Acetic acid	Ethanoic acid	2: 0	CH_3COOH
Propionic acid	n-Propanoic acid	3: 0	CH_3CH_2COOH
Butyric acid	n-Butanoic acid	4: 0	$CH_3(CH_2)_2COOH$
Valeric acid	n-Pentanoic acid	5: 0	$CH_3(CH_2)_3COOH$
Caproic acid	n-Hexanoic acid	6: 0	$CH_3(CH_2)_4COOH$
Caprylic acid	n-Octanoic acid	8: 0	$CH_3(CH_2)_6COOH$

(*Continued...*)

Table 4.1 *Continued*

Common Name	Systemic Name	Abbreviation	Structure
Capric acid	n-Decanoic acid	10: 0	$CH_3(CH_2)_8COOH$
Lauric acid	n-Dodecanoic acid	12: 0	$CH_3(CH_2)_{10}COOH$
Myristic acid	n-Tetradecanoic acid	14: 0	$CH_3(CH_2)_{12}COOH$
Palmitic acid	n-Hexadecanoic acid	16: 0	$CH_3(CH_2)_14COOH$
Stearic acid	n-Octadecanoic acid	18: 0	$CH_3(CH_2)_{16}COOH$
Arachidic acid	n-Psicosonioc acid	20: 0	$CH_3(CH_2)_{18}COOH$

Unsaturated Fatty Acids

These are fatty acids which contain double bonds. Iodine values vary according to the degree of unsaturation. They are generally liquid at room temperature; the general formula is CnH2n-1 COOH. Unsaturated fatty acids are further divided according to the degree of unsaturation.

1. *One double bond:*
   Palmitoleic acid C16: 1; 9
   Oleic acid C18: 1; 9
2. *Two double bonds:*
   Linoleic acid C18: 2; 9; 12
3. *Three double bonds:*
   Linolenic acid C18: 3; 9; 12; 15
4. *Four double bonds:*
   Arachidonic acid C20: 4; 5; 8; 11; 14

Table 4.2 *Unsaturated Fatty Acids*

Common Name	Systemic Name	Abbreviation	Structure
Palmitoleic acid	Hexadecenoic acid	16: 1; 9	$CH_3(CH_2)_5CH{=}CH(CH_2)_7COOH$
Oleic acid	Octadecenoic acid	18: 1; 9	$CH_3(CH_2)_7CH{=}CH(CH_2)_7COOH$
Linoleic acid	Octadecadienoic acid	18: 2; 9; 12	$CH_3(CH_2)_4CH{=}CH_2CH{=}CH-(CH_2)_7COOH$
Linolenic acid	Octadecatrienoic acid	18: 3; 9; 12; 15	$CH_3CH_2CH{=}CHCH_2CH{=}CHCH-CH{=}CH(CH_2)_7COOH$
Arachidonic acid	Eicosatetraenoic acid	20: 4; 5; 8; 11; 14	$CH_3(CH_2)_4CH{=}CH_2CH{=}CH\,CH_2CH + CH\,CH_2CH{=}CH\,CH_2-CH{=}CH(CH_2)_3COOH$

Cyclic Acids

Cyclic acids are fatty acids having a cyclic ring structure containing five carbon atoms. They are unsaturated. Two cyclic acids, hydnocarpic acid and chaulmoogric acids, are of interest because of their presence in chaulmoogric oil, which is used in the treatment of leprosy. The structure of chaulmoorgic acid is shown below:

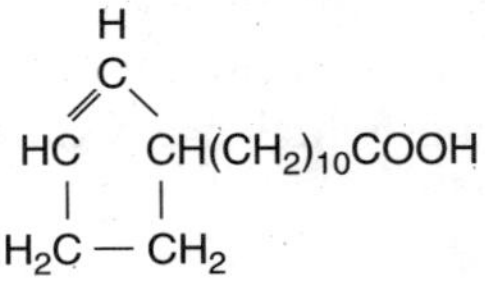

Hydrocarpic acid

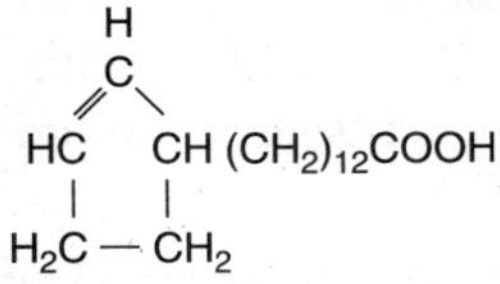

Chaulmoorgic acid

ESSENTIAL FATTY ACIDS

The fatty acids that cannot be synthesised by the body and, therefore, should be supplied in the diet are known as essential fatty acids (EFA). Chemically, they are polyunsaturated fatty acids (PUFA), namely linoelic acid (18: 2; 9, 12) and linolenic acid (18: 3; 9, 12, 15). Arachidonic acid (20: 4; 5, 8, 11, 14) becomes essential. Linolenic and linoleic acids are essential since humans lack the enzyme that can introduce double bonds beyond carbons 9 and 10.

Functions of EFA

Essential fatty acids are required for the membrane structure and function, transport of cholesterol, function of lipoproteins, prevention of fatty liver, and so on. They are also needed for the synthesis of another important group of compounds, namely eicosanoids. They are important for the growth of the cells and the reproductive function. They prolong clotting time and increase the fibrinolytic activity.

Examples:

Linoleic acid: 18: 2; 9, 12

$$CH_3(CH_2)_4CH{=}\underset{12}{CH}CH_2CH{=}\underset{9}{CH}(CH_2)_7COOH$$

Linolenic acid: 18: 3; 9, 12, 15

$$CH_3CH_2CH{=}\underset{15}{CH}CH_2CH{=}\underset{12}{CH}CH_2CH{=}CH(CH_2)_7\underset{9}{COOH}$$

Arachidonic acid:

$$CH_3(CH_2)_4CH{=}\underset{14}{CH}CH_2CH{=}\underset{11}{CH}CH_2CH{=}\underset{8}{CH}CH_2CH{=}CH(CH_2)_3\underset{5}{COOH}$$

Deficiency of EFA

Deficiency in infants and children produces skin lesions like eczema.

Diet rich in EFA: Sunflower oil, corn oil, cottonseed oil, peanut oil, egg, and lard

NOMENCLATURE OF FATTY ACIDS

The naming of fatty acid (systemic name) is based on the hydrocarbon from which it is derived. The saturated fatty acids end with a suffix *-anoic* (e.g. octanoic), while the unsaturated fatty acids end with a suffix *-enoic* (e.g. octadecaenoic acid). In addition to systemic names, fatty acids have common names which are widely used.

NUMBERING OF CARBON ATOMS

It starts from the carboxyl carbon which is taken as one. The carbons adjacent to thin (carboxyl C) are 2, 3, 4, and so on, or alternately, α, β, γ, and so on. The terminal carbon containing methyl group is known as ω carbon.

LENGTH OF HYDROCARBON CHAINS

Depending on the length of carbon chains, fatty acid are categorised into three carbons – short chain with less than six carbons, medium chain with eight to fourteen carbons, and long chain with sixteen to twenty-four carbons.

PROPERTIES OF FATTY ACIDS

Physical Properties

1. *Solubility:* The lower fatty acids like acetic and butyric acids are soluble in water. Higher fatty acids from caproic acid onwards are insoluble in water.
2. *Melting point:* Fatty acids differ in their melting points; generally, the melting point is lowered if
   (a) fatty acids are unsaturated and
   (b) fatty acids have shorter chain lengths.
3. *Spreading:* The phenomenon of spreading is the property of the fatty acids contained in the fat and refers to the formation of a thin and uniform layer of fat or oil when placed over the surface of water.

Chemical Properties

These are due to the carboxyl group and the hydrocarbon chain of fatty acid.

Properties due to carboxyl group

Soap formation: Fatty acids form salts (soaps) with alkalis because of the presence of a terminal carboxyl group.

$$RCOOH + NaOH \longrightarrow RCOONa + H_2O$$

Ester formation

The carboxyl group condenses with alcohol to form esters. Triglyceride is an ester of fatty acid with glycerol. Cholesterol ester is formed by esterification of fatty acid with cholesterol. They are present in blood.

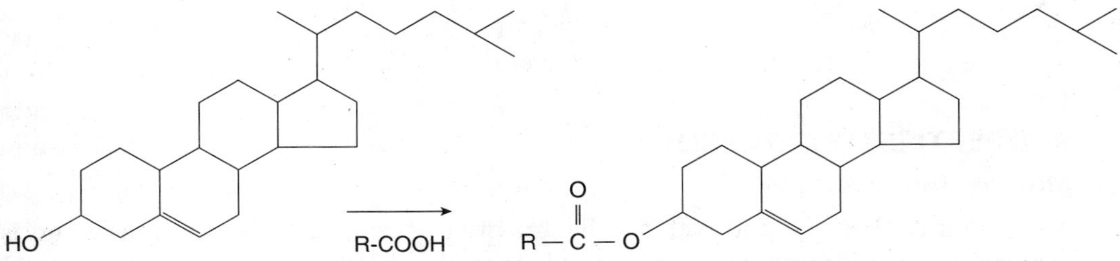

Cholesterol   Fatty acid   Cholesteryl ester

The carboxyl group may be reduced to the corresponding aldehydes and alcohols in the presence of hydrogen at high temperature.

Properties due to hydrocarbon chains

These are mainly due to the double bonds contained in the chain.

Oxidation Reaction

Addition reaction

Isomerism: Unsaturated fatty acids may have isomers, depending upon two factors. One type of isomerism depends upon the position of double bond in the hydrocarbon chain – positional isomerism.

Example: Oleic acid. There are sixteen positions where the double bonds may be placed. Hence, oleic acid forms sixteen isomers.

Geometrical isomerism: It depends on the oxidation of the radicals around the axis of double bond. If the radicals are on the same side of the double bond, the compound is a cis form of isomer. If they are on opposite side of the double bond, it is a trans form of isomer.

$CH_3(CH_2)_7CH$ ‖ $HOOC(CH_2)_7CH$

Oleic acid (in form)

$CH_3(CH_2)_7CH$ ‖ $CH(CH_2)_7COOH$

Elaidic acid - trans form

GLYCEROL

Glycerol is a component of fat. Its properties and reactions are described below. Glycerol is commonly called glycerine and is the alcoholic component of triglycerides. It is closely allied to carbohydrates in its chemical structure and properties, but it is closely associated with fats in its occurrence and function. The structure of glycerol is given below:

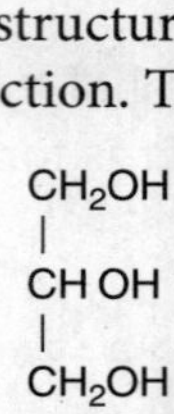

Glycerol

PROPERTIES OF GLYCEROL

Physical Properties

It is colourless heavy liquid with specific gravity of 1.26. It is sweetish in taste and is commonly known as glycerine. It is soluble in water and alcohol but insoluble in lipid solvents like ether, chloroform, and benzene.

Chemical Properties

Oxidation: When glycerol is oxidised with H_2O_2 in slight alkaline solution in the presence of iron salt, a mixture of trioses, glyceric aldehyde, and dihydroxy acetone is formed.

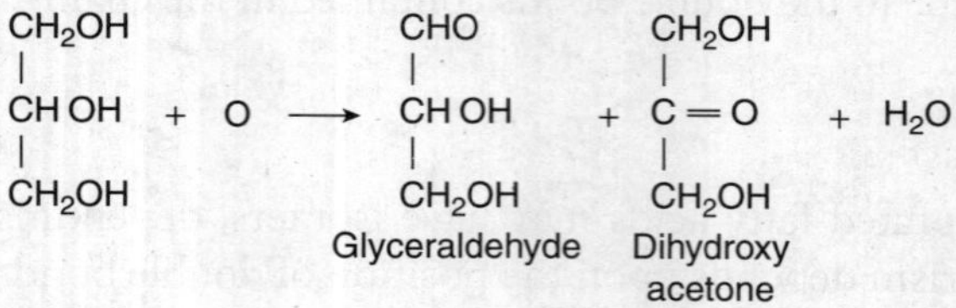

Acrolein formation: When glycerol is heated either alone or in the presence of dehydrating agent like $KHSO_4$, a substance called acrolein, which is an unsaturated aldehyde, is obtained.

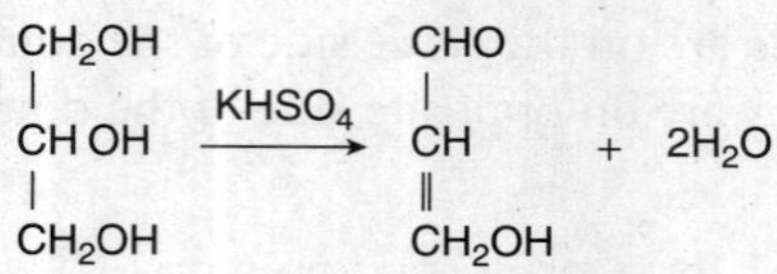

FUNCTIONS

Glycerol is utilised in the body as follows: Glycerol is phosphorylated to form phosphoglycerate (glycerophosphate) action of ATP and specific phosphokinase. Glycerophosphate is oxidised to glyceraldehyde-3-phosphate, which is an intermediate in glycolysis, which is further oxidised to yield energy. Glycerol is widely used as solvent in the manufacturer of cosmetics and in the preparation of medicine. Glycerol trinitrate is used as an explosive. In medicine, nitroglycerine is used as vasodilators in the treatment of coronary heart disease.

COMPOUND LIPIDS

These are the lipids which contain certain chemical groups others than alcohol and fatty acid. Compound lipids include phospholipid, glycolipids, sulpholipids, and lipoproteins.

PHOSPHOLIPIDS

These are also known as phosphatides. They are an important compound, composed of fatty acid, glycerol, phosphoric acid, and nitrogenous base.

Source

They are abundantly present in heart, brain, kidney, egg yolk, and soya bean.

Types of Phospholipids

The important phospholipids are the following:

1. Lecithin
2. Cephalin
3. Sphingomyelin
4. Plasmalogens

Lecithin

Lecithin consists of one molecule of glycerol, esterified with two molecules of fatty acids; one of these fatty acid may be unsaturated and other saturated. It contains phosphoric acid and a nitrogenous base – choline. Two forms of lecithin, alpha lecithin, and beta lecithin are described. Alpha lecithin has the phosphoric acid and choline attached to the third carbon atom (alpha carbon atom) of glycerol; beta lecithin has the phosphoric acid and choline attached to the second carbon atom (beta carbon atom) of glycerol. The structure of lecithin is given below:

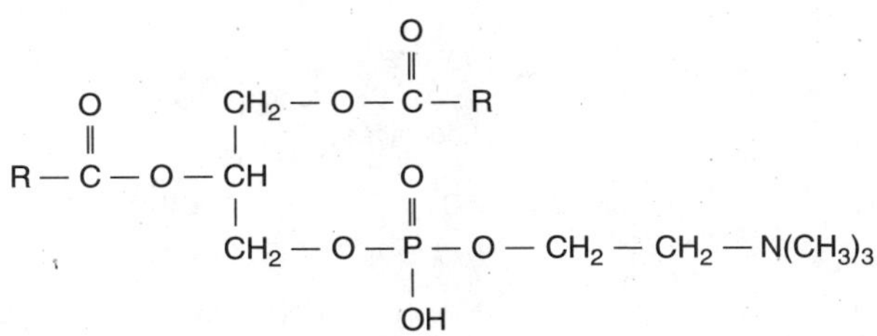

CHOLINE

Choline is the tri ethyl hydroxylethyl ammonium hydroxide. Choline is the component of lecithin and sphingomyelin and hence an important constituent of phospholipid.

$$HO - H_2C - H_2C - N(CH_3)_3$$

SOURCES

Choline is present in egg yolk, liver, heart, kidney, and milk.

FUNCTION

1. Choline is an important factor in the prevention of fatty liver. Absence of choline in the diet leads to accumulation of fat on the liver, leading to fatty liver.
2. Acetylcholine plays an important role in the transmission of nerve impulse.
3. When bacteria acts on tissue choline, water is split from choline to produce poisonous base called neurine.

Physical property

They are white waxy substances, which become brownish when exposed to air and light, owing to autooxidation. This is due to the presence of unsaturated fatty acid in the molecule. They are soluble in alcohol and other fat solvents. They are hygroscopic and mix well with water to form cloudy, colloidal, and slimy solution. They don't have definite melting point. Pure lecithin is a mild emulsifying agent, but when combined with carbohydrates and proteins, they are more effective. They lower the surface tension of aqueous solution.

Chemical property

Lecithin is readily hydrolysed by boiling with acid and alkalis. Lecithin is hydrolysed by enzymes called lecithinase. There are four different types of lecithinase occur in nature: lecithinase A, lecithinase B, lecithinase C, and lecithinase D.

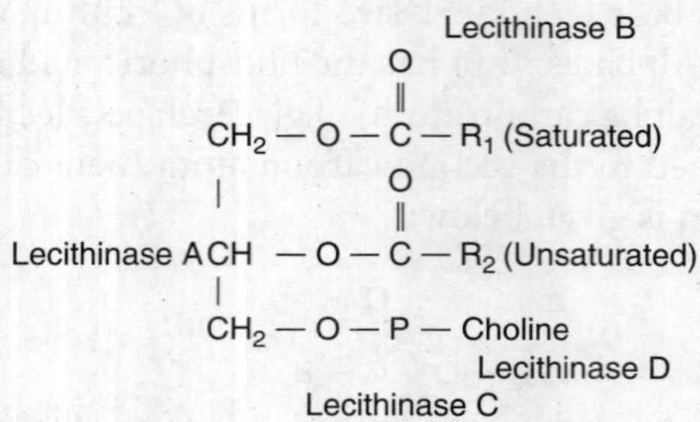

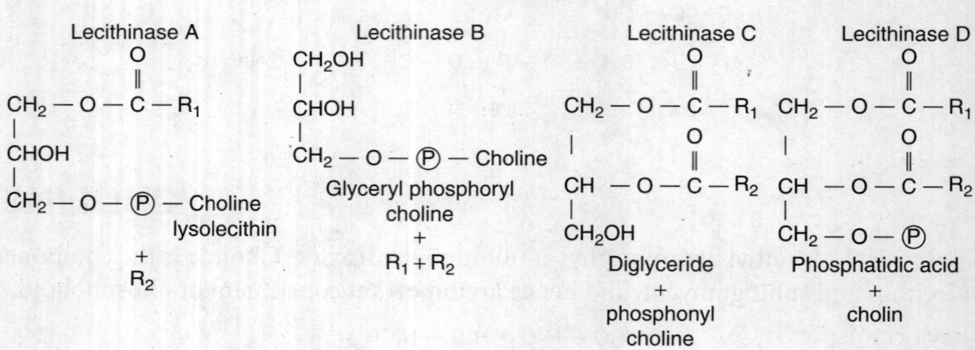

Lecithinase A is found in cobra venom. It is also present in spider venom and poisonous stings. Lecithinase A action is specific; it acts only on the unsaturated fatty acid. Lysolecithin is a powerful haemolytic poison which causes haemolysis of erythrocytes.

Lecithinase B is present in pancreas. It hydrolyses both the saturated and unsaturated fatty acids from lecithin, leaving behind glyceryl phosphoryl choline.

Lecithinase C is found in plants. It hydrolyses lecithin at the point between glycerol and phosphoric acid with the liberation of diglyceride and phosphoryl choline.

Lecithinase D is found in leafy vegetables like cabbage. It hydrolyses the lecithin into choline and phosphatidic acid.

Lecithin forms stable compounds with proteins, carbohydrates, and heavy metals. During absorption of fat, the lecithin content of blood increases as the blood lipids increase.

Biological significance

- Prevention of fatty liver
- Process of blood coagulation

Cephalins

Cephalins occur along with lecithin in all plant and animal cells. They are similar structure as that of lecithin but differ from lecithin in that the base attached to phosphoric acid is not choline. It may be one of the three bases, namely ethanolamine, serine, and inositol. Therefore, the three cephalins are known as follows:

1. Ethanolamine cephalin (phosphatidylethanolamine)
2. Serine cephalin (phosphatidyl serine)
3. Inositol cephalin (phosphatidylinositol)

Ethanolamine cephalin (phosphatidylethanolamine)

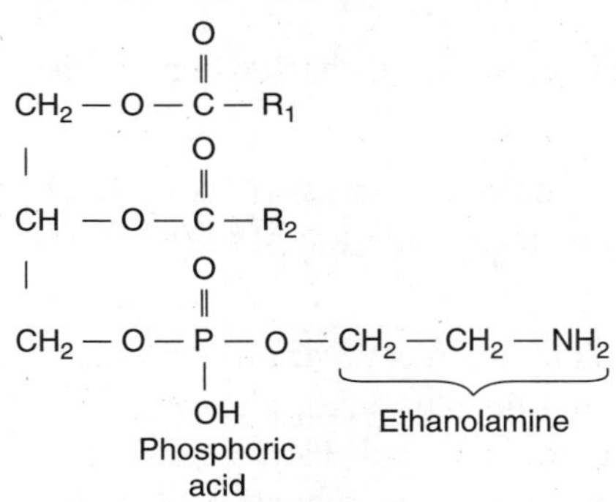

They occur with lecithin, particularly rich in brain and nervous tissue.

Serine cephalin (phosphatidyl serine)

This compound contains amino acid serine instead of ethanolamine. It is found in brain tissues; traces are detected in blood also. Its structure is as follows:

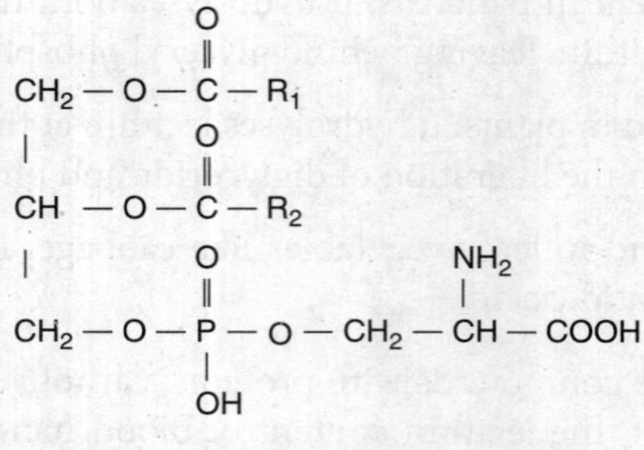

Inositol cephalin – lipositol (phosphatidylinositol)

It is found to occur in brain tissues, nervous tissues, soya bean, and wheatgerm. Inositol is an alcohol and a cyclic compound hexahydroxycyclohexane with molecular formula $C_6H_{12}O_6$.

Its structure is as follows:

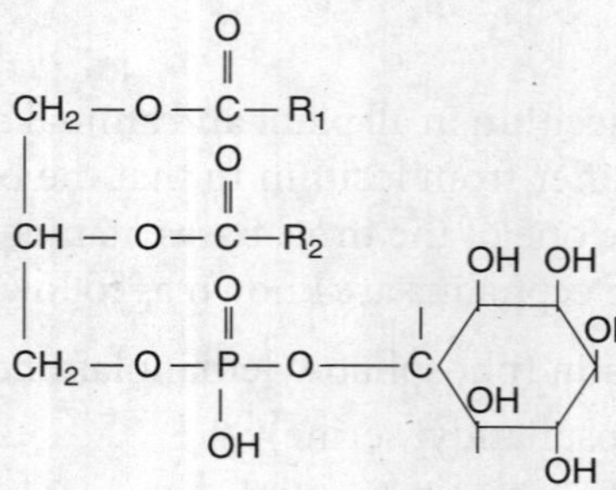

Properties

1. Cephalins are soluble in most fat solvents, except acetone and ethyl and methyl alcohols.
2. They are unstable in air and turn dark when oxidised.

Biological significance

They play an important role in the mechanism of coagulation of blood. Thromboplastin, which is an important factor in coagulation of blood, is a phospholipid.

Sphingomyelin

Sphingomyelins differ radially from cephalin in composition. They are found in brain and nerve tissues. They contain fatty acid, phosphoric acid, choline, and a complex amino alcohol sphingosine or sphingol. Fatty acid is attached to the amino group of sphingosine by amide linkage. The phosphoric acid is attached to one of the hydroxyl group of sphingosine.

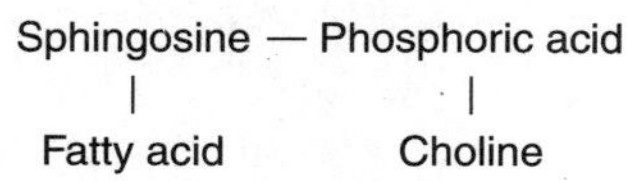

Its structure is given as follows:

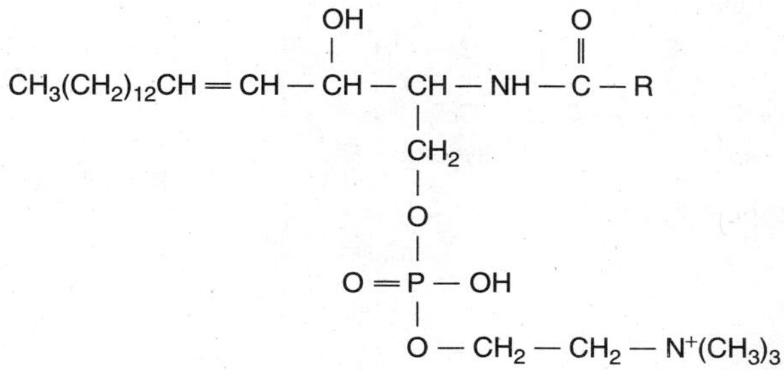

Properties

1. They are white crystalline substances.
2. They form opalescent suspension in water.
3. They are insoluble in fat solvents like ether and acetone.
4. They are optically active.
5. They are stable in air and light.

Clinical aspect of sphingomyelin

Large accumulation of sphingomyelins may occur in brain, liver, and spleen and lead to a disease called Niemann-Pick disease.

Inheritance: Autosomal recessive.

Enzyme defect: Deficiency of the enzyme sphingomyleinase, which cannot degraded sphingomyelin in the body; as a result, large amount of sphingomyelin accumulates in tissues and organs of the body, specially in liver, spleen, and brain.

Plasmalogens

Plasmalogens are present in brain, nervous tissue, muscles, and mitochondria.

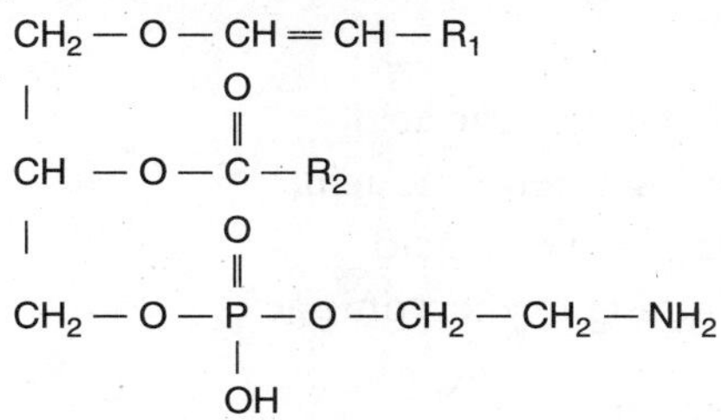

Cardiolipin

It is a phospholipid found in mitochondria (inner membrane) and bacterial wall. It is formed from phosphatidylglycerol. Its structure is as follows:

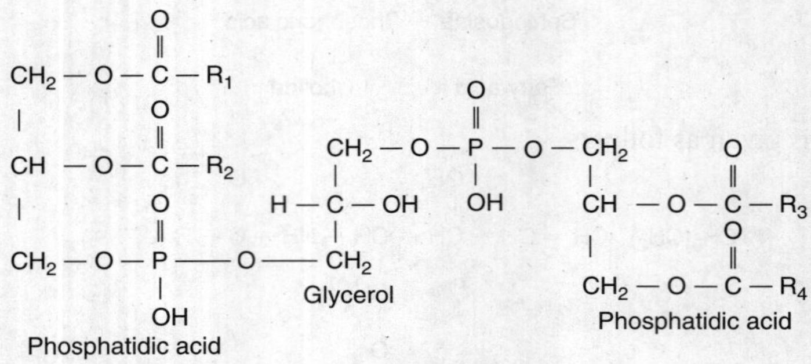

Cardiolipin

GLYCOLIPIDS

These may be further subdivided into cerebrosides and gangliosides

Cerebroside

They are found in brain and myelin sheaths of nerves. They are also known as galactolipids because of their galactose content. They are composed of fatty acid of high molecular weight, sphingosine, and galactose.

Its structure is as follows:

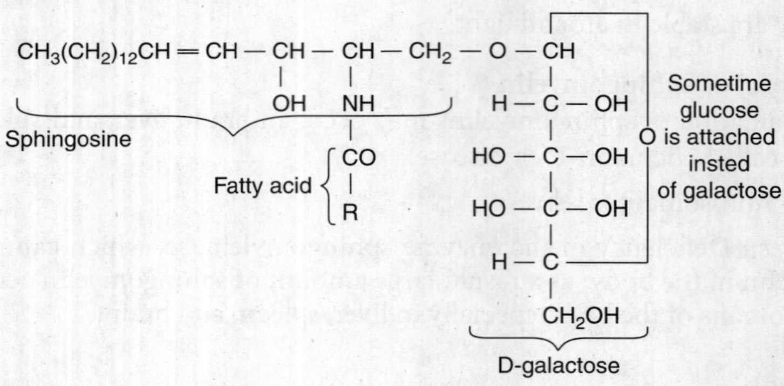

Four cerebrosides have been identified, according to their fatty acid component. They are as follows:

- Kerasin, containing lingoceric acid,
- Cerebron, containing cerebronic acid,
- Nervon, containing nervonic acid
- Oxynervon, containing oxynervonic acid

Clinical aspect of cerebroside

An inherited disorder of cerebroside metabolism. It is autosomal recessive. β-Glucocerebrosidase cleaves the cerebroside to glucose and ceramide. As a result, large amount of cerebroside accumulated in liver spleen, bone marrow, and also brain leads to mental retardation. This type of disease is termed as Gaucher's disease.

GANGLIOSIDES

Gangliosides are glycolipids containing glucose, galactose, and a substance called neuraminic acid and NANA, in addition to fatty acid and sphingosine.

Gal → GalNAC ⟶ Gal ⟶ Glu ⟶ Ceramide (sphingosine + fatty acid)

Gal ↓ NANA

Four important types of gangliosides have been isolated from brain tissue: GM-1, GM-2, GM-3, and GD-3.

Clinical aspect of gangliosides

Tay-Sachs disease–It is an autosomal recessive disorder. It occurs due to the deficiency of the enzyme hexoaminidase A. Due to the deficiency of this enzyme, gangliosides started to accumulate in brain and nervous tissues.

SULPHOLIPIDS

Sulphur containing lipids are present in various tissues such as liver, kidney, testes, brain, and certain tumours. This sulpholipid is also present in plants. It is localised in the chloroplasts.

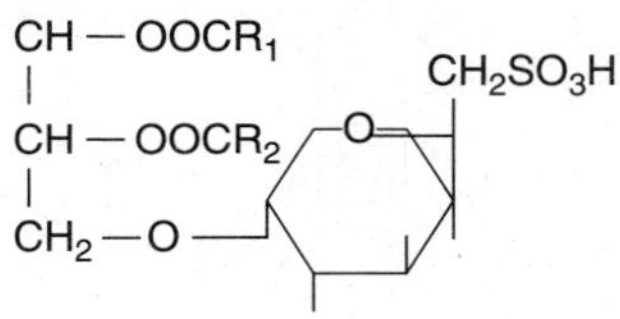

DERIVED LIPIDS

STEROIDS

Steroids represent a large group of compounds which exist in nature, having a common and characteristic structure based on the cyclopentanoperhydrophenanthrene ring (CPPP). It consists of phenanthrene rings (A, B, and C), to which cyclopentane ring (D) is attached.

These compounds vary widely in their occurrence and physiological properties despite their common basic structure. The various substances of physiological interest having the same steroid structure are classified as follows:

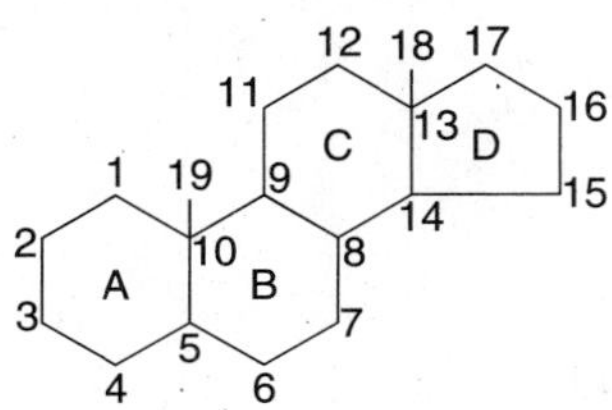

STEROLS

1. Cholesterol
2. 7-dehydrocholesterol
3. Ergosterol
4. Calciferol
5. Coprosterol

STEROID HORMONES

C_{18} STEROIDS

Mammalian ovary contains ovarian folicles and corpus luteal follicles which produce estrogen, estriol, estradiol and corpus luteum produce progesterone which is described in Table 4.3.

Table 4.3 *Steriod Female Sex Hormones*

C_{18} Steroids	C_{21} Steroids	C_{19} Steroids	Progesterone	GLucocorticoids
Estradiol	Mineralocorticoids	Testosterone	Corpus luteum hormone	Cortisone
Estriol	Aldosterone	Androsterone		Cortisol
Estrone	Deoxycorticosterone	Dehydroepiandrosterone		Cotricosterone

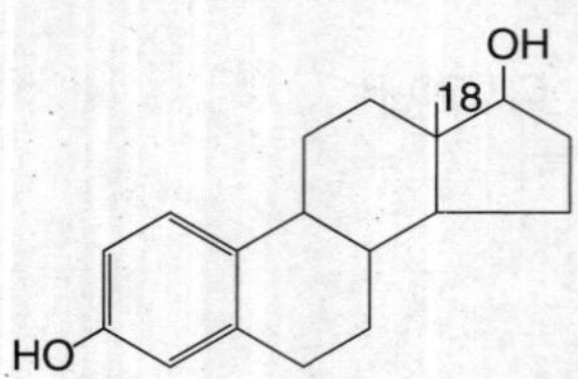

Estradiol

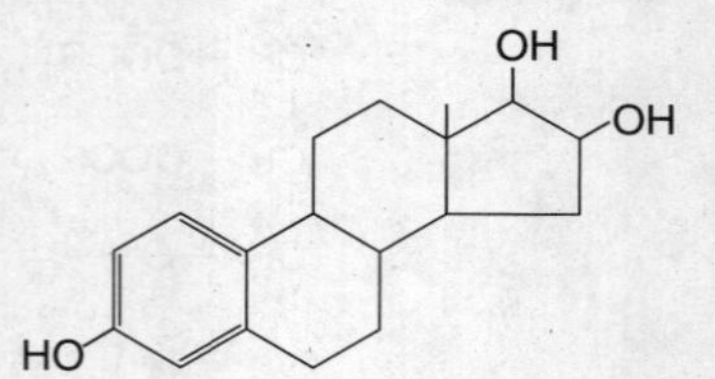

Estriol

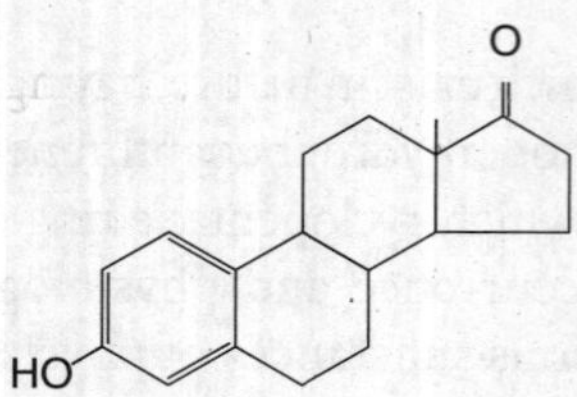

Estrone

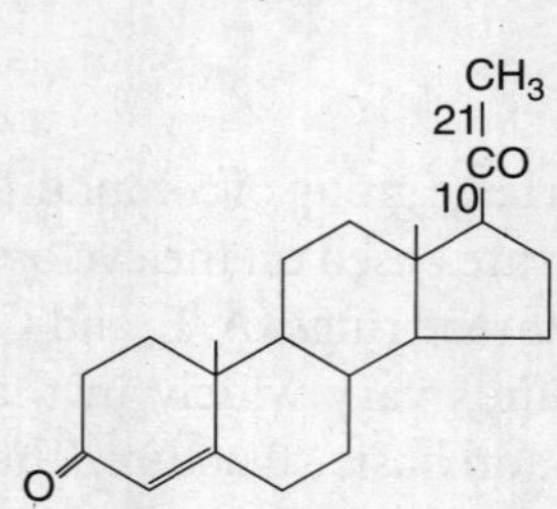

Progesterone

BILE ACIDS AND SALTS

Bile Acids

1. Cholic acid
2. Chenodeoxy cholic acid

3. Lithocholic acid
4. Deoxy cholic acid

Bile Salts

1. Glycocholic acid
2. Taurocholic acid

STEROID HORMONES

Mineralocorticoids C_{21}- aldosterone, deoxycorticosterone, glucocorticoids – cortisone, cortisol, corticosterone

SEX HORMONES

Male: C_{19} – androgen, testosterone, androsterone
Female: C_{18} – estradiol, estriol, estrone, progesterone.

STEROLS

Sterols means solid alcohol. They are widely distributed in animal and plant tissues. They include cholesterol, 7-dehydrocholesterol, ergosterol, calciferol, and coprosterol. Among the sterols, cholesterol forms an important physiological compound.

CHOLESTEROL

Cholesterol means solid bile alcohol. It derives its name from the Greek word *cholesterine* (*chole – bile, sterine* – solid).

Sources

Cholesterol is widely distributed in all the cells of the body. It occurs in large quantities in brain and nerve tissues. It is found in adrenal glands, glandular tissues, plasma, and egg yolk. It may be noted that cholesterol occurs in animal fats and not in plant fats.

Structure

The molecular formula of cholesterol is $C_{27}H_{45}OH$. It is monounsaturated tetracyclic alcohol. The structure of cholesterol, as proposed by Windaus, has the following characteristic features:

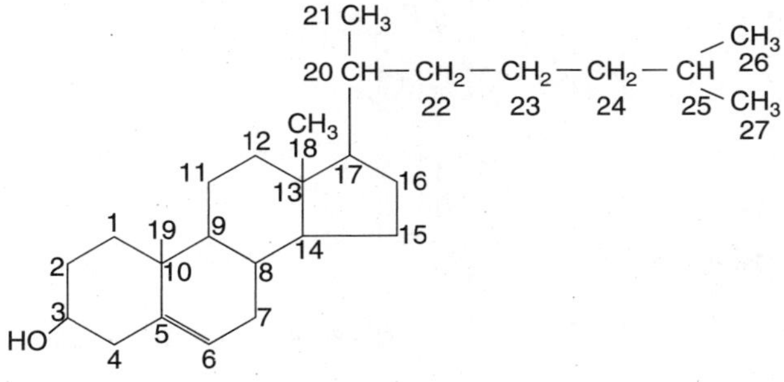

1. Cyclopentanoperhydrophenanthrene ring system
2. An aliphatic side chain attached to C_{17}
3. An OH group attached to C_3
4. A double bond between C_5 and C_6
5. Two methyl groups; one attached at C_{10} and other at C_{13}

Properties

Physical properties

1. They are white shining rhombic plate like crystals.
2. They are tasteless and odourless.
3. Cholesterol has a high melting point of 150°C.
4. Cholesterol is insoluble in water and soluble in fat solvents (chloroform, acetone, and benzene). It is soluble in soap solution and bile salt solution.
5. When mixed with fat or oil, cholesterol has the peculiar property of enabling the fat or oil to absorb water. For example, lanolin is a greasy substance, containing cholesterol, which enables lanolin to absorb water readily. Lanolin is used in pharmacy as a greasy vehicle in the preparation of ointments containing water soluble constituents.
6. Cholesterol is a poor conductor of electricity and serves as an insulator against electric discharge. It appears that cholesterol functions as an insulation cover for the transmission of electrical impulses in the nervous tissue, where it is found abundantly. It act as an insulator against nerve impulses, which are electrical in character.

Chemical properties

The chemical properties are related to the secondary hydroxyl group and the double bond at $C_5 - C_6$.

Cholesterol, when oxidised in suitable conditions, undergoes rapid oxidation to corresponding ketone and cholestenone. The hydroxyl group readily forms esters with acids. The fatty acid esters, namely cholesterol acetate palmitate and propionate, are found in blood and tissues. Cholesterol palmitate, because of the double bond, gives addition and hydrogenation reactions of unsaturated compounds. The iodine value of cholesterol is 65.8.

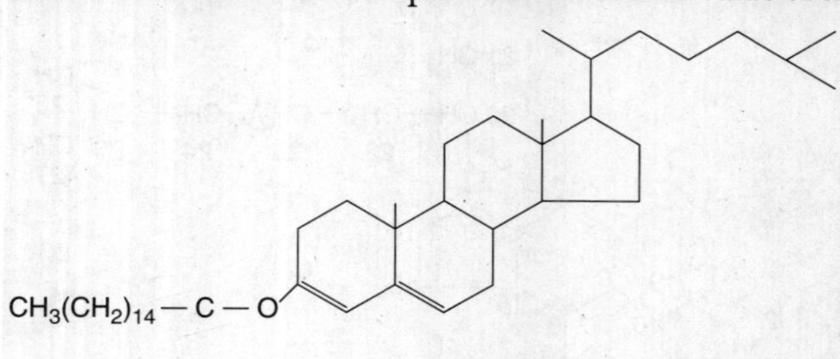

Cholesterol palmitate

Colour Reactions

Liebermann–Burchard reaction

A few drops of acetic anhydride and a few drops of concentrated sulphuric acid and chloroform solution of cholesterol

↓ Changes to

Rose red colour

↓ Quickly changes to

Green colour

The intensity of colour is proportional to the concentration of cholesterol. This reaction is used for the quantitative determination of cholesterol in blood.

Salkowski reaction

Chloroform solution of cholesterol and equal volume of sulphuric acid (equal volume)

↓ Mixed and shaken

Two layers separate

Upper layer – red – contains cholesterol.
Lower layer – yellow colour with green fluorescence – contains sulphuric acid.

Zimmermann reaction

m-dinitrobenzene + sterol containing ketone groups

↓

Violet colour

↓

17-ketosteroids in urine

Cholesterol is the precursor of various physiologically important compounds like bile acids, female sex hormones, and steroid hormones of adrenal cortex.

7-DEHYDROCHOLESTEROL

This is an important sterol present in the skin. It has one more double bond at $C_7 - C_8$. It has only one H atom at $C_7 - C_6$ and none at C_8. It is converted to vitamin D when irradiated with UV. Its structure is as follows:

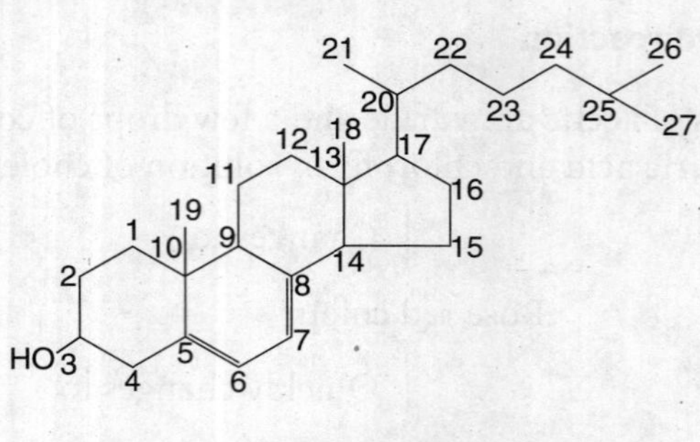

7-dehydrocholesterol

ERGOSTEROL

Ergot is present in ergot and yeast. It's an important compound because it's a precursor of vitamin D, which is converted by irradiation with UV. It has double bond in the side chain. It has an additional CH_3 group in the side chain. Its structure is as follows:

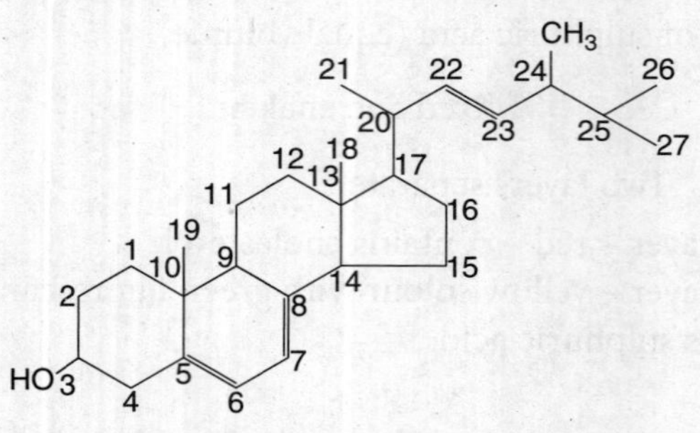

C_{27}

24 — CH_3

24 — double band

Ergosterol

CALCIFEROL

Calciferol is an activated ergosterol.

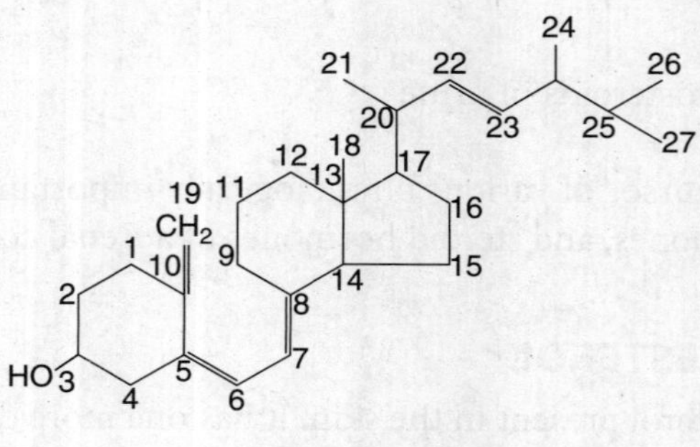

COPROSTEROL

Coprosterol is present in faeces. It is the reduced form of cholesterol, formed as a result of reduction by intestinal bacteria. It does not have the double bond between C_5 and C_6.

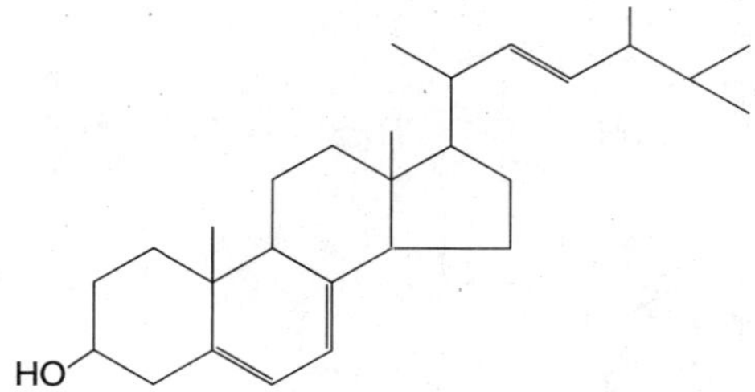

BILE ACID

Bile acids are synthesised from cholesterol. They are important factors in the digestion and absorption of lipids in intestines. Bile acids contain additional hydroxyl groups (1-3) attached to the steroid nucleus. The side chain has a carboxyl group.

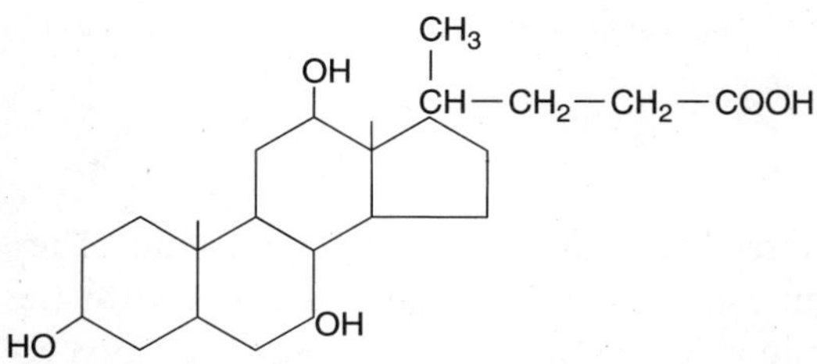

Cholic acid

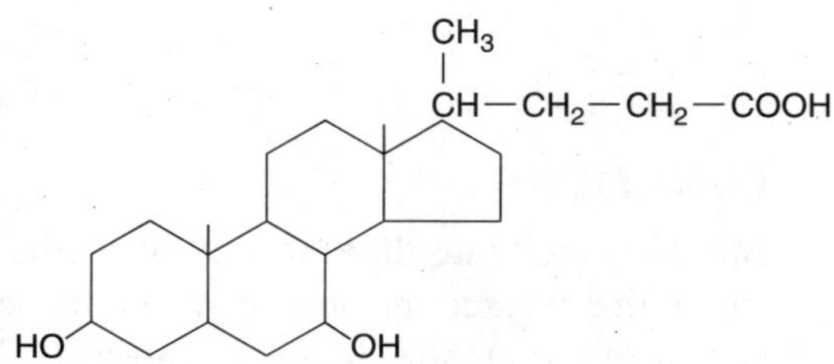

Chenodoxy cholic acid

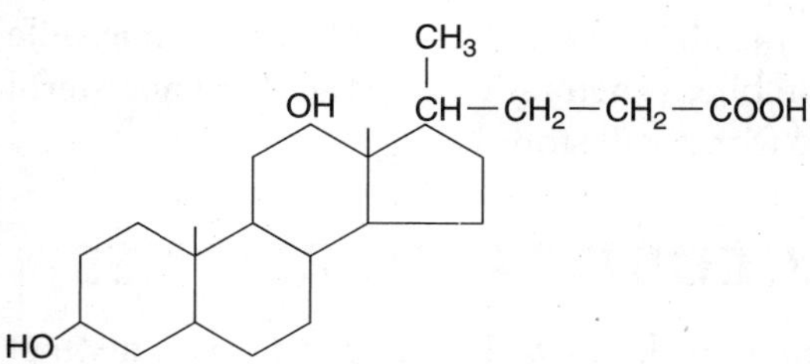

Deoxycholic acid

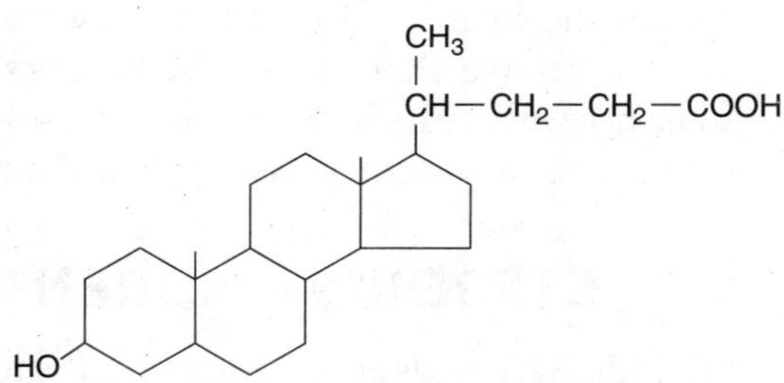

Lithocholic acid

BILE SALT

Cholic acid and chenodeoxycholic acid are conjugated in the liver with either glycine or taurine through peptide linkages, forming the bile salts and glycocholic and taurocholic acids.

Glycocholic acid Taurocholic acid.

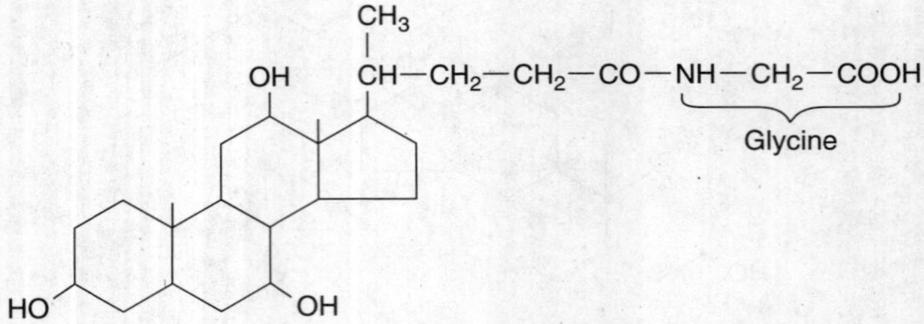

Glycocholic acid

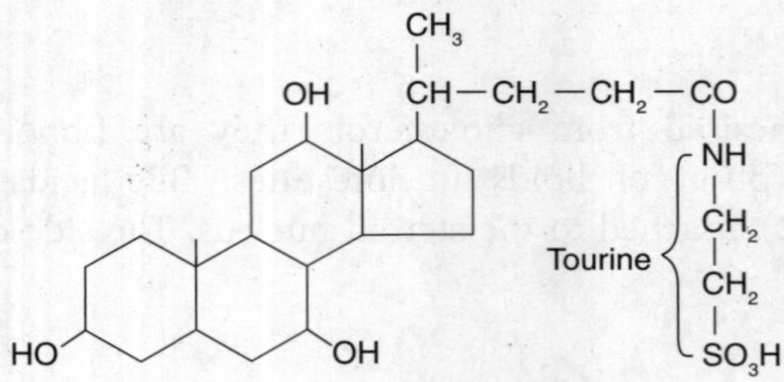

Taurocholic acid

FUNCTIONS

Bile salts facilitate digestion of fat in the intestine by being emulsifying agents. They lower the surface tension of water and emulsify the fats, where it is exposed to the action of pancreatic lipase. In intestine, bile acids combine with mono and digycerides to form tiny particles called micelles, which are about 1/100th of size of usual emulsified triglyceride particle, which facilitates absorption of fat. Bile salts activate pancreatic lipase and cholesterol esterase. They facilitate absorption of cholesterol and fat-soluble vitamins. Bile salts act as stimuli to the secretion of hepatic bile. Bile salts help in maintaining cholesterol level in bile. When ratio of bile acid to cholesterol decreases, cholesterol gets precipitated and forms gall stones.

C_{21} STEROIDS—ADRENAL CORTICAL HORMONES

The adrenal glands are positioned at the top of the kidney. The glands have two distinct parts, namely an inner dark-coloured mass called the medulla and an outer covering layer called cortex.

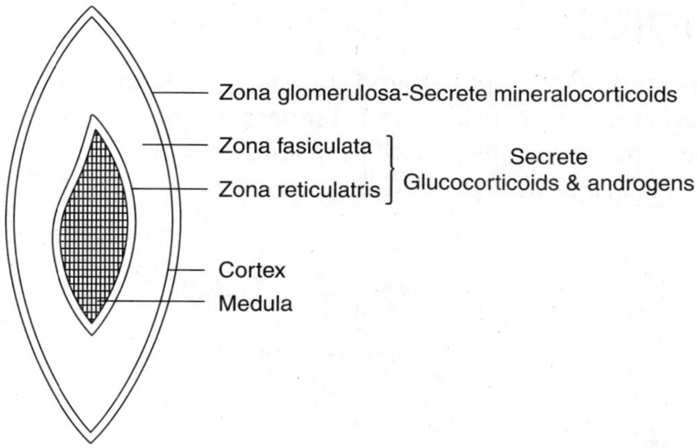

MINERALOCORTICOIDS

Mineralocorticoids are mainly involved in the transport of electrolytes and distribution of water. Examples are aldosterone and deoxycorticosterone. The structure of Aldosterone and Deoxycorticosterone is given below.

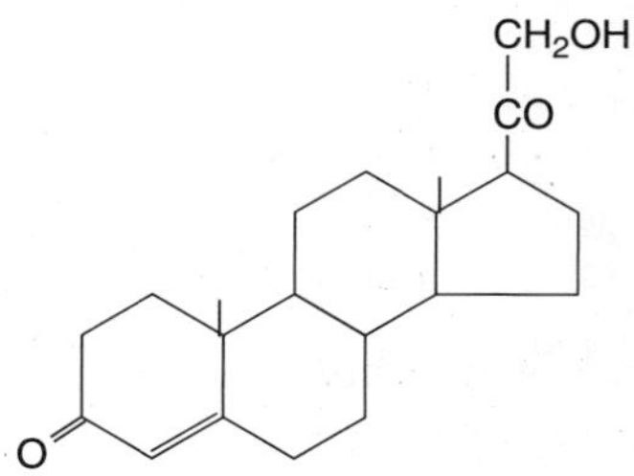

Deoxycorticosterone

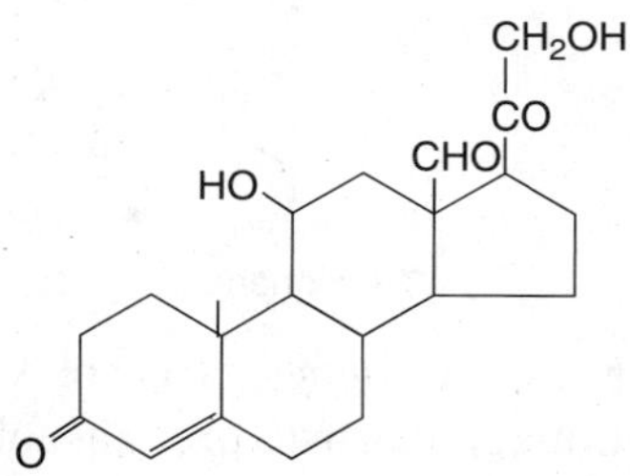

Aldosterone

GLUCOCORTICOIDS

They are concerned primarily with the metabolism of carbohydrates, proteins, and fats. Example: Cortisone, cortisol, and corticosterone. The structure of cortisol is given below:

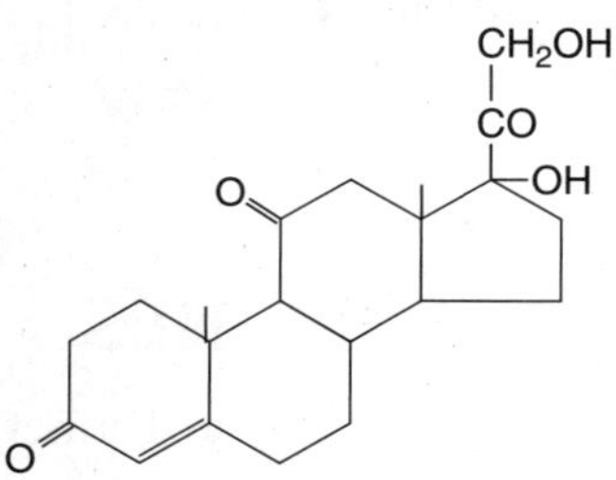

Cortisone

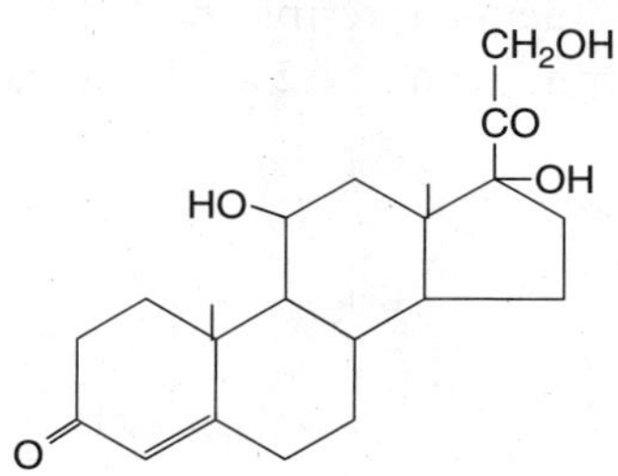

Cortisol

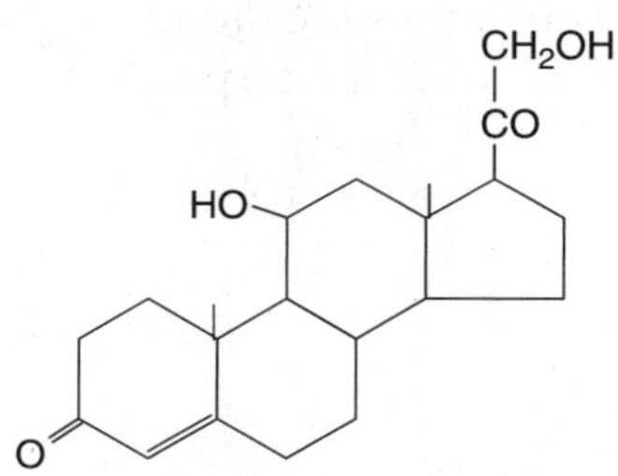

Corticosterone

ISOPRENOIDS

Among the non-saponifiable lipids found in plants are mainly hydrocarbon known as terpenes. The simplest terpenes are called monoterpenes and conform the formula $C_{10}H_{16}$ (equivalent to two isoprene units), sesquiterpenes with the formula $C_{15}H_{24}$, diterpenes with C_2H_{32}, and triterpenes $C_{30}H_{48}$.

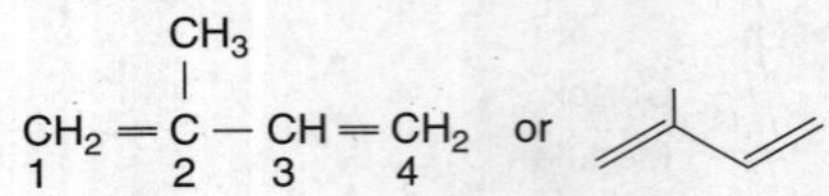

Structure of isoprene

MONOTERPENES

Important monoterpenes are myrcene (from oil of bay), geraniol (from rose oil), limonene (from lemon oil), and menthol (from peppermint oil). Their structures are given below:

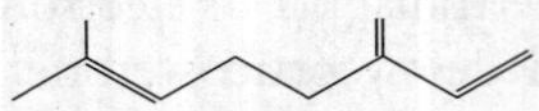

Myrcene

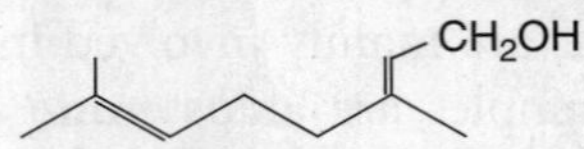

Geranoil

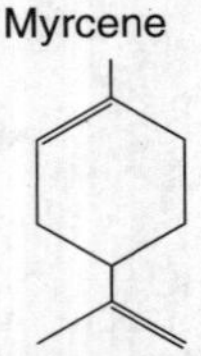

Limonene

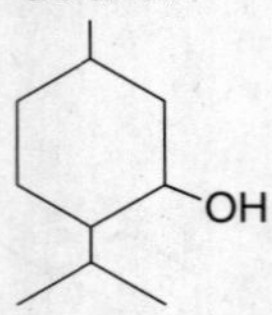

Menthol

Among the sesquiterpenes, farnesol is widely distributed but is present in essential oils in small amounts. Its structure is as follows:

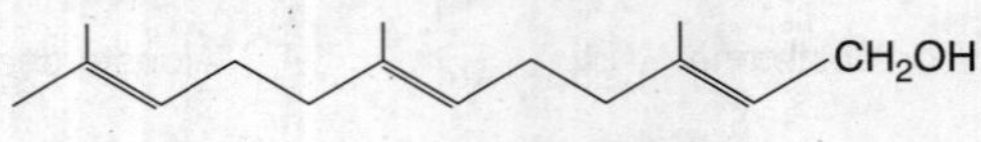

Farnesol

DITERPENES

Diterpenes are found as substituents of resins and balsams. Abeitc and sapietic are the best-known diterpenes. Vitamins A1 and A2 are important monocyclic diterpenes.

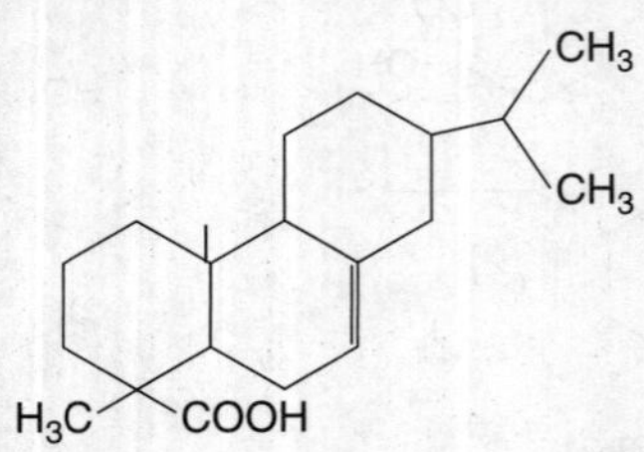

Abietic acid

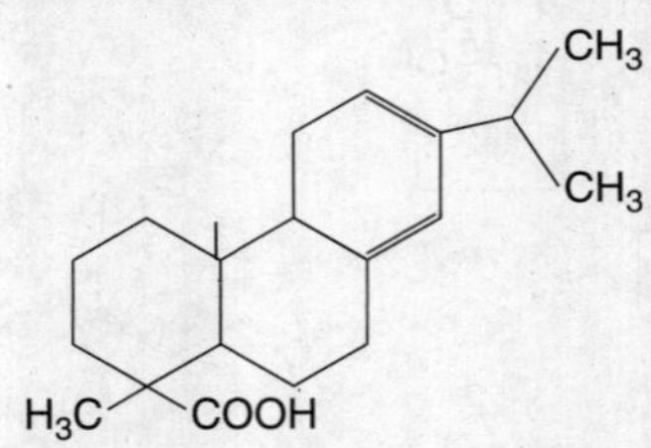

Sapietic acid

Phytol: It's an acyclic diterpenes and is obtained from the hydrolysis of cholorophyll. Given below is its structure:

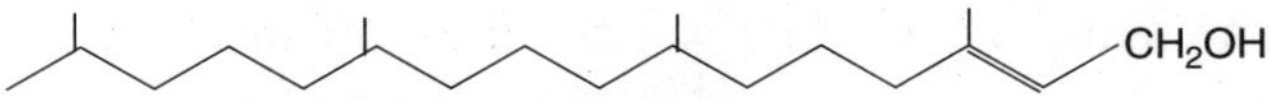

Phytol

TRITERPENES

Triterpenes are mostly the intermediate in the biosynthesis of cholesterol. Two such compounds are squalene and lanosterol. Their structures are as follows:

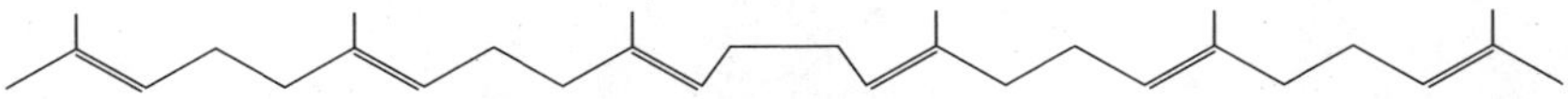

Squalene

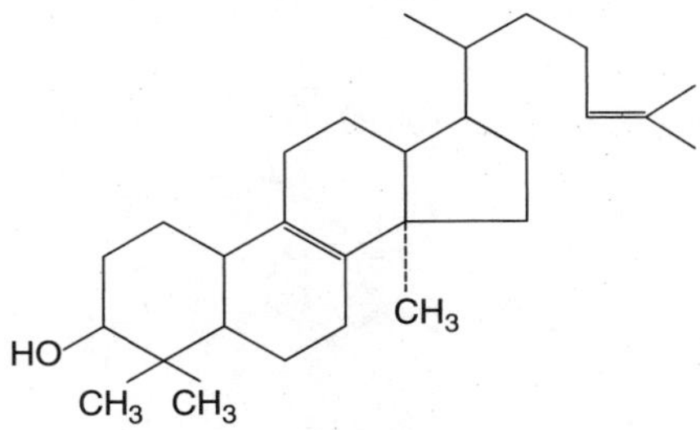

Lanosterol

POLYTERPENE

Rubber is a polyterpene present in the latex of many tropical plants. A molecule of rubber is made of 500–5000 isoprene units joined in a long straight chain.

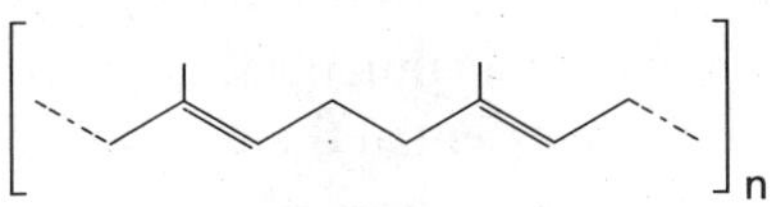

CAROTENOIDS

Carotenoids are tetraterpenes. They are present in both plants and animals. The isoprene derivatives of carotenoids are highly unsaturated. Due to the presence of many conjugated double bonds, they are red or yellow in colour. For example, the pigments of tomato (lycopene) and carrot (α- and β-carotene) are red, whereas oxygen-containing carotenoids are yellow (xanthophyll).

Lycopene

Lycopene is a bright red carotene and carotenoids pigment and phytochemical found in tomatoes and other red fruits and vegetables such as red carrots, red bell peppers, watermelon and pappayas.

Carotene

Carotene is a group of naturally occurring carotenoids with the same molecular formula as that of lycopene. Carotene was first isolated from carrots. Xanthophylls are yellow pigments that form one of two major divisions of the carotenoid group. Xanthophyll is a class of carotenoids with oxygenated carotenes.

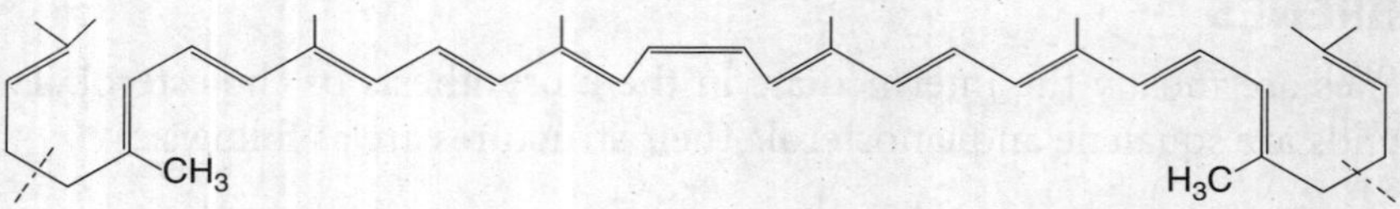

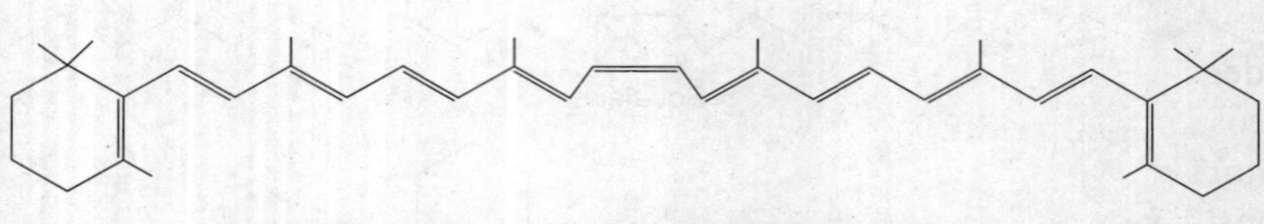

Xanthophyll-oxygen containing carotenicls are yellow

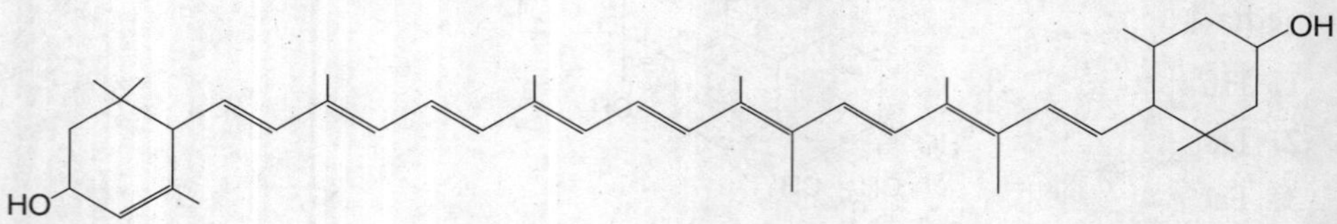

Xanthophyll of leaf (Lutein)

BIOMEDICAL IMPORTANCE OF LIPIDS

1. Lipids are the major dietary constituents that provide energy fuel to the body. Hence, lipids are even more essential than carbohydrates for combustion.
2. One gram of lipids yields 9.5 calories per gram of energy whereas one gram of carbohydrates yields only 4.0 calories.
3. The nervous system is rich in lipids, which are essential for its functioning.
4. Lipoproteins and phospholipids are the essential constituents of cell membranes and cell organelles such as mitochondria.
5. Lipoproteins generally act as the transporter for triglycerides, cholesterol, and phospholipids.
6. Metabolic products of lipids act as the building material for biologically active materials.
7. Lipids also supply essential fatty acids (EFA), which are generally not synthesised in body. Hence, they should be provided through diet and are essential for normal growth and health.

BIOMEDICAL IMPORTANCE OF EMULSIFICATION

The emulsification of dietary fats in the intenstinal canal, brought about by bile salts, is a prerequisite for digestion and absorption of fats.

BIOMEDICAL IMPORTANCE OF EFA

Polyunsaturated fatty acids are the structural elements of mitochondrial membrane. Mitochondrial membrane loses its activity when there is deficiency of EFA, which may cause swelling and reduction of oxidative phosphorylation.

Fats with a high content of polyunsaturated fatty acids lower the serum cholesterol level.

Inadequate EFA may lead to a fatty liver.

Docosahexaenoic acid (22:6n-3) is the common polyenoic acid present in the retinal photoreceptor membrane. It is formed from linolenic acid. Docosahexaenoic acid is responsible for the electrical response of the photoreceptors to illumination. Hence, linolenic acid is essential in the diet for optical vision.

Biomediacl importance of EFA:

1. Human deficiency of some cases are eczma like dermatitis,
2. Degenerative changes in arterial wall and
3. Fatty liver in man may be due to EFA deficiency.
4. There are also some reports that administration of EFA in such cases may produce:
5. Some improvement of eczema in children kept on skimmed milk.
   (a) Prevent fatty liver and
   (b) Lowering of cholesterol level.
   (c) Infant and babies with low fat diet may develop typical skin lesions which has shown to be improved in EFA.

BIOMEDICAL IMPORTANCE OF STEROLS

1. *7-Dehydrocholestrol (provitamin D3):* In the skin, UV rays of the sun convert 7-dehydrocholesterol to cholecalciferol. Hence, 7-dehydrocholesterol is called as provitamin D3. This concept explains the use of sunshine in preventing rickets, a disease caused by vitamin D deficiency.
2. *Ergosterol (provitamin D2):* When irradiated by UV rays, ergosterol is converted to vitamin D2 by the opening of ring B of the sterol. Hence, ergosterol is also called as provitamin D2.
3. *Sitosterol:* Sitosterol decreases the intestinal absorption of exogenous and endogenous cholesterol; this lowers the blood cholesterol level.
4. *Biomedical importance of gangliosides:* Gangliosides are named components of membranes. The sugar units and sialic acid section of the molecule are 'water

soluble (hydrophilic) and negatively charged, whereas the ceramide portion is lipid soluble (hydrophobic). The gangliosides, therefore, can serve as specific membranes' binding sites.

FUNCTION OF PROSTAGLANDINS

1. Anti-hypertensive: Lowers BP
2. Inhibit gastric secretion.
3. Stimulates GI musculature.
4. Bronchodilation used in the treatment of bronchial asthma.
5. Increased uterine contraction can be used as abortifacient.
6. Metabolic effect.
7. Insulin-like effect
8. Luteolysis
9. Thyroid Stimulating Hormone (TSH) like effect
10. Decreases lipolysis
11. Steriodogenesis.

FUNCTIONS OF LEUKOTRIENES

1. Leukotriens (LTS), in general, appear to act as mediators in inflammation and anaphylaxis.
2. Capillary dilatation and vascular permeability.
3. As little as one n-mole of LTC4, D4, and E4 elicits erythema, and wheal formation like histamine also increase vascular permeability.
4. Action on bronchial muscles: Inhalation of LTS causes bronchospasm.
5. SRS-AC slow reacting substance of anaphylactic
6. SRS-A, which is produced by most cells during anaphylactic reaction, has now been shown to have a mixture of LTS like LTS C4,D4, and E4. These leukotrienes are responsible for intense vasoconstriction of bronchial muscles, vasodilation, and increased vascular permeability seen in anaphylactic/allergic reactions.

LIPOPROTEINS

Structure of Lipoproteins

Lipoproteins are produced when proteins combine with lipids such as phospholipids, lecithin, cephalin, fatty acid, cholesterol, glycerides, and fat-soluble vitamins Figure 4.5.

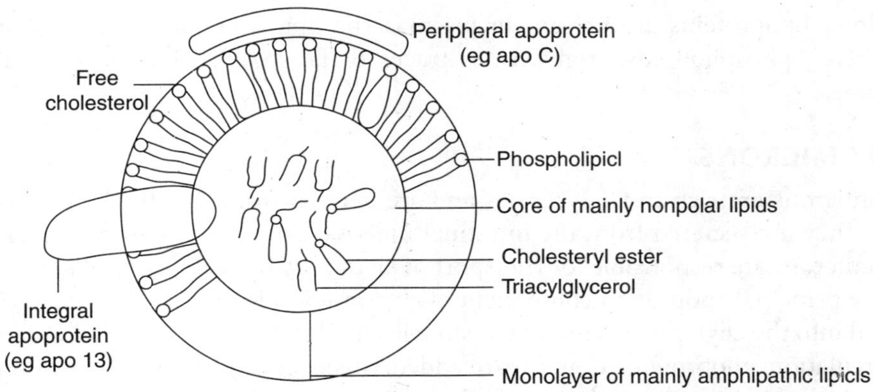

Figure 4.5 *Structure of Lipoprotein*

The density of lipoproteins is lower than that of proteins because of their combination with lipids. Because of this property, the different fractions of lipoprotein can be identified by ultracentrifugation due to different density of each fraction. Figure 4.6 shows the electrophoretic separation of plasma lipoproteins.

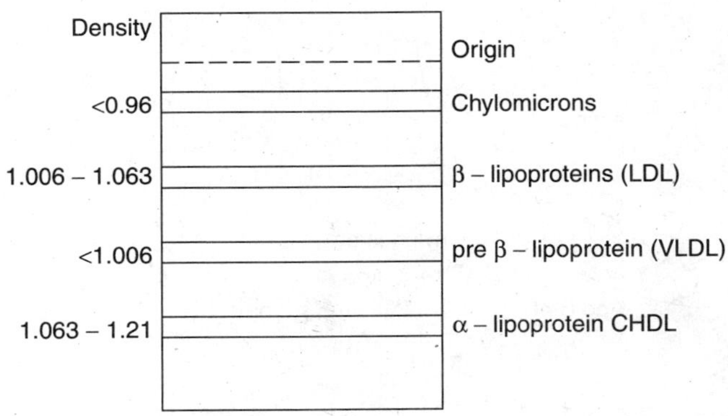

Figure 4.6 *Electrophoretic separation of plasma lipoproteins.*

Lipoproteins are widely distributed in the body tissues. The various sites of distribution are cell membrane and certain internal structures of cells such as nucleus, mitochondria, and microsomes.

Egg yolk contains two lipoproteins known as lipovitellin α and lipovitellin β. The film that stabilises the fat droplets in milk contains lipoprotein.

The rhodopsin or visual purple is a combination of protein opsin and retinene aldehyde of vitamin A.

Blood lipoproteins are present in plasma. The lipid fraction consists mainly of cholesterol, phospholipids, triglycerides traces of fat-soluble vitamins, and steroid hormones.

CHYLOMICRONS

Chylomicrons are rich in triglycerides and are derived from intestinal absorption of lipids. They are secreted from the intestinal mucosal cells into the intracellular fluid. Chylomicrons are responsible for transport of all dietary lipids into circulation.

The principal apoprotein component of chylomicron in apo-B-48, which is incorporated into the chylomicron in the intestine itself. This is the nascent chylomicron. In the circulation, apoproteins C and E are added.This results in the actual chylomicron. Apo-A is also present in chylomicron.The milky appearance (lactescence) of plasma seen after an average fatty meal is due to the presence of chylomicrons, which have the property of scattering light.

The lactescence clears after twelve to fourteen hours as the plasma passes through various organs, and this is due to the clearing action of enzyme lipoprotein lipase, located on the surface of endothelial cells of capillaries and also distributed in adipose tissue. The structure of the formation of chylomicrons and VLDL in the intestine and liver is shown in Figure 4.7.

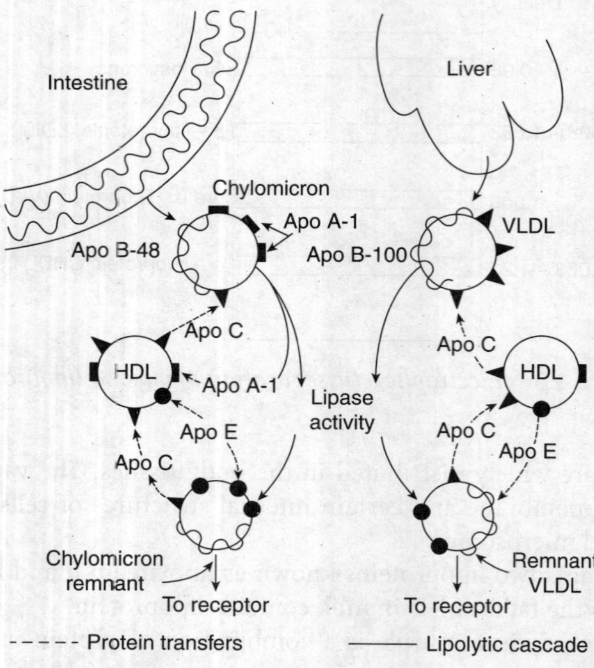

Figure 4.7 *Structure of the Formation of Chylomicrons and VLDL in the Intestine and Liver*

Function

The primary function of chylomicron is transport of dietry fats, mainly triglyderides, to liver and adipose tissues and transport of smaller amount of cholesterol and phospholipids from intestines to liver.

VERY LOW DENSITY LIPOPROTEINS (VLDL)

VLDL are rich in Triglycerides (TG) and are derived mainly from liver and secreted by the intestines in a manner similar to that of chylomicrons.The particle size is smaller than chylomicrons (CM), and the apoproteins of VLDL include apo-B48 and apo-C.

Function

The function of VLDL is transport of Triglycerides from liver to extrahepatic tissue.

LOW DENSITY LIPOPROTEINS (LDL)

They are formed from VLDL in the blood circulation. They are particularly rich in cholesterol. Approximately, 70% of the plasma cholesterol is present in the LDL fraction. The main apoprotein of LDL is apo-B100. The LDL level in plasma is clinically important as high level is indicative of susceptibility to ischemic heart disease (IHD).

Function

The principal function of LDL is transport of cholesterol to extrahepatic tissues.

HIGH DENSITY LIPOPROTEIN (HDL)

HDL is synthesised and secreted from both liver and intestinal mucosa. It contains the maximum amount of protein cholesterol, and phospholipids are the main lipids present in HDL. HDL from intestine contains apo-A and that synthesised in liver contains apo-C.

Function

The principal function of HDL is transport of cholesterol from peripheral cells and transporting it back to the liver from where most of the cholesterol excreted from the body is removed.

CHARACTERISTICS OF LIPOPROTEINS IN HUMAN PLASMA

The comparative characteristic features of different lipoproteins with regard to electrophoretic patterns, size composition, etc., are given in Table 4.4.

Table 4.4 *Characteristics of Human Plasma Lipoproteins*

Characteristics	CM	VLDL	LDL	HDL
Eletrophoretic mobility	Origin	Pre-β	β	α
Density	< 0.96	0.96–1.006	1.006–1.063	1.063–1.21
Apo proteins	A1, A11	B100, C1	B100	A1, A11
	B48	C11, C111E		CI, CII, D, E
Composition				
Protein	2	10	20	40
Lipid	98	90	80	60

APOLIPOPROTEINS (APOPROTEINS)

The protein components of lipoproteins are known as apolipoproteins or, simply, apoproteins. They perform the following functions:

- Act as structural component of lipoproteins.
- Recognise the cell membrane surface receptors.
- Activate enzymes involved in lipoprotein metabolism.

PROSTAGLANDINS

INTRODUCTION

Prostaglandins (PG) are a group of biologically active lipids. They are widely present in mammalian tissues and body fluids. Though they were initially found in the seminal fluid, it is now known that they are present in all tissues. PGs are considered as hormones since their action appears to be confined to the cells of origin.

DEFINITION

Prostaglandins are derivatives of a hypothetical 20 carbon fatty acid, namely prostanoic acid and hence known as prostaglandins or eicosanoids. The structure and biosynthetic pathway of prostaglandins are shown in Figure 4.8.

CLASSIFICATION

Prostaglandins are classified mainly into two groups shown in Table 4.5:

- Prostanoids (PG)
- Leukotrienes (LT)

Prostanoids are further subdivided into three groups:

- Prostaglandins (PG)
- Prostacyclins (PGI)
- Thromboxanes (TX)

Structure of prostaglandins:

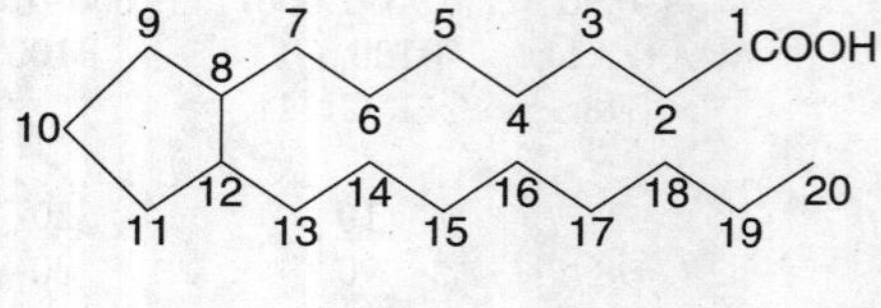

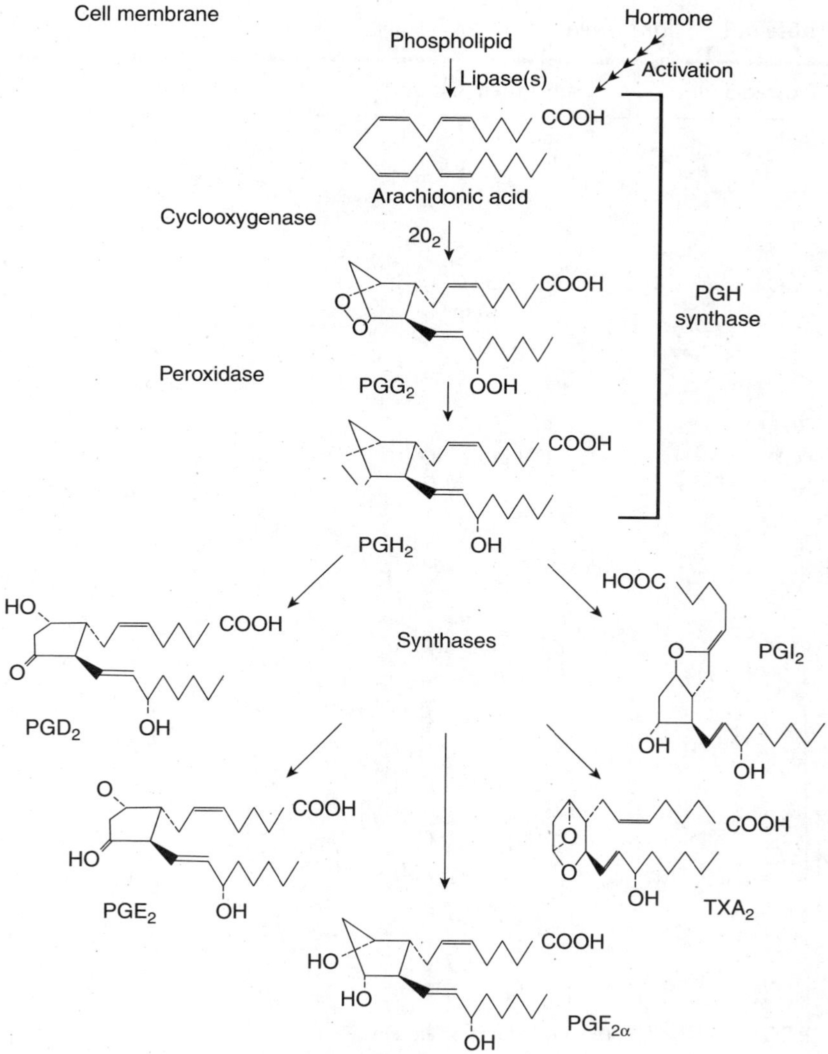

Figure 4.8 *Structure and Biosynthetic Pathway of Prostaglandins*

Table 4.5 *Structure of Prostaglandin*

Prostaglandin	Substituent Group	Structure
PGA	Double bond $C_{10}-C_{11}$ keto group C_9	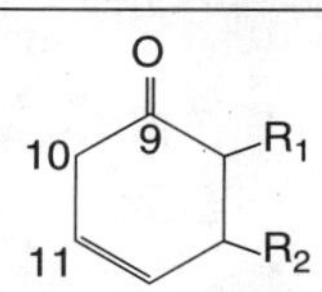

(Continued...)

Table 4.5 *Continued*

Prostaglandin	Substituent Group	Structure
PGB	Double bond C_8-C_{12} keto group C_9	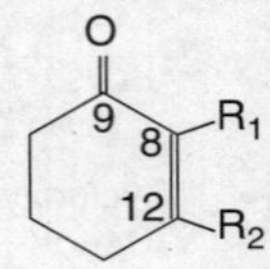
PGC	Double bond $C_{11}-C_{12}$ keto group C_9	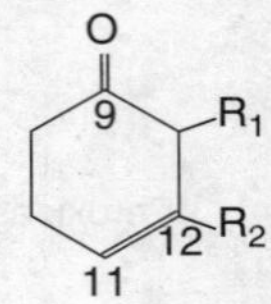
PGD	$-OH$ group at C_9 keto group at C_{11}	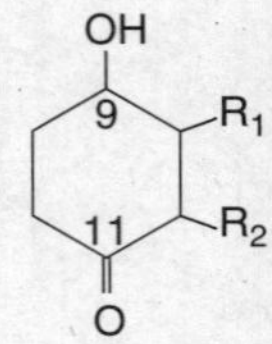
PGE	Keto group at C_9 OH group at C_{11}	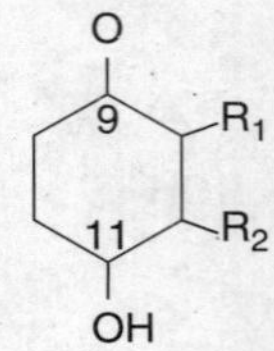
PGF	OH group at C_9 and C_{11}	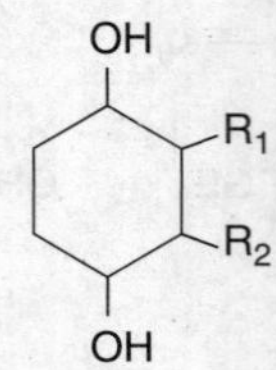
PGG	Two O_2 atoms present at C_9 C_{11} are bonded; hydroperoxide group at C_{15}	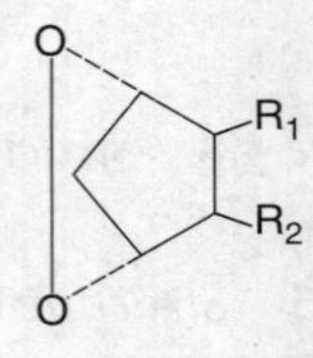
PGH	Ring the same as in PGG; $-OH$ group at C_{15}	
PGI	Attachment of O_2 between C_6 and C_9, forming another 5-carbon ring	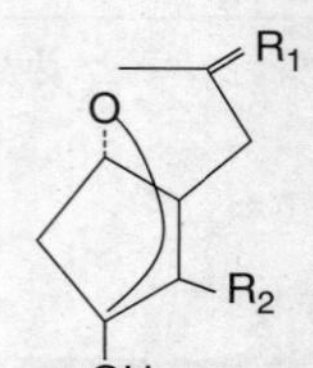

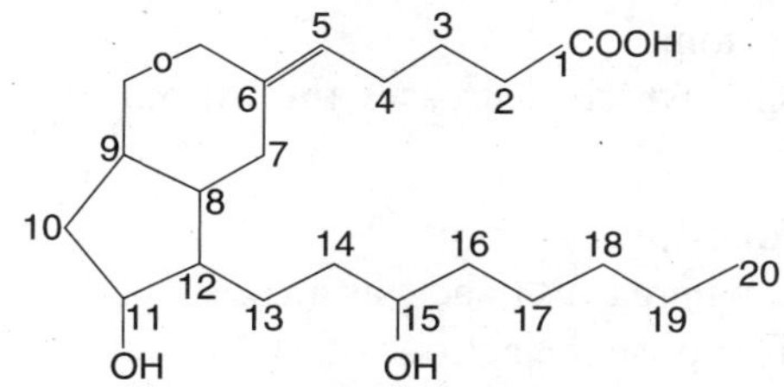

Structure of prostacyclins

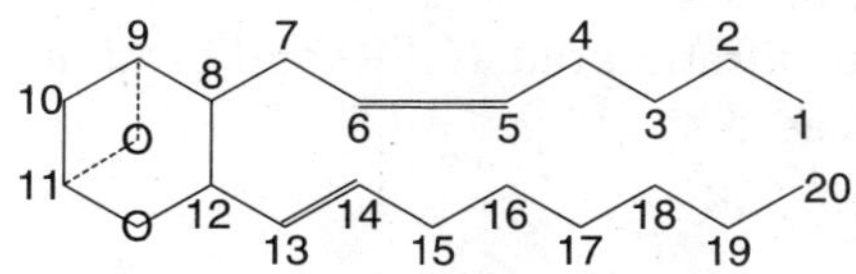

Structure of thromboxanes: TXA_2

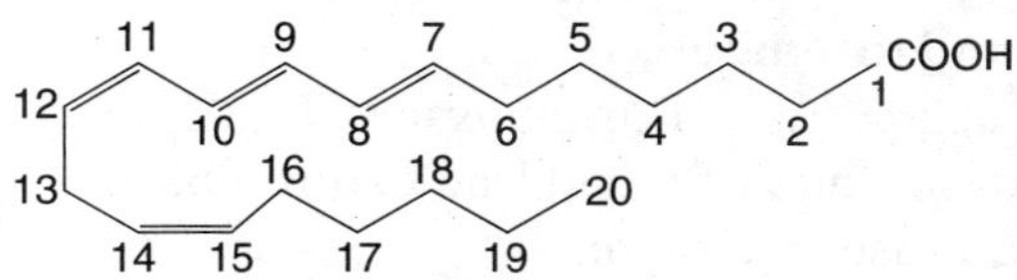

Structure of leucotrienes (LTA_4)

Biochemical Function of Prostaglandins

Prostaglandins have many biochemical functions, which are different in different tissues. PGs also bring about opposing actions in the same tissue.

Regulation of blood pressure

PGE, PGA, and PGI_2 are vasodilators in function. This results in increased blood flow and decreased peripheral resistance to lower the blood pressure. PGS serve as agents in the treatment of hypertension.

Inflammators

PGE_2 induces symptoms of inflammation (redness, swelling, edema, and so on)due to arteriolar vasodilation.

Reproduction

PGE_2 and PGF_2 are used for medical termination of pregnancy and induction of labour.

Pain and fever

Pyrogens (fever-producing agents) promote prostaglandin biosynthesis, leading to formation of PGE_2 in the hypothalamus, the site of regulation of body temperature. PGE_2, along with histamine and bradykinin, causes pain.

Regulation of gastric secretion

PGE inhibits gastric secretion. PGS are used for the treatment of gastric ulcers. However, PGS stimulates pancreatic secretion and increases the motility of intestine which often causes diarrhoea.

Influence on immune system

Macrophages secrete PGE, which decreases the immunological functions of B and T-lympocytes.

Effects on respiratory function

PGE is a bronchodilator, whereas PGF acts as a constrictor of bronchial smooth muscles. Thus, PGE and PGF oppose the actions of each other in the lungs. PGE_1 and PGE_2 are used in the treatment of asthma.

Influence on renal function

PGE increases glomerular filtration rate (GFR) and promotes urine output. Excretion of Na^+ and K^+ is also increased by PGE.

Effects on metabolism

PGE decreases lipolysis, increases glycogen formation, and promotes calcium mobilisation from the bone.

Platelet aggregation and thrombosis

PGI_2 inhibits platelet aggregation. Thromboxanes TX A_2 and PG E_2 promote platelet aggregation and blood clotting that might lead to thrombosis. Thus, PG I_2 acts as vasodilator, while TX A_2 is a vasoconstrictor.

LEUKOTRIENES

Leukotrienes are synthesised by leucocytes, mast cells, lung, heart, spleen, and so on. Leukotriene causes contraction of smooth muscles, broncho constriction, vasoconstriction, adhesion of white blood cells, and release of lysosomal enzymes.

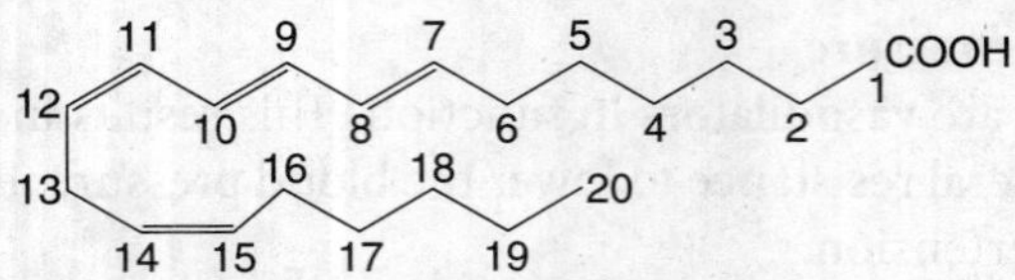

Structure of leucotrienes (LTA_4)

QUESTION BANK

SECTION A

1. What is β-carotene?
2. What is known as cyclic acids?
3. What is terpenes? Give the general formula.
4. Write the structure of sphingomyelin.
5. What are phospholipids? Write the general structure of phospholipids.

6. Define iodine number and give its significance.
7. What is the biological importance of prostaglandins?
8. What is saponification number? What is its significance?
9. Write the structure of cholesterol.
10. What are bile salts and give its uses?
11. Define Reichert–Meissl number of fat.
12. Give two examples of glycolipids.
13. Define phospholipase.
14. What is known as essential fatty acid?
15. Define lipid.
16. Define lipoprotein.
17. What is known as chylomicron?
18. Explain auto-oxidation.
19. What is known as emulsification?
20. Define the term *bile salts*.
21. Write the composition of VLDL.
22. Write the difference between saturated and unsaturated fatty acids.
23. Define derived lipids.
24. What is known as PUFA?
25. Explain triglycerides?
26. Define cephalin.
27. What is known as steroid hormones?
28. Define cerebrosides.
29. Explain Liebermann–Burchard reaction.
30. Define the function of bile salts.
31. Write the structure of mineralocorticoids.
32. Define terpenes.

SECTION B

1. Classify lipoproteins and explain in detail.
2. Give the structure and biological functions of lecithin.
3. Explain the properties and functions of fatty acids.
4. Write a note on derived lipids.

5. What are sphingolipids? Give the structure of any one sphingolipid. Outline the role of shingolipids in biological system.
6. What are phospholipids? How are they classified?
7. Give the structure of lecithin and explain the amphipathic nature of compound.
8. Explain in detail the physical and chemical properties of unsaturated fatty acids.
9. What are fats and oils? Explain the rancidity of fats and oils.
10. What is chylomicrons? Where is it formed and what is its composition?
11. What are soaps and what is the saponification number of fat?

SECTION C

1. How are lipids classified? Give an account of tests to characterise fats.
2. Write a note on glycolipids, their physical and chemical properties, and their functions.
3. Write a note on the occurrence and structure of cerebrosides and gangliosides.
4. Describe the properties of triglycerides.
5. Classify lipids and explain how fats or oils are characterised chemically.
6. Write the detailed classification of fatty acids and write the structure of any two unsaturated fatty acids.

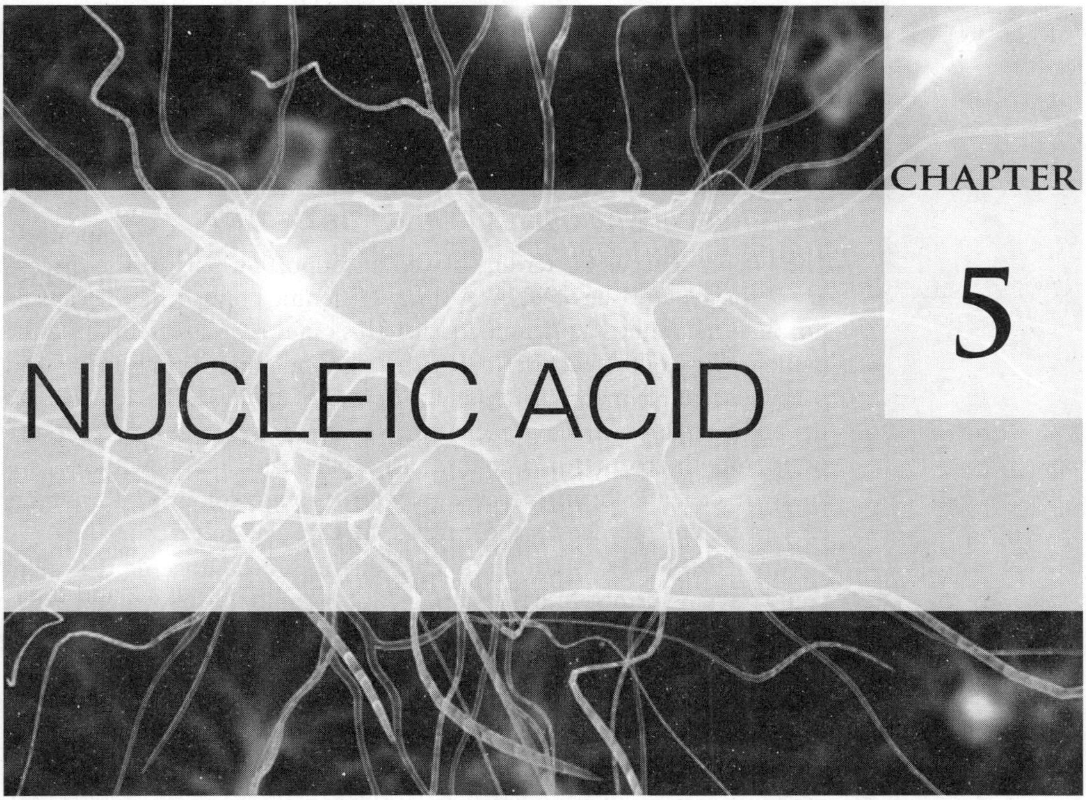

CHAPTER

5

NUCLEIC ACID

Concept Outline

- Nature of Genetic Material
- Definition
- Functions of Nucleic acids
- Types of Nucleic Acids
- Nucleosides
- Nucleotides
- Deoxyribonucleic Acid
- Properties of Nucleic acids
- Ribonucleic Acid

NATURE OF GENETIC MATERIAL

DNA is composed of deoxyribonucleotides. Each deoxyribonucleotide contains a base, a sugar, and a phosphate group. The base in the DNA macromolecule carries genetic information, whereas the sugar and phosphate group support the structural role.

The following experiments prove that DNA is a genetic material:

TRANSFORMATION OF PNEUMOCOCCI BY DNA REVEALED THAT GENES ARE MADE OF DNA

The pneumococcus bacterium played an important part in the discovery of the genetic role of DNA. A slimy, glistening polysaccharide capsule normally surrounds a pneumococcus. This outer layer is essential for the pathogenicity of the bacterium, which causes pneumonia in humans and other susceptible mammals. Mutants devoid of a polysaccharide coat are not pathogenic. The normal bacterium is referred to as the S form (because of its smooth colonies in a culture dish), whereas mutants without capsules are called R forms (because they form rough colonies). R mutants lack an enzyme needed for the synthesis of capsular polysaccharide.

In 1928, Fred Griffith discovered that a non-pathogenic R mutant could be transformed into the pathogenic S form in the following way. He injected some mice with a mixture of live R and heat-killed S pneumococci. The striking finding was that this mixture was lethal to the mice; whereas either the live R or the heat-killed S pneumococci alone are not.

The blood of the dead mice contained live S pneumococci. Thus, the heat-killed S pneumococci had somehow transformed live R pneumococci into live S pneumococci. This change was permanent – the transformed pneumococci yielded pathogenic progeny of the S form. This finding set the stage for the elucidation of chemical nature of the transforming principle.

The cell-free extract of the heat-killed S pneumococci was fractionated, and the transforming activity of its components was assayed in 1994. Oswald Avery, Colin MacLeod, and Maclyn McCarty published their discovery that 'a nucleic acid of the deoxyribose type is the fundamental unit of transforming principle of pneumococcus Type III'. Until 1994, it was believed that the chromosomal proteins carry genetic information and DNA plays a secondary role.

Further support for the genetic role of DNA came from the studies of a virus that infects the bacterium E.coli. The T_2 bacteriophage consists of a core of DNA surrounded by a protein coat. In 1951, Roger Herriot suggested that 'the virus may act like a little hypodermic needle full of transforming principles, the virus as such never enters the cells, only the tail contacts the host and perhaps enzymatically cuts a small hole through the outer membrane, and then nucleic acid of the virus head flows into the cell' as shown in Figure 5.1.

In 1952, Alfred Hershey and Martha Chase tested this idea in the following way: Phage DNA was labelled with the radioisotope ^{32}P,

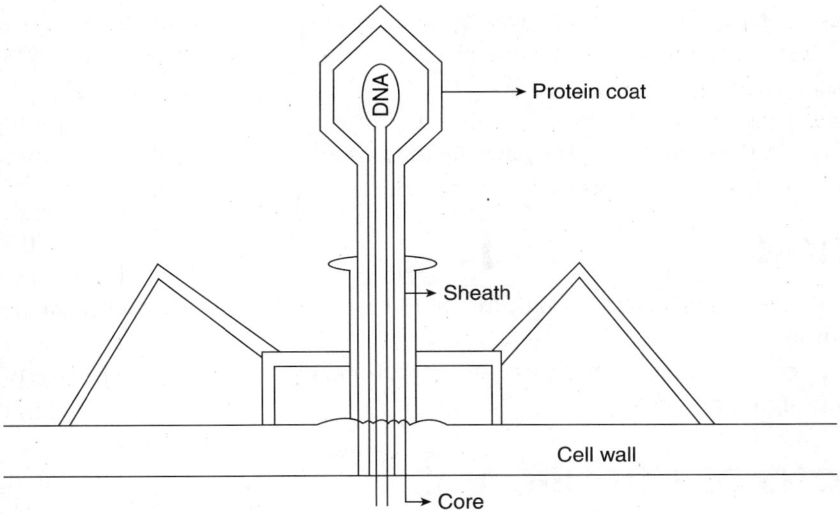

Figure 5.1 *Structural Overview of T_2 Phage*

whereas the protein coat was labelled with ^{35}S. These labels are highly specific because DNA does not contain sulphur, and the protein coat is devoid of phosphorus.

A sample of an E.coli culture was infected with labelled phage, which became attached to the bacteria during a short incubation period. The suspension was spun for a few minutes in a Waring Blendor at 10,000 rpm. This treatment subjected the phage-infected cells to very strong shearing forces, which severed the connections between the viruses and the bacteria. The resulting suspension was centrifuged at a speed sufficient to throw bacteria to the bottom of the tube. Thus, the pellet contained the infected bacteria, whereas the supernatant contained smaller particles. These fractions were analysed for ^{32}P and ^{35}S.

The results of the experiments were as follows:

1. Most of the phage DNA was found in the bacteria.
2. Most of the phage protein was found in the supernatant.
3. The Blendor treatment had almost no effect on the competence of the infected bacteria to produce progeny virus.
4. This experiment also confirms that DNA is the genetic material of life.

The nucleic acids have been the subject of biochemical investigations since 1871, when a twenty-two-year-old Swiss physician and chemist, Friedrich Miescher, isolated a substance from the nuclei of the pus cells and named it as nuclein. This substance was quite different from carbohydrates, proteins, and fats. Later, it was found that the nuclein had acid properties, and hence, it was renamed as nucleic acid.

The nucleic acid is the hereditary determinants of living organisms. They present in nuclei of cells or in the cytoplasm. They are the macromolecules present living either in the free state or bound to proteins as nucleoproteins.

DEFINITION

'The nucleic acids are biopolymers of high molecular weight with mononucleotide as their repeating units.'

The elemental compositions of nucleic acids are carbon, hydrogen, oxygen, nitrogen, and, strangely enough, phosphorus.

FUNCTIONS OF NUCLEIC ACIDS

DNA is considered to be the main source of genetic information. It is exclusively responsible for the identity of different species of organisms. DNA controls all the cellular functions.

The DNA is organised into genes, the fundamental units of genetic information. The genes control the protein synthesis through the mediation of RNA as shown below:

$$\text{DNA} \longrightarrow \text{RNA} \longrightarrow \text{Protein}$$

Besides being present in the nucleus of eukaryotic organisms, DNA is also found in the mitochondria and chloroplasts. DNA in these organisms performs specified functions with regard to protein synthesis.

1. RNA performs several functions, carried out by three different types of RNAs.
2. Messenger RNA (mRNA) specifies the sequence of amino acids in the protein synthesis (translation).
3. Ribosomal RNA (rRNA) is found in combination with proteins and is associated with the structure and function of ribosomes and factories of protein synthesis.
4. Transfer RNA (tRNA) delivers amino acids to ribosomes for protein synthesis.
5. RNA, as ribonucleoprotein, participates in the post-transcriptional modification of other RNAs.
6. RNA (not DNA) is the genetic material carrying hereditary information in many viruses.

TYPES OF NUCLEIC ACIDS

There are two kinds of nucleic acids, deoxyribonucleic acid (DNA) and ribonucleic acid (RNA). DNA is found mainly in the chromatin of the cell nucleus, whereas most of the RNA (90%) is present in the cytoplasm and a little (10%) in the nucleolus. Upon hydrolysis, the two nucleic acids yield three components of phosphoric acid, a pentose sugar, and nitrogenous bases as shown in Table 5.1.

Table 5.1 *Components of Nucleic Acid*

Components	Ribonucleic Acid	Deoxyribonucleic Acid
Acid	Phosphoric acid	Phosphoric acid
Pentose sugar	D-ribose	D-2-deoxyribose
Nitrogenous base		
Purines	Adenine	Adenine
	Guanine	Guanine
Pyrimidines	Cytosine	Cytosine
	Uracil	Thymine

PHOSPHORIC ACID

Phosphoric acid contains three monovalent hydroxyl groups and a divalent oxygen atom, all connected to the pentavalent phosphorus atom. The molecular structure and formula for phosphoric acid (H_3PO_4) is shown here.

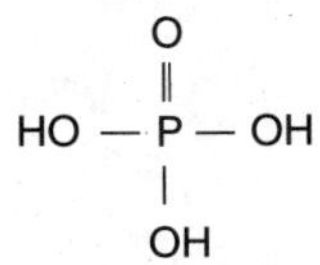

Phosphoric acid

PENTOSE SUGAR

The two types of nucleic acids are distinguished primarily on the basis of the 5-carbon sugar or pentose which they possess. One possesses D-2-deoxyribose, hence the name deoxyribonucleic acid, while the other contains D-ribose, hence the name ribonucleic acid. Both these sugars in nucleic acids are present in the furanose form and are of β configuration.

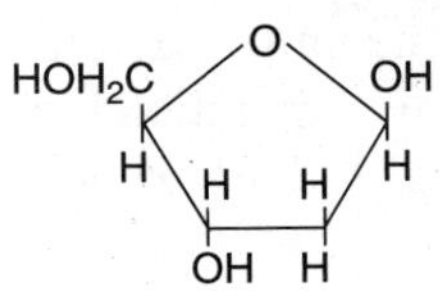

D-2-deoxyribose

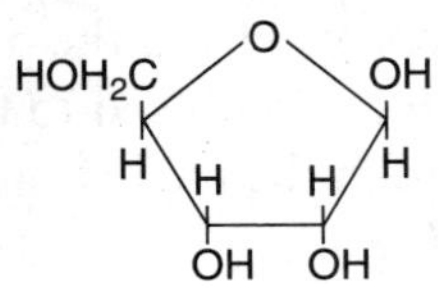

D-ribose

Properties

Two types of sugar are present in the nucleic acids; D-ribose is the parent sugar present in the RNA, whereas D-2-deoxyribose is a derivative in which the OH group on C-2 has been replaced by an H atom. The two sugars are differentiated by specific colour reactions. Ribose reacts with orcinol in hydrochloric acid solution containing ferric chloride. Deoxyribose reacts with diphenylamine in acid solution.

Pentose reacts with phosphoric acid and forms esters. The OH groups of the pentose at C-3 and C-5 are involved in forming 3′-5′-phosphodiester bond between adjacent pentose residues. This phosphodiester bond is an integral part of the structure of nucleic acids.

NITROGENOUS BASES

Two types of nitrogenous bases are found in all nucleic acids. These are derivatives of pyrimidine and purine.

Pyrimidine Derivatives

These are all derived from their parent heterocyclic compound, pyrimidine.

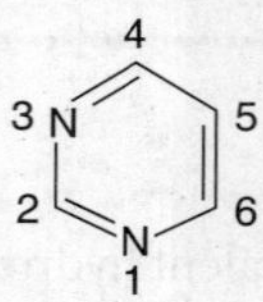

The common pyrimidine derivatives found in nucleic acids are uracil, thymine, and cytosine. Their structures are given below:

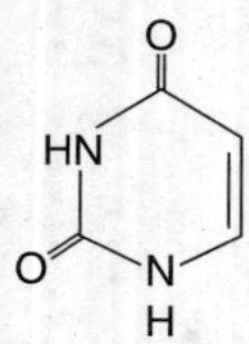

Uracil
[2-4,dioxy pyrimidine]

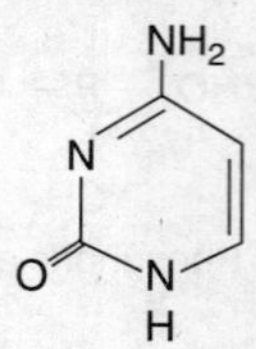

Cytosine
[2-oxy, 4-amino pyrimidine]

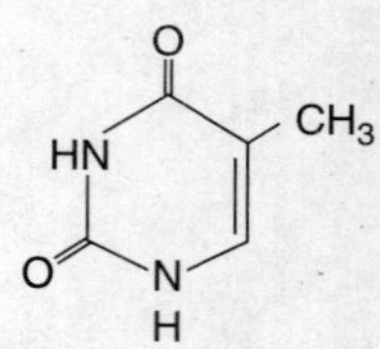

Thymine
[5-methyl uracil
or
2-4,dioxy-5-methyl
pyrimidine]

Purine Derivatives

Purine derivatives are derived from the parent compound purine, which is made up of six-membered pyrimidine ring fused to the five-membered imidazole ring. The main purine derivatives found in the nucleic acids are adenine and guanine. Their structures are given below:

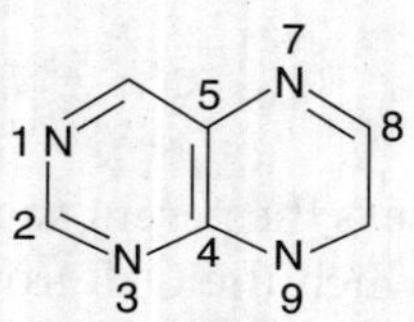

Purine

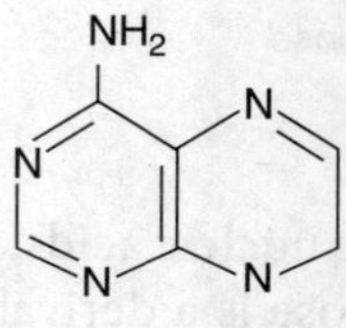

Adenine
[6-amino purine]

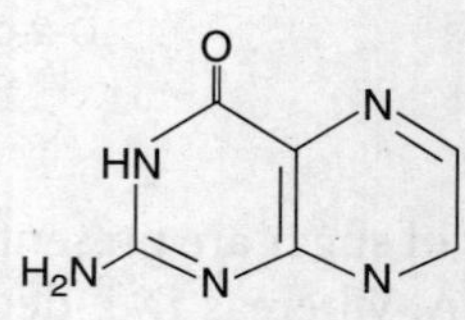

Guanine
[2-amino 6-oxy purine]

Tautomer Forms of Purines and pyrimidines

The existence of a molecule in keto (lactam) and enol (lactim) form is known as tuatomerism. The heterocyclic rings of purines and pyrimidines with oxo functional groups exhibit tuatomerism as simplified below:

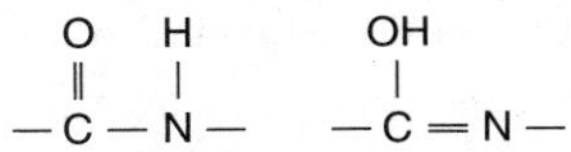

Lactam form Lactim form

NUCLEOSIDES

DEFINITION

The nucleotides are compounds in which nitrogenous bases (purine and pyrimidine) are conjugated to the pentose sugars (ribose or deoxyribose) by a β-glycosidic linkage. The β-glycosidic linkage involves C-1 of sugar and the hydrogen atom of N-9 (in the case of purines) or N-1 (in the case of pyrimidine).

STRUCTURE OF NUCLEOSIDES

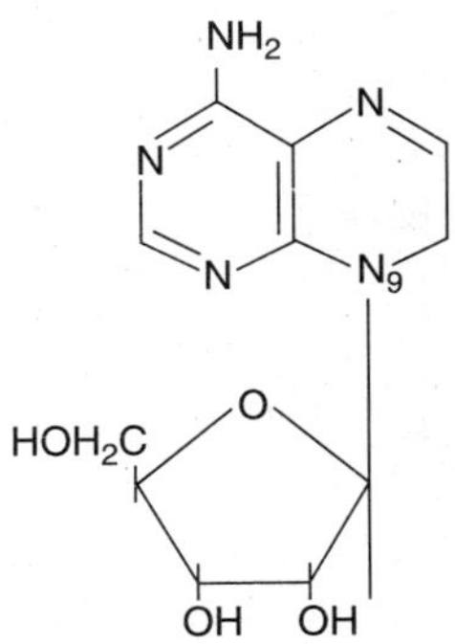

Adenosine

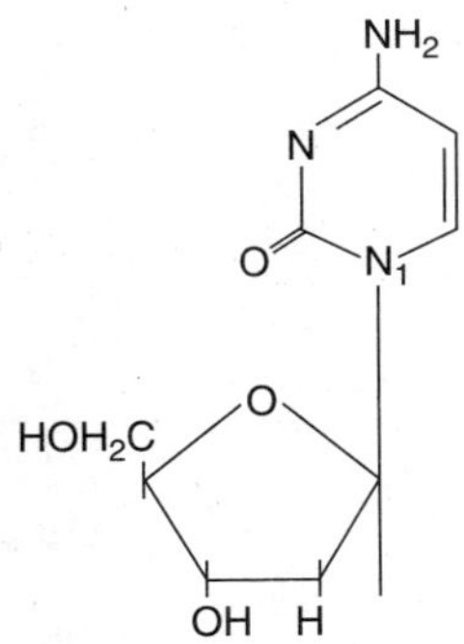

Deoxycytidine

NOMENCLATURE

The nucleosides are generally named for the particular purine or pyrimidine present. Nucleosides containing ribose are ribonucleoside, while those possessing deoxyribose are deoxyribonucleoside.

The common nucleosides are given in Table 5.2.

Table 5.2 *Common Nucleosides*

Base	Sugar	Nucleoside	Trivial name	Abb
Adenine	Ribose	Adenine ribonucleoside	Adenosine	AR
Guanine	Ribose	Guanine ribonucleoside	Guanosine	GR

(Continued…)

Table 5.2 *Continued*

Base	Sugar	Nucleoside	Trivial name	Abb
Cytosine	Ribose	Cytosine ribonucleoside	Cytidine	CR
Uracil	Ribose	Uracil ribonucleoside	Uridine	UR
Adenine	Deoxyribose	Adenine deoxyribonucleoside	Deoxyadenosine	AdR
Guanine	Deoxyribose	Guanine deoxyribonucleoside	Deoxyguanosine	GdR
Cytosine	Deoxyribose	Cytosine deoxyribonucleoside	Deoxycytidine	CdR
Thymine	Deoxyribose	Thymine deoxyribonucleoside	Deoxythymidine	TdR

NUCLEOTIDES

Nucleotides are the phosphoric acid esters of nucleosides.

In ribose moiety of a ribonucleoside, phosphorylation is possible only at three position (C2′, C3′, and C5′) since C1′ and C4′ are involved in the furanose ring formation. In deoxyribosenucleoside, only two positions (C3′ and C5′) are available for phosphorylation.

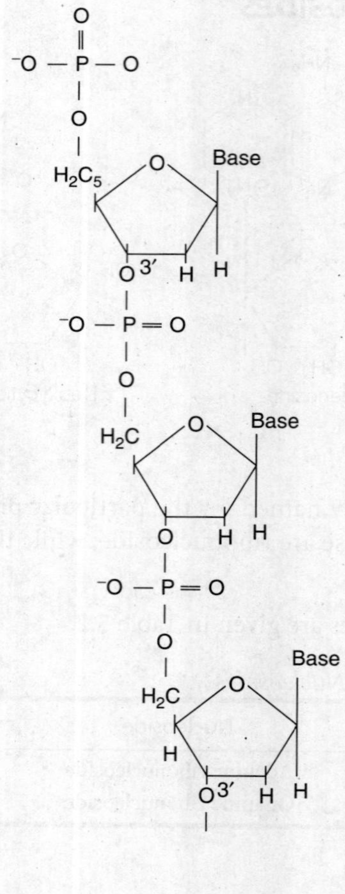

NOMENCLATURE

When a single phosphate moiety is added to a nucleoside, it is termed as a mononucleotide. For example, adenosine monophosphate (AMP) contains adenine + ribose + phosphate.

The common nucleotides are given in Table 5.3.

Table 5.3 *Common Nucleotides*

Base	Sugar	Nucleotide	Trivial name	Abb
Adenine	Ribose	Adenosine 5′ monophosphate	Adenylic acid	AMP
Guanine	Ribose	Guanosine 5′ monophosphate	Guanylic acid	GMP
Cytosine	Ribose	Cytidine 5′ monophosphate	Cytidylic acid	CMP
Uracil	Ribose	Uridine 5′ monophosphate	Uridylic acid	UMP
Adenine	Deoxyribose	Deoxyadenosine 5′ monophosphate	Deoxyadenylic acid	dAMP
Guanine	Deoxyribose	Deoxyguanosine 5′ monophosphate	Deoxyguanylic acid	dGMP
Cytosine	Deoxyribose	Deoxycytidine 5′ monophosphate	Deoxycytidylic acid	dCMP
Thymine	Deoxyribose	Deoxythymidine 5′ monophosphate	Deoxythymidylic acid	dTMP

FUNCTIONS OF NUCLEOTIDES

They are the building blocks or monomeric units in the nucleic acid (DNA and RNA) structure. Nucleotides are the structural components of several coenzymes of B complex vitamin (examples: FAD, NAD+, NADP+, and coenzyme A). They serve as carriers of high-energy intermediates in the biosynthesis of carbohydrates, lipids, and proteins (example: UDP-glucose, CDP- acylglycerol, and S-adenosylmethionine). Nucleotides are intimately involved in the energy reactions of the cell (example: ATP is the energy currency of the cell). Nucleotides control several metabolic reactions by their action as allosteric regulators. Cyclic AMP and cyclic GMP are the second messengers in hormonal function.

NUCLEOSIDE DI- AND TRI-PHOSPHATE

Nucleoside monophosphate possesses only one phosphate moiety (AMP and TMP). The addition of second or third phosphate to the nucleoside results in nucleoside diphosphate (e.g. ATP). The additional phosphates are linked by acid anhydride bonds possessing high-energy group potential. These bonds yield about 7 Cal/mol on hydrolysis.

DEOXYRIBONUCLEIC ACID

DNA is a polymer of deoxyribonucleotides and is found in chromosomes, mitochondria, and chloroplasts. The nuclear DNA is bound to basic proteins called histones. DNA is present in every nucleated cell and carries the genetic information. The human body contains 0.5 gm of DNA.

INTERNUCLEOTIDE LINKAGES

The backbone of a DNA molecule consists of deoxyribose linked by phosphodiester bridges. The 3′ hydroxyl of the adjacent sugar moiety of one deoxyribonucleotides is joined to the 5′ hydroxyl of the adjacent sugar by an internucleotide linkage called a phosphodiester bond.

SCHEMATIC REPRESENTATION OF DNA

The monomeric deoxynucleotides in DNA are held together by 3′-5′-phosphodiester bond. DNA structure is often represented in a shorthand form as shown in Figure 5.2. A DNA molecule is made of two chains of several thousands of linked nucleotides. Each nucleotide consists of a phosphate group, a pentose (= a sugar with five carbon atoms and a base containing nitrogen (adenine, thymine, guanine and cytosine).

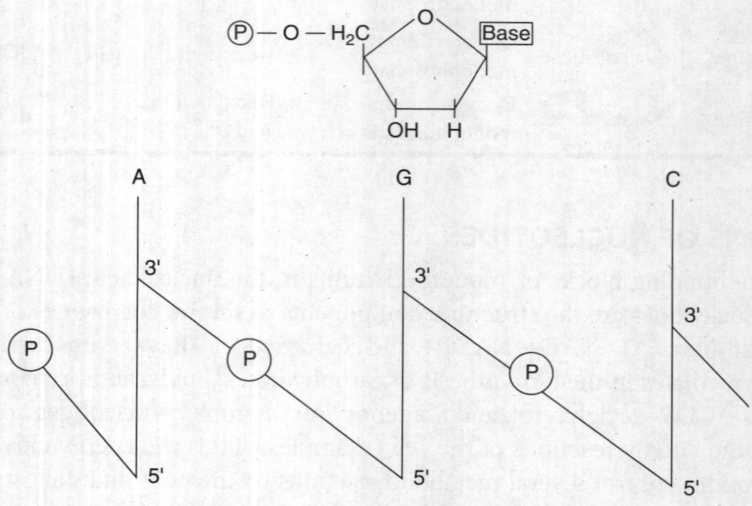

Figure 5.2 *Shorthand Form of the Structure of DNA*

PRIMARY STRUCTURE OF DNA

1. Chromosomal DNA is made up of long DNA molecules with the size MW 1.6×10^6 to 2×10^9.
2. A DNA polymer generally contains 10^{10} deoxynucleotides.

3. The four different types of bases that are present in the DNA molecules are adenine deoxynucleotides (dA), thymine deoxynucleotides (dT), guanine deoxynucleotides (dG), and cytosine deoxynucleotides (dC).
4. Nucleotides are connected to one another by covalent 3′-5′-phosphodiester linkage. The ester linkage of a single phosphate residue is with the 3′—OH of one nucleotide with 5′—OH group of ribose of the next molecule.
5. Phosphodiester linkage gives rise to a linear polydeoxynucleotide strand with two free ends on both sides.
6. Each strand has one free 5′phosphate group without phosphodiester linkage at one end, which is called the 5′ end. The opposite group is called the 3′ end.
7. The primary structure mainly consists of different deoxynucleotides in one strand joined together by phosphodiester linkages.
8. The backbone of the primary structure is sugar phosphate residues, while the purine and pyrimidine bases connected with the sugar residue project laterally from the backbone.

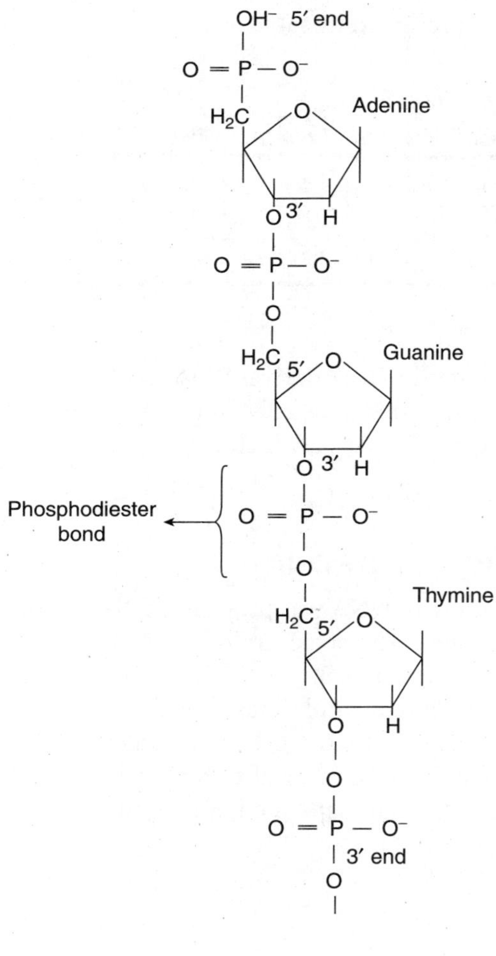

CHARGAFF'S RULE OF DNA COMPOSITION

Erwin Chargaff, in late 1940s, quantitatively analysed the DNA hydrolysates from different species. The important conclusions drawn by him are the following:

1. The sum of purines is equal to the sum of pyrimidines; that is, pu/py = 1. In other words, A + G = T + C.
2. The ratio of adenine to thymine is also one, A/T = 1.
3. The ratio of guanine to cytosine is also one, G/C = 1.
4. Bases with 6-amino groups are equal to bases with 6-keto groups; that is, A + C = G + T.
5. The ratio of A + T/G + C, known as dissymmetry ratio, varies greatly from one species of DNA to the other and is characteristic of that species. When the dissymmetry ratio exceeds one, such a DNA is called AT type. When the value is less than one, such a DNA is designated as GC type. In most animals, the value ranges from 1.3 to 3.2 and in higher plants, from 1.1 to 1.7. DNA composition of human liver is presented in Table 5.4.

Table 5.4 *DNA Composition of Human Liver*

Species	Base proportions, moles %				Pu / Py ratio A+G/T+C	A/T	G/C	A+T/G+C
	A	G	T	C				
Human liver	30.3	19.5	30.8	19.9	0.99	1.00	0.98	1.54

The double helical structure of DNA derives its strength from Chargaff's rule.

Single-stranded DNA, which is usually single stranded, does not obey Chargaff's rule. However, double-stranded RNA, which is the genetic material in certain viruses, satisfies Chargaff's rule.

SECONDARY STRUCTURE OF DNA

The double helical structure of DNA was proposed by James Watson and Francis Crick in 1953 (Noble Prize, 1962). The structure of DNA double helix is comparable to a twisted ladder as shown in Figure 5.3.

The salient features of Watson and Crick model of DNA are given below.

DNA molecule consists of two helical polynucleotide chains, which are coiled around a common axis in the form of a right-handed double helix. Such double helices cannot be pulled apart and can be separated only by an unwinding process.

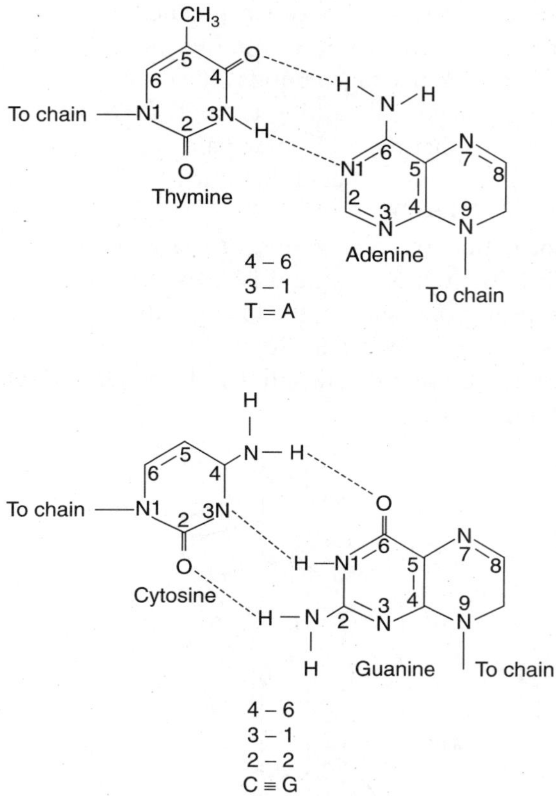

Figure 5.3 *Complementary Base Pairing in DNA*

They are called plectonemic coils, that is, coils that are interlocked about the same axis.

The two strands are antiparallel, that is, one strand runs from 5′ to 3′ direction, while the other from 3′ to 5′ direction. This is comparable to two parallel adjacent roads carrying traffic in opposite direction.

The width (or diametre) of a double helix is 20 Å(2 nm).

Each turn of the helix is 34 Å (3.4 nm) with 10 pairs of nucleotides, with each pair at a distance of about 3.4 Å.

Each DNA strand has a hydrophilic deoxyribose phosphate backbone (3′-5′-phosphodiester bond) facing outwards, while the hydrophobic bases are stacked inside.

The two nucleotides are complementary to each other due to the base pairing.

The two strands are held together by hydrogen bonds formed by complementary base pairs. The A−T pair has two hydrogen bonds while G−C pair has three hydrogen bonds. The G≡C is stronger by about 50% than A=T.

The hydrogen bonds are formed between purines and pyrimidines only. If two purines face each other, they would not fit into the allowable space. And two pyrimidines would be too far to form hydrogen bonds. The only base arrangement possible in DNA structure, from special considerations, is A−T, T−A, G−C, and C−G.

Chargaff's rule explains that the DNA base pairs are complementary to each other. So, the number of adenine equals to that of thymine (A = T) and the number of guanine is equal to that of cytosine (G = C).

The genetic information resides on one of the two strands known as template strand or sense strand. The opposite strand is antisense strand. The double helix has major grooves and minor grooves along the phosphodiester backbone. Proteins interact with DNA at these grooves, without disrupting the base pairs and double helix. The diagrammatic representation of Watson–Crick model of double helical structure of DNA is given in Figure 5.4.

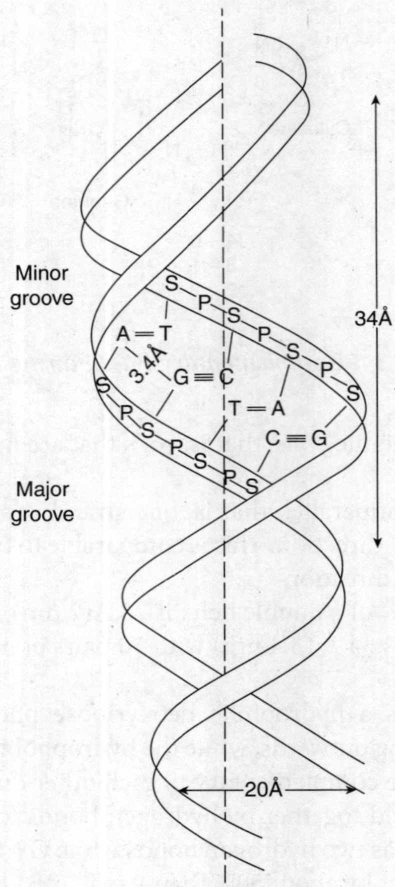

Figure 5.4 *Watson–Crick Model of Double Helical Structure of DNA*

MOLECULAR WEIGHT OF DNA

DNA is among the largest known. It is difficult to isolate DNA without fragmentation. The entire DNA of bacteria, such as Escherichia coli, has a molecular weight of 2.6×10^9. Viral DNAs range in size from about 1 to 350×10^6. The residue weight of a single nucleotide is 300 to 350. Thus, there are about three thousand nucleotides per million molecular weight of DNA.

SHAPE AND SIZE OF DNA MOLECULE

DNA molecules are huge in size. On an average, a base pair of B-DNA with a thickness of .34 nm has a molecular weight of 660. The term kilobase pair (Kb = 1,000 base pairs) is common in the DNA structure. The number of base pairs and the length of DNA (contour length) varies from species to species.

DNA molecules are characteristically highly elongate structures. The Escherichia coli chromosome is a single molecule of double helical DNA, containing about 4 million base pairs. The DNA contains 4.6 million base pairs. Its contour length is about 3 mm and 1 mm in diameter.

The largest chromosomes of Drosophila melanogaster contain a single DNA molecule with 6.2×10^7 base pairs.

For instance, the DNA from polyomavirus contains 4,600 base pairs and has a contour length of 1.6 μm. Furthermore, tropocollagen, the basic structural unit of collagen, has the shape of a rod, which is 3,000 Å in length and 15 Å in diametre.

DNA molecule from the T7 bacteriophage is, however, linear in shape.

In case of humans, the twenty-three haploid chromosomes have 2,900,000 KB with a total contour length of 990 mm.

TYPES OF DNAs

The double helical structure of DNA exists in at least six different types A to E and Z. Among these, B, A, and Z forms are important.

A-DNA

The number of base pairs per turn is about 11, and the pitch of the helix of A-DNA is 2.46 nm.

In both A-DNA and B-DNA, the sugar moiety and base are on the opposite sides of the glycosidic bond. In the presence of high concentrations of cations, the nucleotides will rotate into syn-conformations. In these conditions, a different DNA conformation can exist. The chain zigzag of Z-DNA results in syn- and anti-conformations.

A-DNA appears when the DNA fibre (B-DNA) is dehydrated; that is, relative humidity is reduced from 92% to 75% and Na^+, K^+, and Cs^+ ions are present in the medium. In other words, in solution, DNA assumes the B form and under conditions of dehydration, the A form.

Like B-DNA, A-DNA is a right-handed double helix, made of anti-parallel strands held together by Watson–Crick base pairing. But the vertical rise base pair is 2.3 Å, and the number of base pair per helical is 11, relative to the 3.4 Å rise and 10.4 base pairs per turn found in B-DNA. Such helix in A-DNA moves a total distance of (2.3 × 11) or 25.30 Å per turn, which is called the helix pitch. Obviously, the rotation per base pair A-DNA is 360/11 or 32.72 Å. The diametre of A-DNA helix is 25.5 Å, whereas that of B-DNA helix is 23.7 Å. The helix of A-DNA is, therefore, wider and shorter than that of B-DNA.

B-DNA

The B-type of DNA double helix described by Watson and Crick is the most predominant type under physiological conditions.

Adjacent nucleotides in each chain are rotated by 34.60 relative to each other. Double helix completes one turn approximately every 10.4 base pairs.

A single turn of a double helix spans a distance of 3.4 nm. Therefore, a single base pair increases the length of a double helix by 0.33 nm.

Diametre of the double helix is 2.37 nm.

When the B-DNA crystals are dried and the salt content of the crystal is lowered, B-DNA becomes the short and stubby molecule of A-DNA.

C-DNA

C-DNA is produced at 66% relative humidity and in the presence of Li^+ ions. This form of DNA is known as the right-handed helix with an axial rise of 3.32 Å per base pair. Per turn of the helix contains 9.33 base pairs and the pitch of the helix is, therefore, 3.32 × 9.33 or 30.97 Å. The rotation per base in C-DNA is 360/9 or 38.58 Å. The C-helix has a diameter of 19 Å and the tilt of the base pair is 7.8 Å.

D-DNA

D-DNA is rarely present and has only eight base pairs per helical turn. This form is present in all DNA molecules irrespective of their base sequence. The axial rise per base pair is 3.03 Å with a tilt of 16.7° from the axis of the helix.

H-DNA

H-DNA is found in polypyrimidine or polypurine segments that contain a minor repeat. An example is a long stretch of alternating T and C residues. The main feature of H-DNA is the pairing and interwinding of three strands of DNA to form a triple helix. H-DNA in triple helix contains pyrimidines in two strands and the third contains purines.

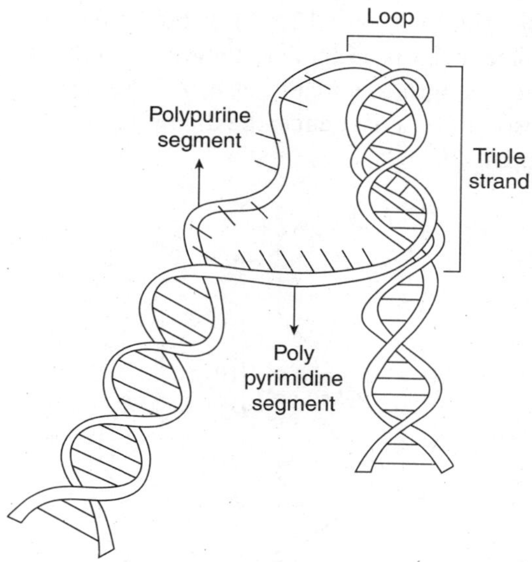

Z-DNA

Z-DNA is longer and thinner than B-DNA. It is a left-handed helix. A single turn of Z-DNA has twelve base pairs, and the pitch of the helix is 4.56 nm. The diameter of the double helix is 1.84 nm. The major groove of Z-DNA is not a groove but it simply forms a convex surface. The minor groove of Z-DNA is like a deep cleft and it spirals around the structure. Almost all the cells contain B-DNA, though regions with guanine and cytosine base pairs assume the Z conformation.

Z-DNA is the more radical departure from B-DNA and is characterised by a left-handed helical rotation. Rick, Nordheim, and Wang discovered it in 1984. They found that a hexanucleotide, CGCGCG, forms a duplex of anti-parallel strands held together by Watson-Crick base pairing as excepted. Surprisingly, they found that this double helix was left-handed and the phosphates in the backbone were in a zigzag manner; hence they termed this new form as Z-DNA. Another remarkable characteristic of Z-DNA is that in the adjacency, regular sugar residues have alternating orientation, and it is because of this reason that in Z-DNA, the repeating unit is a dinucleotide as against the B-DNA, where the adjacent sugar residues have some orientation so that the repeating unit in B-DNA is a mononucleotide. Unlike A- and B-DNAs, Z-DNA contains only one deep helical turn, with an axial groove. There are twelve base pairs (or six repeating dinucleotide units) per helical turn, with an axial of 3.8 Å per base pair; the bases are inclined at 9° with the axis of the helix. Because twelve base pairs are accommodated in one helix Z-DNA, as against 10.4 in B-DNA, the angle of twist per repeating unit (dinucleotide) is (360 / 12 × 2) or 60° as against 34.61° per nucleotide in B-DNA. One complete helix is 45.60 Å in length in contrast to 35.36 Å in B-DNA. Since the base gets more length to spread out in Z-DNA and since the angle of tilt is 60°, they are more closer to the axis; hence, the diametre of Z-DNA molecule is 18.4 Å, whereas it is 23.7 Å in B-DNA.

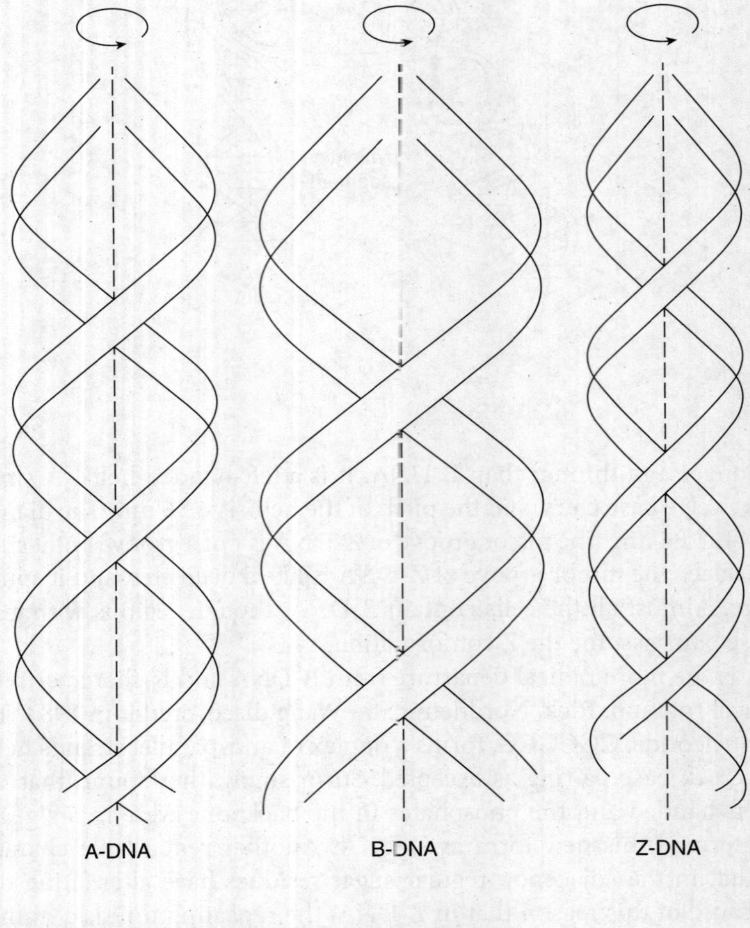

PROPERTIES OF NUCLEIC ACIDS

DENATURATION

The two strands of DNA helix are held together by hydrogen bonds (change in pH or increase in temperature) results in the separation of polynucleotide strands. This phenomenon of loss of helical structure of DNA is known as denaturation. The phosphodiester bonds are not broken by denaturation. Loss of helical structure can be measured by increase in absorbance at 260 nm. The transition from native to a denatured form is usually very abrupt and is accelerated by reagents like urea and formamide, which enhance the aqueous solubility of the purine and pyrimidine groups. Denaturation involves the following changes:

Increase in absorption of ultraviolet light (= hyperchromic effect). All nucleic acids show a strong absorption in the ultraviolet with a maximum near 260 mμ. When the native DNA is denatured, there occurs a marked hyperchromicity, that is, an increase in the optical absorbency of the purine and pyrimidine bases. This change reflects a decrease in hydrogen bonding.

Decrease in specific optical rotation. Native DNA shows positive rotation, which is decreased upon denaturation. This change is analogous to the change observed in protein denaturation.

Decrease in viscosity. Due to the rigid double helical structure and rod-like character of DNA, the native DNA possesses high viscosity.

EFFECT OF TEMPERATURE ON DENATURATION

The DNA double helix, although stabilised by hydrogen bonding, can be denatured by heat. This occurs at characteristic temperature called denaturation temperature or melting temperature T_m.

T_m is defined as the temperature at which the helical structure of DNA is lost.

Since G−C base pairs are more stable than A-T base pairs, the T_m is greater for DNAs with higher G−C content. Thus, the T_m is 650°C → 65°C for 35% G−C content, while it is 700°C → 70°C for 50% G−C content. Formamide destabilises hydrogen bonds and, therefore, lowers T_m.

EFFECT OF pH ON DENATURATION

Denaturation of DNA helix also occurs at acid and alkaline pH values, at which, changes of the substituents on the purine and pyrimidine bases can occur. In acid solutions near pH 2 to 3, at which amino groups bind protons, the DNA helix is disrupted. Similarly, in alkaline solutions near pH 12, the enolic hydroxyl groups ionise, thus preventing the amino group hydrogen bonding.

SYNTHESIS OF POLYNUCLEOTIDE BY KORNBERG ENZYME

Arthur Kornberg in 1955 discovered the enzyme DNA polymerase from E. Coli for the synthesis of nucleotides as shown in Figure 5.5.

DNA polymerase is 109 Kdal single polypeptide chain. It catalyses the step-by-step addition of polynucleotide units to a DNA chain.

$$(\text{DNA})_n \text{ residues} + \text{dNTP} \rightleftharpoons (\text{DNA})_{n+1} + \text{PPi}$$

DNA polymerase requires the following components to synthesis DNA chain.

All four deoxyribonucleoside 5′-triphosphate dATP, dGTP, dTTP, and dCTP must be present. DNTP used to refer these deoxynucleoside triphosphate-Mg^{2+} is also required.

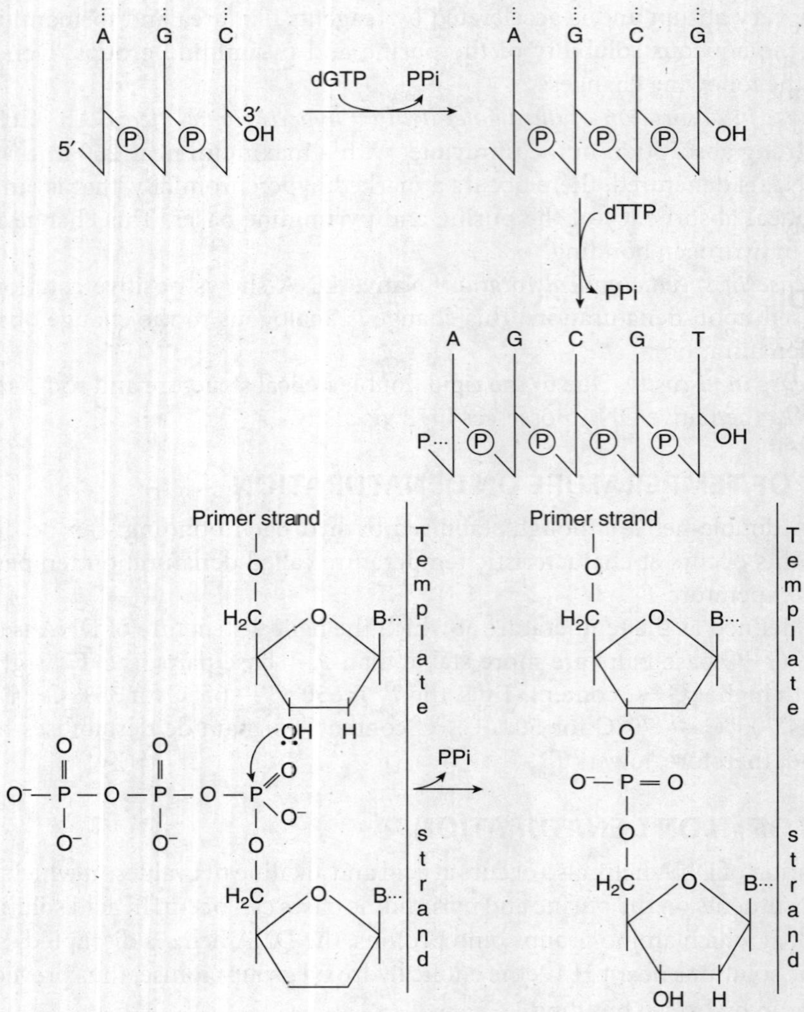

Figure 5.5 *DNA Polymerase Catalyses Elongation of DNA Chain in the 5′-3′ Direction*

DNA polymerase adds deoxyribonucleotides to the 3′ hydroxyl terminus of a pre-existing DNA or RNA. A primer chain with 3′-OH group is required.

A DNA template is essential. The template can be single- or double-stranded DNA.

Chain elongation reaction catalysed by DNA polymer occurs by means of a nucleophilic attack of the 3′-OH terminus of the primer on the innermost phosphate atom of the incoming deoxynucleoside triphosphate.

A phosphodiester bond is formed, and pyrophosphate is concomitantly released. The subsequent hydrolysis of pyrophosphate drives the polymerisation forward. This displacement of the overall equilibrium could not occur if the activated intermediate were nucleoside diphosphate.

The elongation of DNA chain proceeds from 5′ to 3′ direction. About ten nucleotides are added per second per molecule of DNA polymerase l.

Polymerisation is processive; that is, many nucleotides are added without the release of the enzyme from the template.

DNA polymerase catalyses the formation of phosphodiester bond only if the base on the incoming nucleotide is complementary to the base on the template strand. Thus, DNA polymerase is a template-directed enzyme.

HYBRIDISATION

A new technique called hybridisation was developed in 1961 by Sol Spiegelman. The concept lying behind this technique is that the ability of the complementary DNA strands to pair with one another can be used to detect similar DNA sequence into two different species or with genome of single species as shown in Figure 5.6.

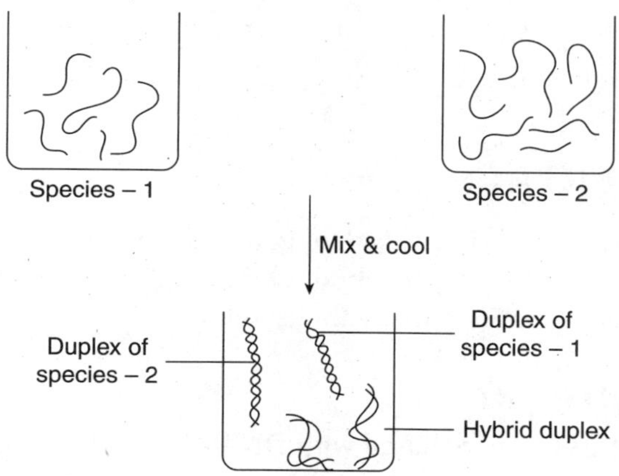

Figure 5.6 *Diagrammatic Representation of Hybridisation*

If the duplex DNA of human and mouse cells is denatured by heating, then mixed and kept at 65°C for many hours, most of the DNA starts to anneal. Almost all the mouse DNA strands anneal with complementary mouse DNA strands and the human DNA strands anneal with complementary human DNA strands. However, some strands of the human DNA will also anneal those of the mouse DNA to form hybrid duplexes. This explains the fact that different organisms have some common evolutionary heritage; they also have some proteins and RNAs with the same functions and often similar structures. Hybridisation techniques can be used to detect a specific RNA rather than DNA. The isolation and identification of specific genes and RNAs relies on these techniques, and new applications are making it possible to accurately identify an individual on the basis of a crime or predict the onset of some diseases in an individual before symptoms appear.

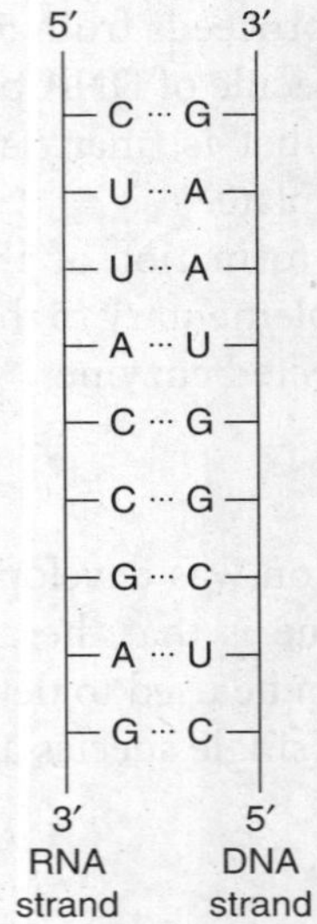

An RNA–DNA hybrid can be formed if RNA and DNA have complementary sequences.

RIBONUCLEIC ACID

Ribonucleic acid (RNA), like DNA, is a long, unbranched macromolecule, consisting of nucleotides joined by 3′ → 5′ phosphodiester bond. The number of ribonucelotides RNA ranged from as few as 75 to many thousands.

DIFFERENCES WITH DNA

Although many features can be shared with DNA, the RNA molecules possess several specific differences.

As apparent from its name, the sugar moiety in RNA, to which the phosphate and the nitrogen bases are attached, is ribose rather than the 2′deoxyribose of DNA. Ribose contains a 2′ hydroxyl group not present in deoxyribose.

RNA contains the pyrimidine uracil (U) in place of thymine, which is characteristic of DNA molecule. Uracil, like thymine, can form a base pair with adenine by two hydrogen bonds. However, it lacks the methyl group present in thymine. It may be noted that in one case, however, RNA possesses thymine.

The native RNA is single stranded rather than a double-stranded helical structure characteristic of DNA. However, given the double-stranded pattern, in the region of hairpin loops, A pairs U, and G pairs with C. Guanine can also form a base pair with uracil but is less stronger than G−C base pairs with C. Guanine can also form a base pair with uracil but is less stronger than G−C base pair. The base pairing in RNA hairpins is frequently imperfect. The apposing bases may not be complementary, and one or more bases of a single strand may be complementary and one or more bases along a single strand may be looped out to facilitate the pairing of others. The proportion of helical regions in various types of RNA varies over a wide range, and 50% is typical.

Since the RNA molecule is a single-stranded and complementary for only one of the two stands of a gene, it need not have complementary base ratios. In other words, its adenine content does not necessarily equal its uracil content, not does its guanine content necessarily equal its cytosine content.

RNA can be hydrolysed by weak alkali (pH 9) at 100°C) to 2′ 3′cyclic diesters of mononucleotides via as intermediate compound called 2′-3′-5′-triester. This intermediate, however, cannot be formed in alkali-treated DNA because of the absence of 2′ hydroxyl group in its molecule. Thus, RNA is alkali labile, whereas DNA is alkali stable.

TYPES OF RNA

In all prokaryotic and eukaryotic organisms, three general types of RNAs are found: ribosomal, transfer, and messenger RNAs. Each of these polymeric forms serves as extremely important informational links between DNA, the master carrier of information and proteins. The three types of RNA molecules differ from each other by size, function, and general stability. Table 5.5 shows RNA molecules in E. coli.

Table 5.5 *RNA Molecules in Escherichia Coli*

Type	Relative amount	Sedimentation	Molecular weight	Number of nucleotides
Ribosomal RNA (rRNA)	80%	23S 16S 5S	1.26×10^6 0.55×10^6 3.6×10^4	3,7001,700120
Transfer RNA (tRNA)	15%	4S	2.5×10^3	75
Messenger RNA (mRNA)	5%	Heterogeneous		

Ribosomal RNA (rRNA)

It is the most stable form of RNA and is found in ribosomes. It has the highest molecular weight. In bacterium Escherichia Coli, there are three kinds of RNA called 23S, 16S, and 5S RNA because of sedimentation behaviour. One molecule of each of these three types of rRNA is present in each ribosome. Ribosomal RNA is abundant of all types of RNAs and makes up about 80% of the total RNA of a cell.

Ribosomal RNA represents about 40%–60% of the total weight of ribosomes. The ribosomes of prokaryotes are found in plastids and mitochondria. Prokaryotes contain three types of RNA: 23S and 5S RNA in the larger (50S) subunit and 16S RNA in the smaller (30S) subunit. Eukaryotes also contain three kinds of RNA: 28S and 5S RNA in the larger (60S) subunit and 18S RNA in the smaller (40S) subunit.

The mammalian ribosomes deserve special rRNA. They contain two major nucleoprotein subunits, a larger one (60S), and a smaller one (40S). The 60S subunit contains a 5S, 5.8S, and a 28S rRNA. There are also probably more than fifty polypeptides. The

40S subunit contains a single 18S rRNA and about thirty polypeptide chains. All types of rRNA, except the 5S rRNA, are processed from a single 45S precursor RNA molecule in the nucleolus.

It is remarkable to note that rRNA from all sources has G–C contents more than 50%. The rRNA molecule appears as a single unbranched polynucleotide strand (primary structure). In low ionic strength, the molecules are like a rod with random coiling, but at high ionic strength, the molecules exhibit a compact helical structure with complementary base pairing and looped outer region. The helical structure results from the folding back of a single-stranded polymer at areas where hydrogen bonding is possible. The function of rRNA is to donate a specific sequence to mRNA where it has to be translated.

Transfer RNA (tRNA) or Soluble RNA (sRNA)

Transfer RNA is the smallest polymeric form of RNA. In abundance, the tRNA comes next to rRNA and amounts to about 15% of total RNA of the cell. The tRNA molecule serves a number of functions; the most important of which is to act as specific carriers of activated amino acids to specific sites on the protein-synthesising templates. As the function of tRNA is to bind the specific amino acids, there must be at least one for each amino acid. There are, thus, at least twenty tRNA molecules in every cell. Since the code is degenerate (i.e. there is more than one codon for an amino acid), there may also be more than one tRNA for a specific amino acid.

Primary structure (or base sequence)of a tRNA molecule

Robert W. Holley and coworkers determined the base sequence of a tRNA molecule specific for the amino acid alanine in 1965. The molecular structure of rRNA is shown in Figure 5.7.

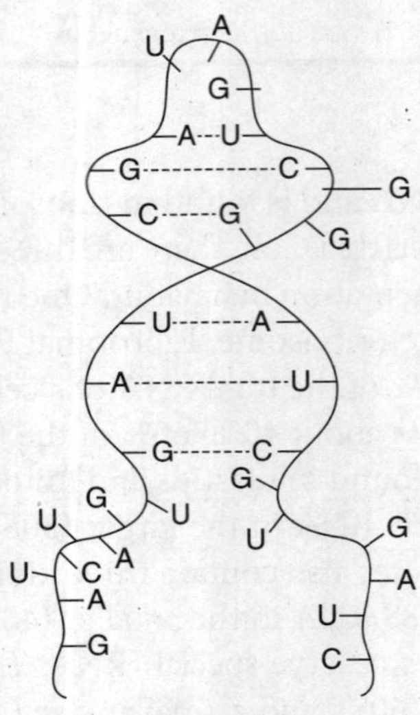

Figure 5.7 *Molecular Structure of rRNA*

Ribosomes and their RNAs are given in Table 5.6.

Table 5.6 *Ribosomes and their RNAs*

Ribosomes	rRNA
Prokaryotic ribosomes	
30S	16S
60S	5S, 23S
Eukaryotic ribosomes	
40S	18S
60S	5S, 28S

Holley first suggested the Cloverleaf model (i.e. consisting of three folds), which is based on a secondary structure, which would contain a maximum of intermolecular hydrogen bonding.

The alanyl tRNA of yeast consists of an unbranched chain of seventy-seven ribonucelotides, including 8a, 11u, 25g, 23C, and 10 unusual nucleotides, which are inosine ribothymidine, dihydrouridine, pseuodouridine, and methylated, derivatives of inosine and guanosine. The 5′ terminus is phosphorylated, whereas the 3′ terminus has free hydroxyl group. The attachment site for the amino acid alanine is 3′-hydroxyl group of the adenosine residue at 3′ terminus molecule. In the sequence inosine--guanine-cytosine, the middle of the molecule is anticodon. It is complementary to guanine-cytosine-cytosine, one of the codon for alanine.

```
             Anticodon
     3′                          5′
-----C----------G--------I-----


-----G----------C-------C---
     5′                     3′
              Codon
```

Common structural features of tRNAs

1. Almost all tRNA molecules have a cloverleaf shape, consisting of three folds.
2. All tRNA molecules are unbranched chains, containing from seventy-three to ninety-three ribonucleotides; tyrosyl-tRNA contains seventy-eight nucleotides, and seryl-tRNA contains eighty-five nucelotides. Their molecular weight approximates 25,000.
3. They contain from seven to fifteen unusual bases. Many of these unusual bases are methylated or dimethylated derivatives of A, U, G, and C. These include nucleotides of pseudouridine, various methylated adenines and guanines, methylated pyrimidines such as thymine and 5-methylcytosine and others.

Not all these are present in any one source of tRNA, but pseudouridne is the most abundant and universally distributed.

Although the role of these bases is uncertain, two roles seem certain:

(a) Methylation prevents the formation of certain base pairs so that some of the bases become accessible for other interactions.

(b) Methylation imparts hydrophobic character to some portions of tRNA molecules, which may be important for their interaction with synthetases and ribosomal proteins.

4. The 5′ end of tRNA is phosphorylated. The 5′ terminal residue is usually Pg.

5. The base sequence at the 3′ end of all tRNA is CCA. All amino acids bind to this terminal adenosine via the 3′-OH group of its ribose.

   (5′ end) G --------------- tRNA --------------- CCA (3′ end)

6. About 50% of the nucleotides in tRNAs are base paired to form double helices.

7. There are, however, five groups of bases, which are not base paired. These five groups, of which four form loops, are the following:

   (a) The 3′ CCA terminal region

   (b) The ribothymine-pseudouracil-cytosine (= TψC) loop

   (c) The extra arm or little loop, which contains a variable number of residues

   (d) The dihydrouracil (=DHU) loop, which contains several dihydrouracil residues

   (e) The anticodon loop, which consists of seven bases with the sequence

   5′ ------------ py-py-X-Y-X-modified purine – variable base --- 3′

   This loop contains a triplet of bases, which allows the tRNA to hydrogen bond to a complementary sequence on mRNA attached to a ribosome.

8. The four loops are recognition sites. Each tRNA must have at least two such recognition sites: one for the activated amino acid – enzyme complex, with which it must react to form the aminoacyl-tRNA and another for the site of messenger RNA molecule, which contains the code (codon) for that particular amino acid. It is interesting to note that the former involves recognition by bases of amino acid residues (either of the activated amino acid or of a site on the enzyme molecule), whereas the latter involves recognition by bases of bases (hydrogen bonding).

9. An unique similarity among all tRNA molecules is that the overall distance from CCA at one end to the anticodon at the other end is constant. The difference in nucleotide numbers in various tRNA molecules is, in fact compensated for by the size of the extra arm, which is located between the anticodon loop and TψC loop.

The tertiary structure (or three-dimensional structure) of a tRNA molecule is given in Figure 5.8.

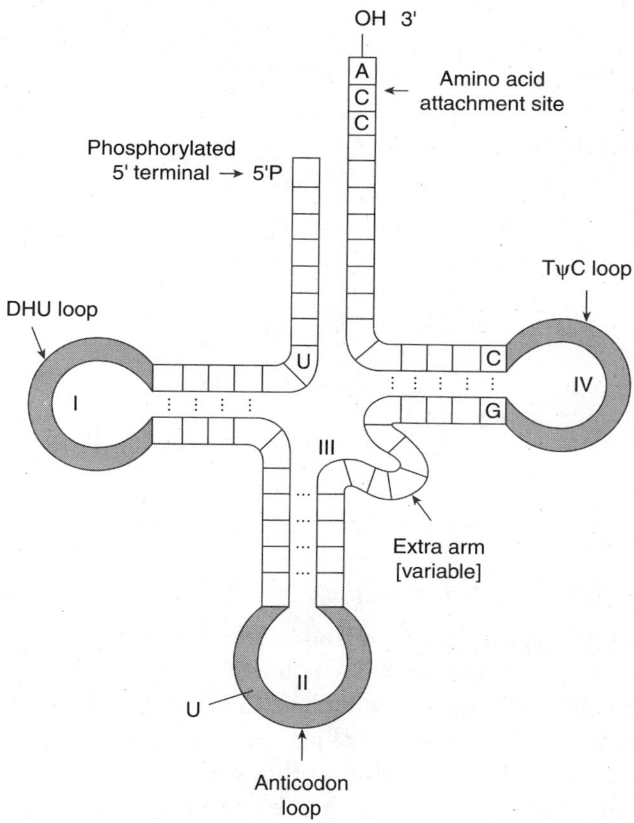

Figure 5.8 *Structure of tRNA*

Alexander Rich and Aaron Klug on the basis of their X-ray crystallographic studies have elucidated the 3-D structure of the phenylalanine-accepting tRNA from yeast as shown in Figure 5.9. The important features observed by them are the following:

The molecule is L-shaped.

The two segments of the double helix contain about 10 base pairs per turn of helix. The helix is perpendicular to each other and forms an L shape.

Amino acids are attached to the CCA terminus, which is at one end of the L. The other end of L contains an anticodon loop. DHU and TψC loops occupy the corners of the L.

The CCA terminus and the helical region of the adjacent group do not react with the remaining amino acids and thus induce a change in conformation during amino acid activation and protein synthesis.

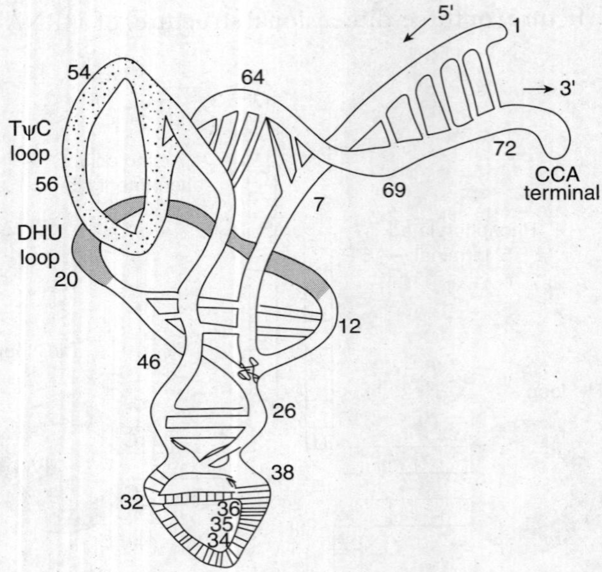

Figure 5.9 *Tertiary structure (Three dimensional structure)*

Messenger RNA (mRNA) or Template RNA

The abundance of RNA n the cytoplasm and its role in protein synthesis suggested that the genetic information of nuclear DNA is transmitted to an RNA which functions at the sites of protein synthesis. In 1961, the two Nobel laureates François Jacob and Jacques Monod postulated that control of protein formation, at least in certain micro-organisms, is determined by the rate of synthesis of templates. This requires that the templates do not accumulate in contrast to the constant presence of DNA, rRNA, and tRNA. They proposed that the messenger should have the following properties:

The mRNA is a polynucleotide. The base composition of the messenger shows the base composition of the DNA that codes for it. The mRNA is heterogeneous in size and the molecular weight of an mRNA will be at least half million.

It is associated with ribosomes for a short period and should be synthesised and degraded very rapidly.

Messenger RNA is the most heterogeneous in size and stability among all the types of RNAs. It has large molecular weight approaching 2×10^6 and amounts to about 5% of the total RNA of a cell. It synthesised on the surface of DNA template. Thus, it has base sequence complementary to DNA and carries genetic information or message for the assembly of amino acids from DNA to ribosomes, the site of protein synthesis. In prokaryotic cells, mRNA is metabolically unstable with a high turnover rate, whereas it is rather stable in eukaryotic. It is synthesised by DNA-dependent RNA polymerase.

On account of its heterogeneity, mRNA varies greatly in chain length. Since a few proteins contain less than hundred amino acids, the mRNA coding for these proteins must have at least 100 × 3 or 300 nucleotides residues. In E.coli, the average size of mRNA is 900 to 1,500 nucleotides units.

If mRNA carries the codes for the synthesis of simple protein molecule, it is called monocistronic type, and if it has codes for more than one kind of protein, it is known as polycistronic types as in E.coli.

mRNAs are unstable in the bacterial systems with a half life from a few seconds to about two minutes. In mammalian systems, however, mRNA molecules are more stable with a half life ranging from a few hours to one day.

mRNAs are single strands and complementary to the sense strand of their respective structural genes. The mRNA molecules, especially those of mammals, have some peculiar characteristics. The 5′ end of mRNA is capped by a 7-methylguanosine triphosphate, which is linked to an adjacent 2′-0-methylribonucleoside as it's 5′-hydroxyl through the three phosphates. Yet the function of this capping of mRNA is to protect the mRNA from the digestion of ribonuclease. The translation of mRNA into proteins begins at the capped 5′ end. In most mRNAs, the other end, that is, the 3′—OH end, has a polymer of 20–250 nucleotides in length. The specific function of the poly A tail at the 3′ hydroxyl end of mRNA is also avoided from the digestion of enzyme. It probably serves to maintain the intracellular stability of the specific mRNA.

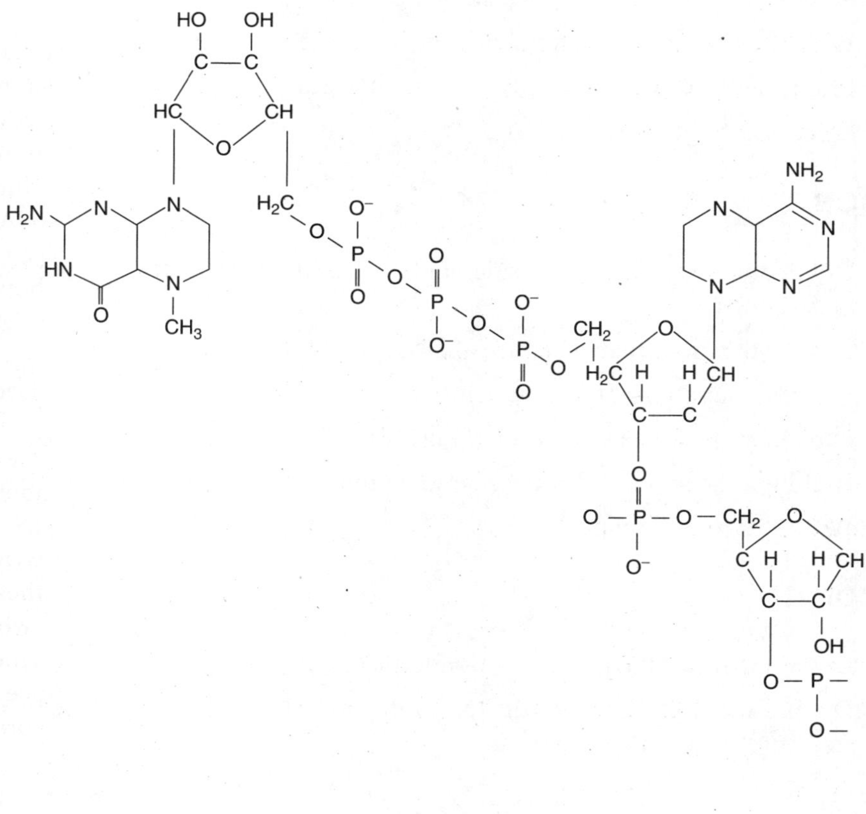

Heterogeneous Nuclear RNA (hnRNA)

In mammalian cells, including those of human beings, DNA-dependent RNA polymerase first synthesises precursor RNA in the nucleoplasm. This precursor is then degraded by a nuclear nuclease to mRNA that is then translocated to the cytoplasm, where it becomes associated to the ribosomal system. This precursor RNA constitutes the fourth class of RNA molecules and is designated as heterogeneous nuclear RNA (hnRNA). Molecules may have molecular weights exceeding 10^7 daltons, whereas the mRNA molecules are generally smaller than 2×10^6 daltons. Most mammalian mRNA molecules are 400–4,000 nucleotides in length, whereas an hnRNA molecule possesses 5,000–50,000 nucleotides. Thus, the hnRNA molecule appears to be processed to generate the mRNA molecules, which then enter the cytoplasm to serve as templates for protein synthesis.

QUESTION BANK

SECTION A

1. Define nucleic acids and mention their elemental compositions.
2. Discuss the properties of pentose sugar.
3. What do you mean by internucleotide linkages?
4. Briefly describe tautomerism with a suitable example.
5. Discuss in brief about hnRNA.

SECTION B

1. Differentiate between nucleosides and nucleotides with relevant examples and illustrations.
2. Explain the molecular structure of DNA in detail.
3. Enlist the functions of nucleic acids.
4. Explain in detail the structures of purines and pyrimidines.
5. Highlight the major differences between the A, B and Z forms of DNA.
6. Write short notes on mRNA.

SECTION C

1. Explain in detail the properties of nucleic acids.
2. Discuss the different types of RNA with special mention about the primary structure of the tRNA molecule.

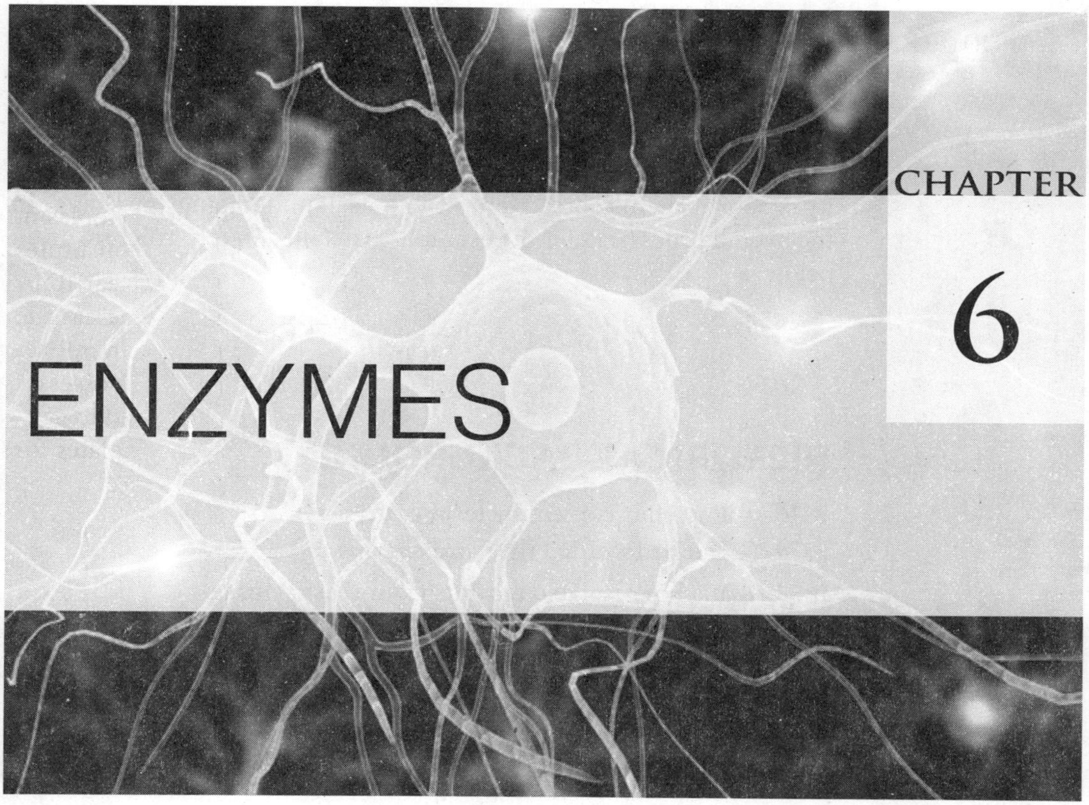

CHAPTER 6

ENZYMES

Concept Outline

- Enzymes
- Historical Background
- Chemical Nature of Enzymes
- Characteristics of Enzymes
- Enzymes are Highly Specific
- Proteolytic Enzymes
- Regulation of Enzymes
- Thermo Stability – Heat Sensitivity
- Active Site
- Enzyme Kinetics of Single Substrate Reactions
- Enzyme Inhibition
- Examples of Inhibition
- Regulation of Enzyme Activity
- Modifiers of Enzyme Activity
- Enzyme Activators
- Metal-Activated Enzymes and Metalloenzymes
- Coenzymes
- Nomenclature and Classification

ENZYMES

Enzymes are biocatalysts defined as a substance that increases the velocity or rate of chemical reaction without itself undergoing any change in the overall process. Refer the example as detailed in the structure given below:

$$CO_2 + H_2O \xrightarrow[\text{Anhydrase}]{\text{Carbonic}} H_2CO_3$$

HISTORICAL BACKGROUND

1836 – Berzelius coined the term *catalysis.*
1878 – Kuehne worded the word *enzyme.*
1883 – Buchner isolated enzyme from yeast zymase.
1926 – James Sumner isolated from Jack Beans. He got noble prize in 1940.
1930 – Northrop crystallised proteolytic enzyme from gastro-intestinal tract.

CHEMICAL NATURE OF ENZYMES

Enzymes are sometimes considered under two broad categories:

1. Endozymes or intracellular enzymes – act within the cells in which they are produced.
   Example: All metabolic enzymes.
2. Exozymes or extracellular enzymes – produced in the cells and act outside the cell.
   Example: Digestive enzymes.

Chemically, the enzymes may be divided into two categories:

1. *Simple protein enzymes:* They contain simple proteins only.
2. *Complex protein enzymes:* They contain conjugated proteins.

The functional unit of enzyme is known as *holoenzyme*, which is often made up of *apoenzyme* (the protein part) and coenzyme (non-protein part).

$$\underset{\text{(Active enzyme)}}{\text{Holoenzyme}} \longrightarrow \underset{\text{(Protein part)}}{\text{Apoenzyme}} + \underset{\text{(Non-protein part)}}{\text{Coenzyme}}$$

The term *prosthetic group* is used when the non-protein part (tightly) covalently binds with the apoenzyme. The coenzyme or cofactors can be easily separated from apoenzyme.

Coenzyme: Examples: NAD, NADP, FAD, FMN, BIOTIN, and TPP
Cofactors: Examples: Mg^{2+}, Mn, Ca, Cu, Co, and Fe

The word ***monomeric enzyme*** is used when if it's made up of single polypeptide. Examples: ribonuclease and trypsin.

Oligomeric enzymes possess more than one polypeptide unit. Examples: Lactate dehydrogenase and aspartate transcarbamoylase.

Multienzyme complex possesses specific sites to catalyse different reactions in sequence. Examples: Pyruvate dehydrogenase and fatty acid synthase.

CHARACTERISTICS OF ENZYMES

1. Enzymes have immense catalytic power.
2. The enzymes accelerate (activate) the rate of chemical reaction and do not normally participate in these reactions, if they do so, at the end of the reaction they are recovered as such without undergoing any quantitative or qualitative change.
3. The catalytic power of an enzyme is measured by the turnover number.

Turnover number is defined as the number of substrate molecules converted into product per unit time when the enzyme is fully saturated with substrate. It depends upon its molecular weight.

Example: Transfer of Carbon dioxide from tissues into blood and then to the alveolar air would be less complete in the absence of an enzyme. Carbonic anhydrase is one of the fastest enzymes known.

Each enzyme can hydrate 10^5 molecules of CO_2/sec.

For most enzymes, the turnover numbers fall between $1–10^4$ per second as shown in Table 6.1.

Table 6.1 *Turnover Numbers for Enzymes*

Enzymes	Turnover number
Lysozyme	0.5
DNA polymerase I	0.5
Lactate dehydrogenase	1000
Carbonic anhydrase	600,000

ENZYMES ARE HIGHLY SPECIFIC

Enzymes are highly specific in two ways—first in the reaction catalysed and second in the choice of reactants called substrates.

ABSOLUTE SPECIFICITY

Enzymes are capable of acting on only one substrate.

Example: Urease acts only on urea to produce ammonia and CO_2.

$$\text{Urea} \xrightarrow{\text{Urease}} NH_3 + CO_2$$

$$H_2O + CO_2 \xrightarrow[\text{Anhydrase}]{\text{Carbonic}} (H_2CO_3)$$

GROUP SPECIFICITY

Enzymes are capable of catalysing the reaction of a structurally related group of compounds.

Example: Lactate dehydrogenase catalyses the inter-conversion of pyruvic acid and lactic acids and also a number of other related compounds.

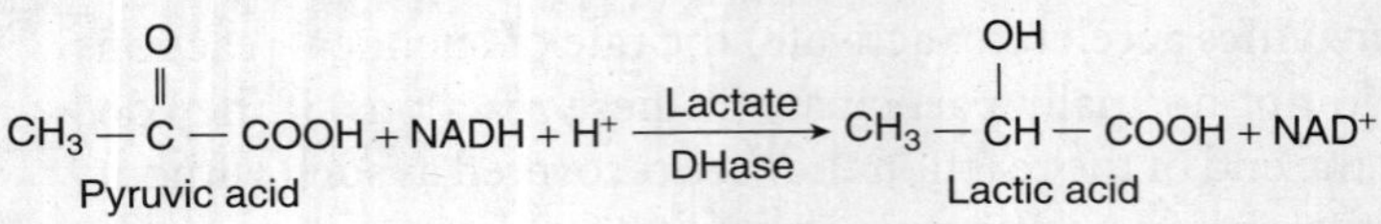

OPTICAL SPECIFICITY

Enzymes are capable to react with only one of two optical isomers.

Example:

$$\text{L-Arginine} \xrightarrow{\text{Arginase}}$$

$$\text{D-Amino acid} \xrightarrow[\text{Oxidase}]{\text{D-Amino acid}} \text{Keto acid}$$

GEOMETRICAL SPECIFICITY

Some enzymes exhibit specificity toward the cis-trans forms.

Example: Fumarase, the inter-conversion of fumaric and malic acid.

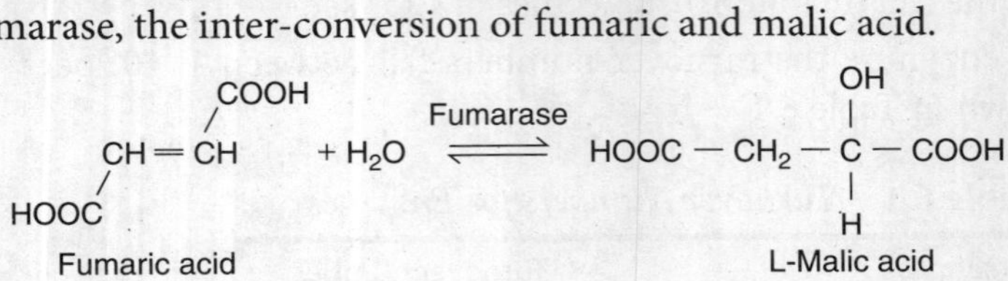

PROTEOLYTIC ENZYMES

The degree of specificity of enzymes for substrate is usually high and sometimes virtually absolute. Proteolytic enzymes, for instance, catalyse the hydrolysis of peptide bond.

1. Specificity of pepsin:

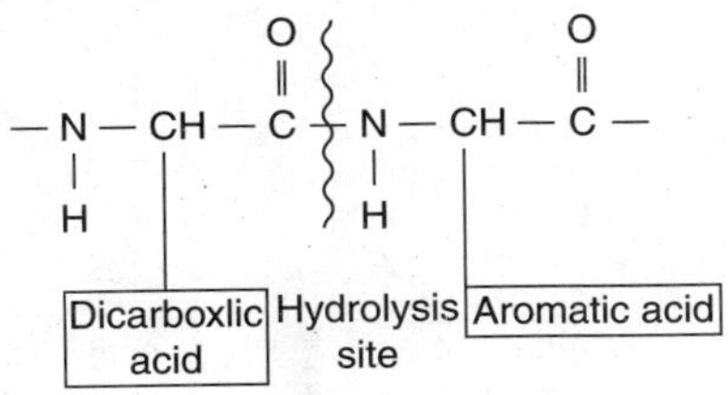

2. Specificity of trypsin:

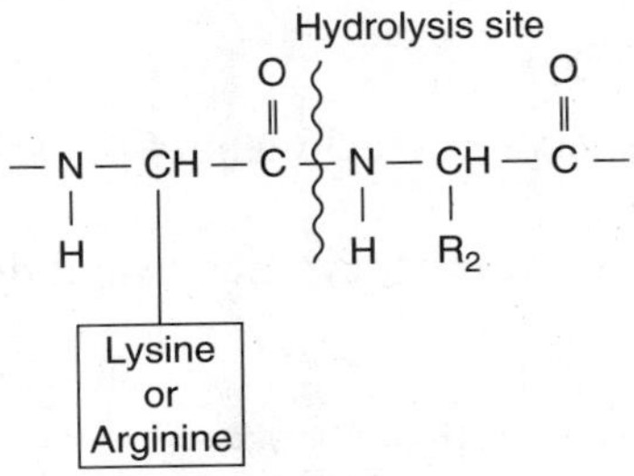

3. Specificity of chymotrypsin:

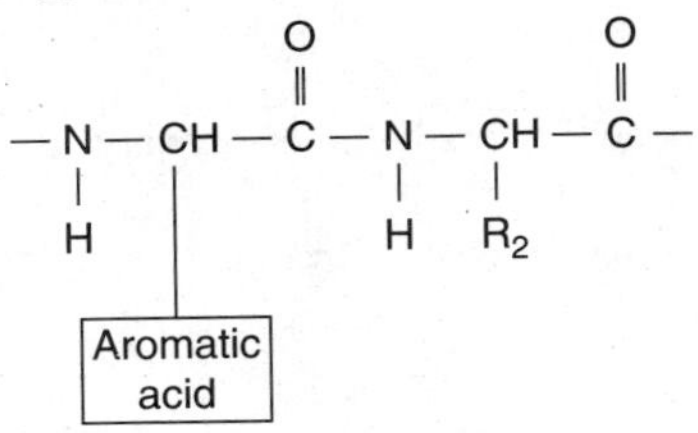

SPECIFICITY OF THROMBIN

Thrombin is an enzyme involved in blood coagulation is even more specific in that the side chain on the carboxyl of susceptible peptide bond must be arginine, whereas the one on the amino must be glycine.

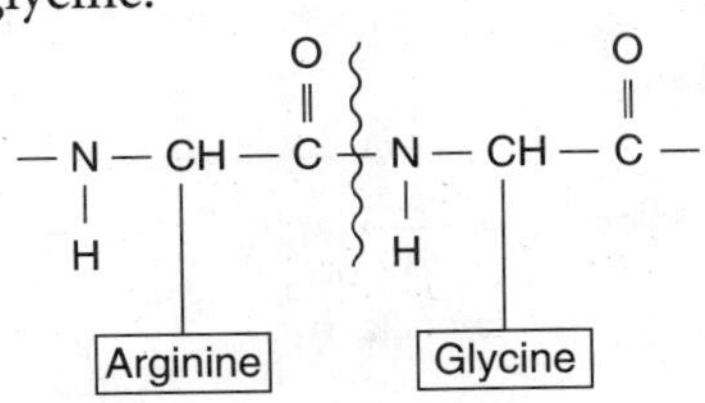

REGULATION OF ENZYMES

ZYMOGEN ACTIVATION

Enzymes are usually synthesised in an inactive precursor form known as zymogen, and this gets activated at a physiologically appropriate time and place. Digestive enzymes are controlled in this way, which is called as proteolytic activation or zymogen activation.

Example: Trypsinogen is synthesised in pancreas and activated in small intestine to form active enzyme trypsin.

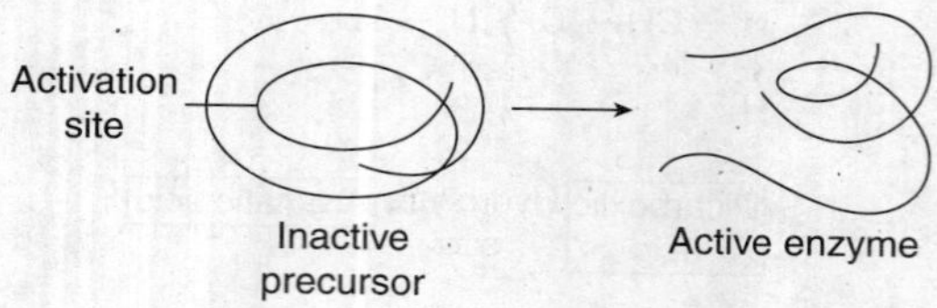

COVALENT MODIFICATION

Enzymes are controlled by reversible attachment of phosphoryl groups to specific serine and threonine residues.
Example: The activities of enzymes that synthesise and degrade glycogen are coordinate in this way.

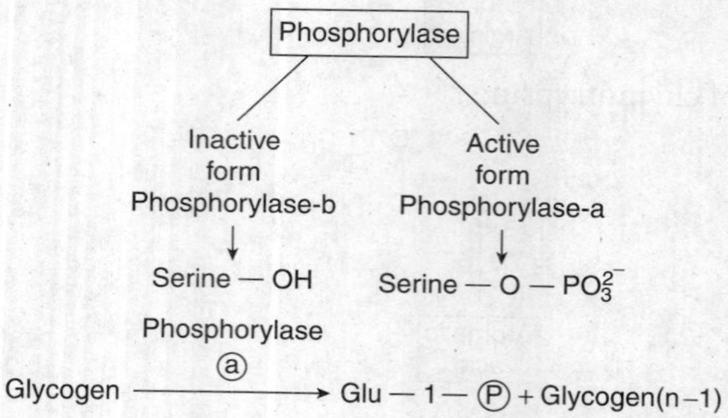

FEEDBACK INHIBITION

The enzyme that catalyses the first step in a biosynthetic pathway is controlled by ultimate product.
Example: HMG-CoA reductase.

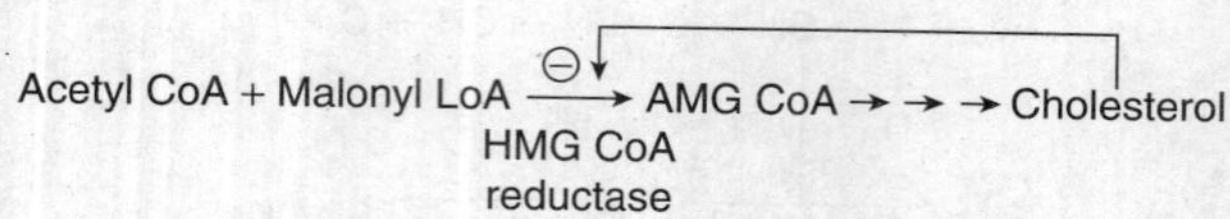

THE SPECIFICITY OF SOME ENZYMES IS UNDER PHYSIOLOGICAL CONTROL

The synthesis of lactose by mammary gland is a particular example. Lactose synthetase catalyses the synthesis of lactose and consists of a catalytic subunit. The catalytic subunit (refer Figure 6.1) by its own cannot synthesise lactose; it helps or catalyses the attachment of galactose to a protein–carbohydrate (glycoprotein).

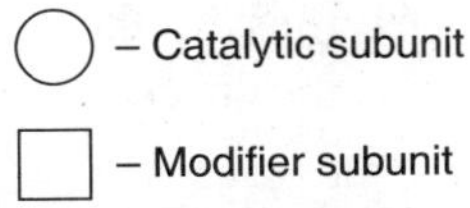

Figure 6.1 *Structure of Catalytic and Modifier Subunit of Lactose Synthetase*

Modifier subunit (refer Figure 6.2) converts the specificity of the catalytic subunit in order to link galactose to glucose to form lactose. This subunit is controlled by hormones. During pregnancy, catalytic subunit is formed in abundance in the mammary gland, but little modifier subunit is formed. At the time of birth or parturition, modifier subunit is synthesised in large amount. The modifier subunit that binds to catalytic subunit and synthesise lactose in large amount during the period of parturition.

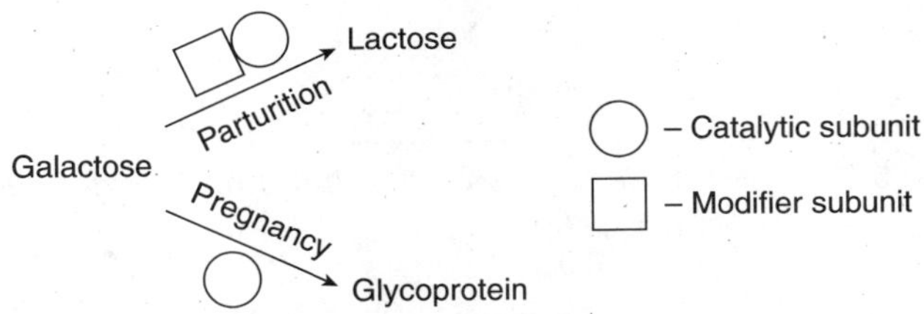

Figure 6.2 *Modifier Subunit*

ENZYMES TRANSFORM DIFFERENT KINDS OF ENERGY

In many biochemical reactions, the energy of reactant is converted into a different form with high efficiency.

Example: Light energy $\xrightarrow{\text{Photosynthesis}}$ Chemical bond energy

Free energy food $\xrightarrow{\text{Mitochondria}}$ Chemical bond energy currency-ATP

Chemical bond energy ATP $\xrightarrow[\text{Contraction}]{\text{Muscular}}$ Mechanical energy

ATP $\xrightarrow{\text{Cells and organelles}}$ Transport ions

High energy bond $\xrightarrow{\text{Nerves}}$ Electric impulse

ENZYMES DO NOT ALTER REACTION EQUILIBRIUM

An enzyme is a catalyst, and it cannot alter the equilibrium of a chemical reaction. This means that an enzyme accelerates the forward and reverse reaction by some factors precisely.

$$A \underset{K_2}{\overset{K_1}{\rightleftharpoons}} B$$

To attain this equilibrium without enzymes, it takes several hours, but with enzymes, it is attained within a second. Thus, enzymes accelerate the attachment of equilibria but do not shift their portions.

ENZYMES DECREASE THE ACTIVATION ENERGY

The reactants or substrates require energy before they start to react. For this reaction, it takes time if enzymes are not present, but in the presence of enzyme, it reduces the activation or transition energy and favours the reaction as shown in Figure 6.3.

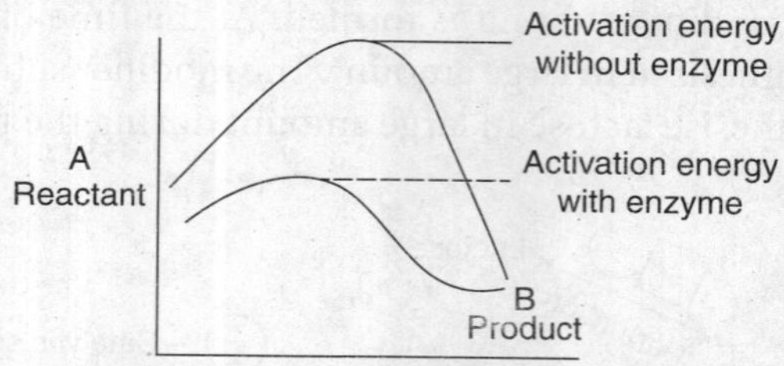

Figure 6.3 *Effect of Enzyme on Activation Energy of a Reaction (A is the reactant and B is the product)*

THERMO STABILITY – HEAT SENSITIVITY

The rate of an enzyme action increase with the rise in temperature of 10°C is called temperature quotient or Q10 (which means enzyme activity is increased two to three times for a rise in temperature of 10°C).

The optimum temperature for most of the enzymes varies from 30°C–40°C.

Example: Catalase optimum temperature is 30°C. The enhanced enzyme activity with the rise in temperature is due to the fact that the energy of molecules becomes greater, which, in turn, enhances the inherent reactivity of molecules and the frequency of their collisions.

Decrease in temperature 0°C or below inactivates the enzyme, but this is a reversible type of change, and enzyme regains its catalysing power upon increasing the temperature to optimum. Increase in temperature 60°C or above inactivates the denaturation, and it is irreversible in the change of their chemical structure. It is because of high rate of enzyme destruction at increasing temperature that an enzyme is stable as per listed in Table 6.2.

Table 6.2 *Effect of Temperature on Maintaining the Enzyme Activity*

0°C	–	Weeks
10°C	–	Days
30°C	–	Hours
70°C	–	Seconds

Figure 6.4 is a graph which represents the reactions between the temperature and the rate of enzyme activity.

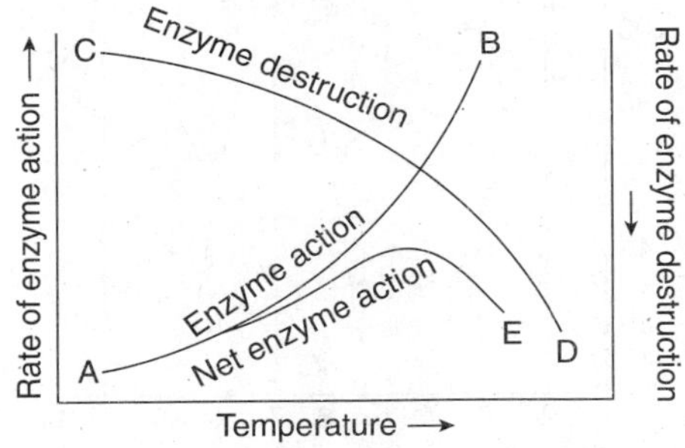

Figure 6.4 *Graph Representing the Relations Between the Temperature and the Rate of Enzyme Activity*

pH SENSITIVITY

The pH value or the hydrogen ion concentration of the medium (buffers) controls the activity of an enzyme to a great extent. This is mainly related to the degree of dissociation, the electric charge of enzyme, and through this, the formation of enzyme substrate complex. Each enzyme acts best in a certain pH, which is optimum to it. The approximate optimum pH value for most enzymes is nearly neutral. This value depends on the following:

1. Nature of buffer system
2. Age of cell tissues
3. Nature of substrate

Table 6.3 is an example to validate this fact.

Table 6.3 *Enzyme with its Optimum pH*

Enzyme	Optimum pH	Nature of the medium
Pepsin	1.5–1.6	Acidic
Lipase (pancreas)	8	Basic
Lipase (stomach)	4–5	Acidic
Catalase	7	Neutral
Urease	7	Neutral
Trypsin	7.8–8.7	Basic
Arginase	10	Basic

Maximum enzyme activity is obtained at or near the isoelectric point (pl) of enzyme. The effect of pH on enzyme action is given in Figure 6.5.

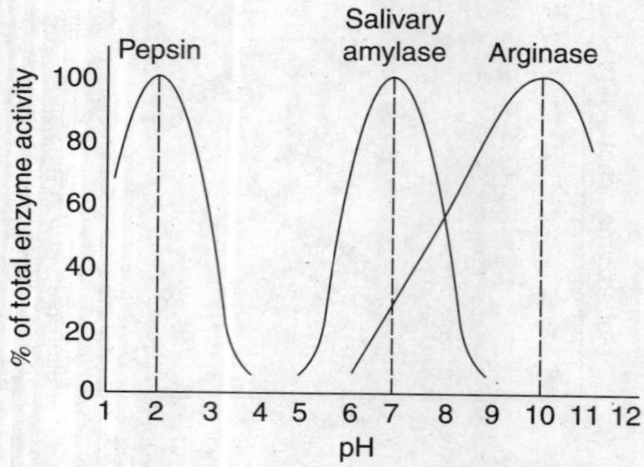

Figure 6.5 *Effect of pH on Enzyme Action*

ACTIVE SITE

As the substrate molecules are comparatively much smaller than the enzyme molecules, there should be some specific regions or sites on the enzyme for binding with substrate. Such sites of attachment are variously called *active sites or catalytic sites or substrate sites.* Active site of an enzyme as shown in Figure 6.6.

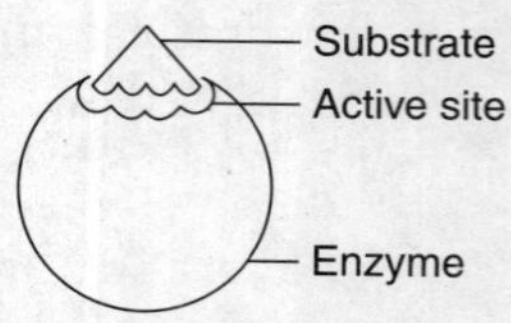

Figure 6.6 *Diagrammatic Representation of Active Site*

The active site or (active centre) of an enzyme is defined as the small region at which the substrate(s) binds and participates in catalysis. The salient features of active site are given below:

1. The active site takes up relatively small part of total volume of an enzyme.
2. The active site is a three-dimensional entity.
3. The active site of an enzyme is not a point or a line or even a plane. It is an intricate three dimensional form made up of groups that come from different parts of linear amino acid sequences.

   Example: Lysozyme. The important groups in the active sites are numbered 35, 52, 59, 62, 63, and 101 in the linear sequence of amino acids as shown in Figure 6.7.

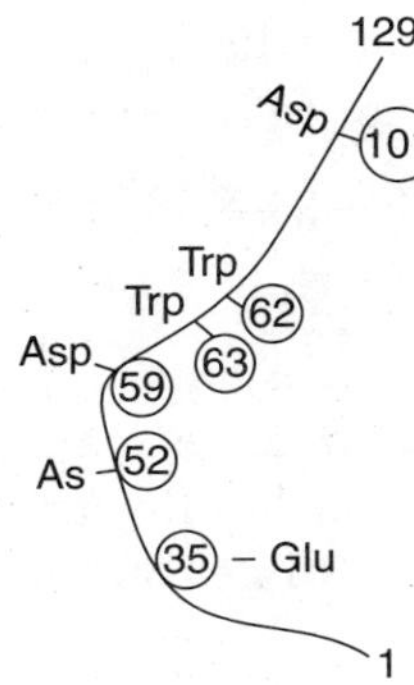

Figure 6.7 *Diagrammatic Representation of Active Site of Enzyme Lysozyme*

4. Substrates are bound to enzymes by relatively weak forces.
5. Active sites are clefts or crevices from which water is largely excluded. It contains several polar residues that are essential for binding and catalysis. The non-polar character of cleft enhances the binding of substrate.
6. Polar residues are aspartic acid, glutamic acid, lysine, serine, and so on.
7. The specificity of binding depends on the precisely defined arrangement of atoms in an active site.
8. Active site changes its configuration in order to bind a substrate which is only slightly different in structure from its own substrate.

ENZYME KINETICS OF SINGLE SUBSTRATE REACTIONS

INTRODUCTION

Enzyme kinetics is the most powerful technique for elucidating the catalytic mechanisms of enzymes. Almost all the enzyme-catalysed reactions involve more than one substrate. Although the single-substrate enzyme reaction is something of a 'rarity' in biochemical systems, it nevertheless represents a very useful simple model for the development and discussion of theories of enzyme kinetics, and the majority of the conclusions reached are directly applicable to more complicated enzyme systems.

MICHAELIS–MENTEN EQUATION

It was Michaelis and Menten, in 1913, who proposed a successful explanation for the effect of substrate concentration on the enzyme activity. According to them, enzyme (E) and substrate (S) combine rapidly to form a complex, the enzyme substrate complex (ES) or Michaelis constant. This complex then breaks down relatively slowly to form the product (P) of the reaction. (The formation of ES complex is central to all current thinking on the mechanism of enzyme action.)

$$\text{First step (rapid): } E + S \leftrightarrow ES \tag{1}$$
$$\text{Eleventh step (slowest): } ES \leftrightarrow E + P \tag{2}$$

Where K_1, K_2, K_3, and K_4 are the rate constants for each reaction.

When the substrate concentration reaches maximum, all the E will be converted to ES form. Now the reaction becomes rate limiting and the overall reaction rate becomes "insensitive" to further increase in substrate concentration.

Assumptions

Michaelis–Menten (M–M) theory is based on two important assumptions:

1. Assumption of Equilibrium (Leonor Michaelis and Maud Menten's assumption) The reaction between the enzyme and its substrate (equation 1) remains in equilibrium, any effect of the equation 2 on its equilibrium being assumed to be negligible.

   These conditions will be achieved when the rate of breakdown of the ES complex is back to the free enzyme and substrate is much greater than the rate of its breakdown to free enzyme and products; that is, $K_2 >>> K_3$.

   It implies that the equilibrium between enzyme and substrate (equation 1) is attained so rapidly in comparison with the breakdown of ES (equation 2), that ES remains in equilibrium with E and S always while the enzyme action is proceeding. At equilibrium, the rate of forward reaction (V_f of equation 1) is equal to the rate of backward reaction (V_b).

   $$V_f = K_1[E]\,[S]$$
   $$V_b = K_2[ES] \text{ (by law of mass action)}$$

   At equilibrium,

   $$K_1[E]\,[S] = K_2[ES]$$
   $$K_2/K_1 = [E][S]/[ES] = K_s$$

   Here K_s is the dissociation constant of the first step in the enzymatic reaction. $K_s = K_m$ only when $K_2 >>> K_3$ under equilibrium condition otherwise noted.

   $$K_m = K_2 + K_3/K_1$$
   $$K_m = K_2/K_1 = K_s \text{ as } K_2 \text{ is negligible.}$$

   If we have used the equation assumption, we can say equivalent $K_m = K_s$ in M–M equation.

2. Assumption of steady state (Briggs and Haldane's assumption)

   The above assumption that the reaction between E and its S remains in equilibrium is not always correct. In view of the very high catalytic activities shown by many enzymes (the mixing of E and S is over in milliseconds), it is probable that this may not always be the case that the concentration of ES will differ from its equilibrium value.

An alternate treatment, applicable to such cases, was put forward by Briggs and Haldane.

This is based on the postulate that at any moment, the rates of formation and breakdown of the ES complex are essentially equal, so that its concentration [ES] can be regarded as a constant or in 'steady state' (until the S is entirely exhausted) over the short period of time necessary for a velocity measurement. (Over a longer period, of course, [ES] will change as the reaction proceeds.)

3. During the initial period, the concentration of free substrate remains unchanged so that in equations 1 and 2, the concentration of substrate [S] is equal to the total substrate concentration $[S_T]$.

   These conditions are usually achieved when the total substrate concentration is much greater than the total enzyme concentration, as is usually the case in kinetics studies.

$$[S] >>> [E]$$

4. By Michaelis–Menten equation, we measure only the initial velocity; at that condition, the amount of P is very less. So the formation of [ES] from E + P is negligible. Therefore, K_4 is ignored.

Derivation of Michalis–Menten Equation

According to Michaelis and Menten, the rate determining the step (slowest) of the reaction is the breakdown of the enzyme substrate complex [ES]. Thus, the velocity V of the overall reaction, according to the law of mass action, is given by

$$V_o = K_3[ES] \tag{3}$$

Determination of ES value

Rate of formation of [ES] complex

Let ET be the total concentration of enzyme both free [E] and bound [ES], and then the concentration of free enzyme [E] will be

$$[E] = [ET] - [ES]$$

Let [S] be the concentration of substrate in free form; then the rate of formation of [ES] is given by

$$d[ES]/dt = K_1[E]\ [S] \tag{4}$$

$$= K_1\{[ET] - [ES]\}\ [S] \tag{5}$$

Rate of breakdown of [ES] complex

The rate equation for the breakdown of ES may be given by the sum of two reaction – one is the backward reaction of equation 1 and the other is the forward reaction of equation 2. Thus,

$$-d[ES]/dt = K_2[ES] + K_3\,[ES] \tag{6}$$

Assume the rate of formation of ES from E + P is very small (K_4), hence can be ignored.

When the reaction system enters the 'steady state' (defined as the condition in which the concentration of ES remains constant), the rate of formation of [ES] is equal to the rate of breakdown of [ES].

(5) = (6)

$$K_1\{[ET] - [ES]\}\,[S] = K_2\,[ES] + K_3\,[ES] \tag{7}$$

$$K_2 + K_3/\,K_1 = \{[ET] - [ES]\}\,[S]/[ES] \tag{8}$$

$K_2 + K_3/K_1 = K_m$ where K_m is called Michaelis–Menten constant.

Thus,

$$K_m = \{[ET] - [ES]\}\,[S]/[ES] \tag{9}$$

Solving for [ES], we have

$$K_m\,[ES] = [ET]\,[S] - [ES]\,[S]$$

$$K_m\,[ES] + [ES]\,[S] = [ET]\,[S]$$

$$[ES]\,\{K_m + [S]\} = [ET]\,[S]$$

$$[ES] = [ET]\,[S]/K_m + [S] \tag{10}$$

Substituting (10) in equation (3); $V_o = K_3\,[ES]$

$$V_o = K_3\,[ET]\,[S]/K_m + [S] \tag{11}$$

Now, if E is completely saturated with the substrate, V will increase to the maximum velocity V_{max}, which according to law of mass action, will be given by,

$$V_{max} = K_3\,[ET] \tag{12}$$

Substituting (12) in (11), we get

$$V_o = V_{max}\,[S]/K_m + [S]$$

This is the Michalis–Menten equation for a single substrate E catalysed reaction (Or equation for rectangular hyperbola) refer Figure 6.8. This equation adequately describes the kinetics of the most enzyme catalysed reactions (although the assumptions of these theories are not always correct, some exceptions exist).

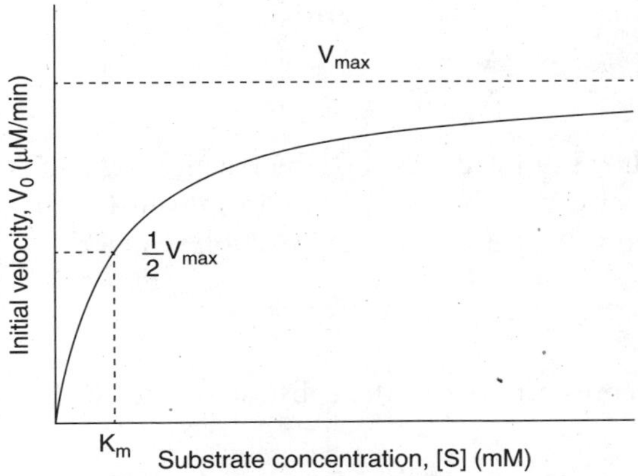

Figure 6.8 *Michaklis Menten Plot (Hyperbola curve)*

Verification of M–M equation

The three situation at A, B, and C on hyperbola curve can be verified from the M–M equation.

1. At point A

   Substrate concentration is much less than K_m. If [S] is added, its value changes very little; hence [S] can be ignored from the denominator of M–M equation.

   $$V_o = V_{max}\ [S]/K_m + [S]$$
   $$V_o = V_{max}\ [S]/K_m \text{ as [S] is negligible.}$$

   Since V_{max} and K_m are both constants, V_{max}/K_m can be replaced by a new constant K, so we have

   $$V_o = K[S]$$

   V_o therefore depends on [S] concentration

2. At point B

   [S] is equal to K_m, hence

   $$V_o = V_{max}\ [S]/K_m + [S]$$
   $$V_o = V_{max}\ [S]/[S] + [S] \text{ as } K_m = [S] \text{ at B}$$
   $$V_o = V_{max}\ [S]/2[S]$$
   $$V_o = V_{max}\ /2$$

   K_m, therefore, is the substrate concentration, at which V_{max} is half.

3. At point C

   The substrate concentration is far greater than K_m, so by adding K_m to [S], the value of [S] does not change much; hence, K_m may be dropped. Then,

$$V_o = V_{max}\,[S]/K_m + [S]$$
$$V_o = V_{max}\,[S]/[S] \text{ as } [S] >>> K_m$$
$$V_o = V_{max}$$

When E is fully saturated with [S], the reaction velocity V will increase to the maximal velocity, V_{max}; that is, at this point, any further increase in the substrate concentration will have no effect on reaction velocity.

Michaelis Constant (K_m)

Definition

It is defined as the concentration of the substrate at which the initial rate is half the maximal rate.

Units

K_m is expresses in moles per litre.

Properties

It is independent of enzyme concentration but dependent on substrate conc.

It varies with pH and temperature.

Typically, the K_m values lie between 10^{-2} to 10^{-5} moles/litre for various enzymes as given in Table 6.4.

Table 6.4 *K_m Value for Different Enzymes*

Enzyme	Substrate	K_m	$K_{cat}(s^{-1})$	$K_{cat}/K_m(m^{-1}s^{-1})$
Acetylcholineesterase	Acetylcholine	9.5×10^{-5}	1.4×10^4	1.5×10^8
Carbonic anhydrase	Co_2	1.2×10^{-2}	1.0×10^6	8.3×10^7
Catalase				
Fumarase	H_2O_2	2.5×10^{-2}	1.0×10^7	4.0×10^8
Urease	Fumarate	5.0×10^{-6}	8.0×10^2	1.6×10^8
Chymotrypsin	Urea	2.5×10^{-2}	1.0×10^4	4.0×10^5
	N-acetyl tyrosine ethyl ester	6.6×10^{-4}	1.9×10^2	2.9×10^5

Determination(measurement)

The K_m value can be determined from Michaelis plot, LB plot, and Eadie Hofstee's plot.

Significance

1. K_m is a measure of the affinity of an enzyme for its substrate, and they are inversely related.

   Low K_m value → E has high affinity for [S]
   High K_m value → E has low affinity for [S]

$$K_m = K_2 + K_3/K_1$$

$$K_m = K_2/K_1 + K_3/K_1$$
$$K_m = K_S + K_3/K_1$$

Where K_S = dissociation constant for equation 1

$K_m = K_S$, provided K_3/K_1 is small compared to K_S or $K_2 >> K_3$.

Under these conditions, K_m is the reciprocal of the affinity of E for its substrate. As K_S decreases ($\rightarrow K_m \downarrow K_S$), the enzymes' affinity for the substrate increases.

Example: Hexokinase and glucokinase (refer Table 6.5) are the enzymes that catalyse the phosphorylation of glucose.

Glucose → glucose-6-phosphatase. (Hexokinase or glucokinase enzyme)

Table 6.5 *Characteristics of Hexokinase and Glucokinase Towards Glucose*

Hexokinase (Brain)	Glucokinase (Liver)
Low K_m value	High K_m value
High affinity for [S[i.e. glucose	Low affinity for [S] i.e. glucose
Hence takes up glucose even when blood glucose levels are low	Hence takes up glucose only when blood glucose level is high

2. K_{cat}/K_m is a measure of catalytic efficiency

We may define the catalytic constant K_{cat} of an enzyme as

$$K_{cat} = V_{max}/(E_X)$$

$$\text{Unit of } K_{cat} = V_{max}/(E_T) = \text{conc}^N \cdot \text{time}^{-1}/\text{conc}^N$$

$$\text{Unit of } K_{cat} = \text{time}^{-1}.$$

This quantity is also known as the turnover number of an enzyme because it is the number of substrate molecules converted to product per unit time per active site of an enzyme (or per E molecule when E is fully saturated with [S]).

For Michaelis–Menten equation,

$$V_{max} = K_3(E_t)$$

(when E is fully saturated with S, V_o becomes V_{max})

$$V_{max}/(E_T) = K_3$$

$$K_3 = K_{cat}; \text{ only if saturating [S] conc.}$$

Under physiological conditions, E are not fully saturated with substrate. In vivo, we have only low or optimum concentration of [S] but not high concentration. Then we have another term K_{cat}/K_m, which gives the catalytic efficiency.

$$V_{max} = K_3[E_T]$$

$$\text{Or } \quad V_{max} = K_{cat}[E_T]$$

K_{cat} is sometimes known as the turnover number of the enzyme.

When [S] << <<K_M

a. $K_m + [S] K_m$

b. $E_T = E$ free, at low [S] conc; as the amounts of bound E are very low, all are free forms.

Therefore, we have

$$V = V_{max} [S]/K_m + [S]$$

$$\text{Or} \quad V = K_{cat} [E_t][S]/K_m + [S]$$

When $[S] \ll \ll K_m$

$$V = K_{cat}/K_m [E][S]$$

Rate of the reaction varies directly with how often enzyme and substrate encounter one another in solution. The quantity K_{cat}/K_m is, therefore, a measure of an enzyme catalytic efficiency.

Applications

1. It is useful in evaluating the affinity of the enzyme; for the substrate, the low K_m value of an E indicates its high affinity for[S].
2. The K_m value also gives an idea regarding the type of inhibition of an E caused by an inhibitor (double-reciprocal treatments).
3. K_m value is a useful index during the purification of enzymes.
4. It has got biomedical importance in the treatment of leukaemia.
5. It is also a useful index in the kinetic studies.
6. It is used in the assay of enzymes.

Application of M–M equation

M–M equation is sufficient to describe most of the enzyme-catalysed reaction.

It can be utilised to determine K_m and V_{max} at various substrate concentration, and the values can be used in predicting rate-limiting steps.

Limitation of M–M equation

According to M–M equation, at very high values of [S], the initial velocity V_o asymptotically approaches V_{max}. In practice, however, it is very difficult to accurately assess V_{max} from direct plots of V_o versus[S] (from hyperbola curve). Only when the graph is linear, we can determine the V_{max}. Therefore, the M–M equation is algebraically transformed into other forms that are more useful in plotting experimental data or in determining the parameters like V_{max} and K_m.

Transfrmation of M–M equation

1. Lineweaver Burk plot (1/V vs 1/S)

   Hans Lineweaver and Deon Burk took the reciprocal of both the sides of M–M equation.

$$V_o = V_{max}[S]/KS + [S]$$

$$1/V_o = K_m + [S]/V_{max} [S]$$

$$1/V_o = K_{m/}V_{max}[S] + [S]/V_{max}[S]$$
$$1/V_o = K_{m/}V_{max} \cdot 1/[S] + 1/V_{max}$$

This is the Lineweaver Burk (LB) equation.
LB plot is then a double-reciprocal plot as shown in Figure 6.9.
The plot is a straight line unlike M–M plot, which is hyperbolic. The plot is obtained when 1/V is plotted against 1/S.

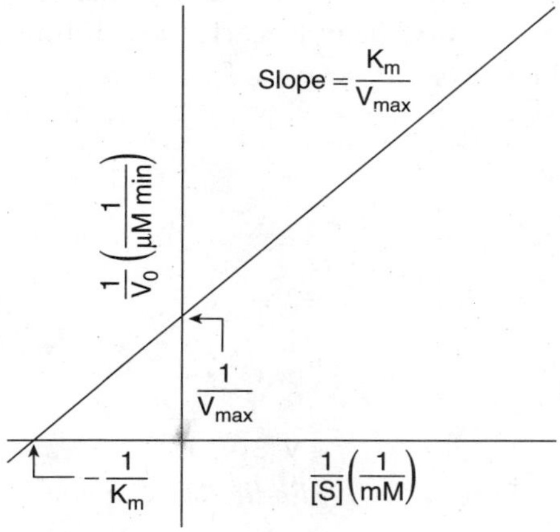

Figure 6.9 *A Double-Reciprocal Plot or Lineweaver Burk Plot*

Advantages
An advantage of LB plot over other plot is that accurate determination of V_{max} is possible. One of the major applications of LB is to study enzyme inhibitor.
Disadvantage
For high substrate concentration, when we insert it, we get less value and, therefore, a lot of errors will take place while determining the K_m.

2. Eadie–Hofstee plot (V vs V/S)

$$V = V_{max}[S]/K_m + [S] \text{ (M–M equation)}$$

Taking reciprocal on both sides,

$$1/V_o = K_{m/}V_{max}[S] + 1/V_{max} \text{ (LB equation)}$$

Multiply by [S] on both sides.

$$[S]/V = K_m[S]/V_{max}[S] + [S]/V_{max}$$

Multiply by V_{max} on both sides.

$$[S]V_{max}/V_o = K_mV_{max}/V_{max} + [S]/V_{max} \cdot V_{max}$$
$$[S]/V_o \cdot V_{max} = K_m + [S]$$
$$[S]V_{max} = V_oK_m + V_o[S]$$

$$V_o[S] = -V_oK_m + V_{max}[S]$$

Divide by [S]

$$V_o = -K_m \cdot V_o/[S] + V_{max}$$

This is Eadie–Hofstee's equation.

Advantages

EH plot is slightly superior to LB plot as it yields accurate V_{max} and K_m values and magnifies the deviation from linearity which might not be apparent in LB plot as shown in Figure 6.10.

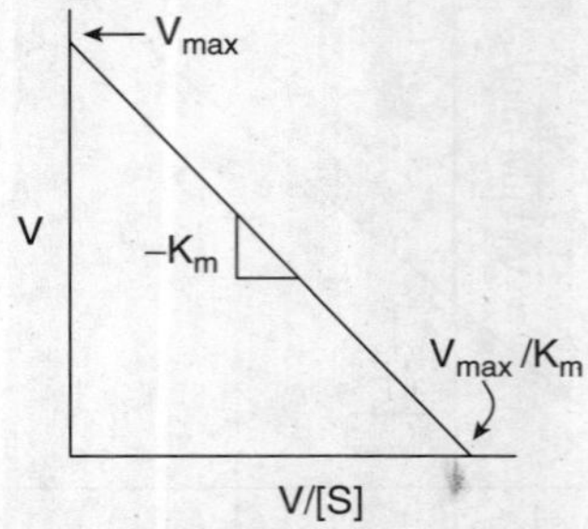

Figure 6.10 *Eadie–Hofstee plot*

Disadvantage

Here, there is still reciprocal of [S].

3. Augustsson plot [[S]/V vs [S]] or Hanes Wolf plot as shown in Figure 6.11.

$$V_o = V_{max}[S]/K_m + [S] \text{ (M–M equation)}$$

Taking reciprocal,

$$1/V_o = K_m/V_{max}[S] + 1/V_{max} \text{ (LB plot)}$$

Multiply by[S]

$$[S]/V_o = K_m[S]/V_{max}[S] + [S]/V_{max}$$

This is Augustsson equation.

$$[S]/V_o = 1/V_{max} \cdot [S] + K_m/V_{max}$$

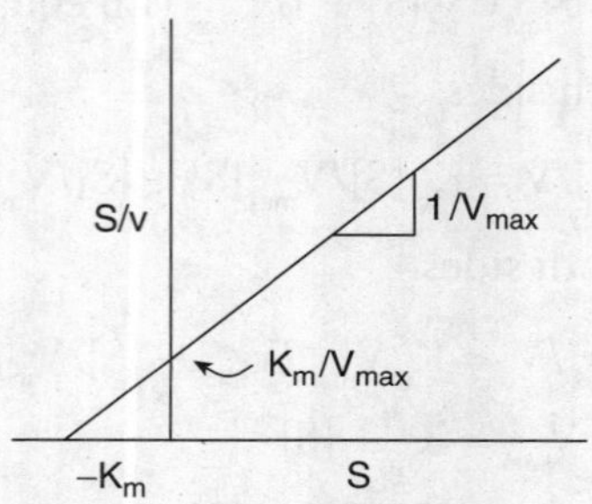

Figure 6.11 *Hanes–Wolf Plot*

Transformation of M–M equation is given in Table 6.6.

Table 6.6 *Transformation of M–M Equation*

S.No	Plot	Equation	Slope (m)	Y intercept	X intercept
1.	Lineweaver Burk plot	$1/V_o = K_m/V_{max}.1/[S] + 1/V_{max}$	K_m/V_{max}	$1/V_{max}$	$-1/K_m$
2.	Eadie Hofstee plot	$V_o = -K_m V_o/[S] + V_{max}$	$-K_m$	V_{max}	V_{max}/K_m
3.	Augustsson plot	$S/V_o = 1/V_{max} \cdot [S] + K_m/V_{max}$	$1/V_{max}$	K_m/V_{max}	$-K_m$

Important conclusions about MichaeLIs-Menten Kinetics

Characteristics of K_m

The Michaelis–Menten constant is characteristics of an enzyme and a substrate and reflects the affinity of the enzyme for that substrate. K_m is numerically equal to the substrate concentration, of which the reaction velocity is equal to ½V_{max}. K_m does not vary with the concentration of enzyme.

Small K_m

A numerically small (low) K_m reflects a high affinity of enzyme for substrate (e.g. glucokinare) as shown in Figure 6.12. Because low concentration of substrate is needed to half the substrate of the enzyme, that is, to reach a velocity that is ½V_{max}.

Large K_m

A numerically high K_m reflects a low affinity for substrate (e.g. hexokinase) as shown in Figure 6.12. Because a high concentration of substrate is needed to half the substrate of the enzyme.

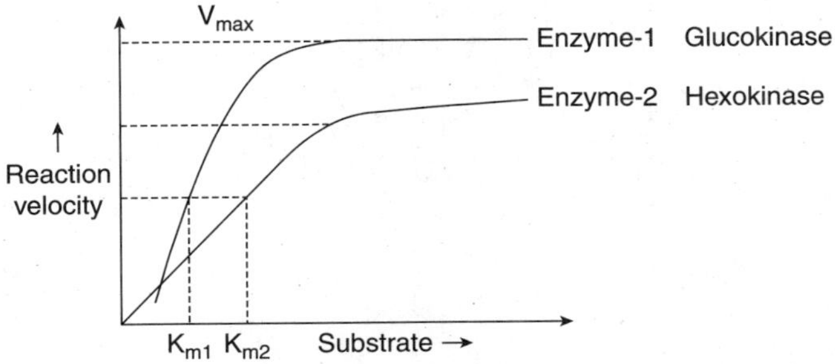

Figure 6.12 *Graphical Representation of K_m for Different Enzymes with Substrate Affinity*

Order of reaction

When [S] is lesser than K_m, the velocity of the reaction is roughly proportional to the substrate concentration. The rate of the reaction is said to be first order with respect to substrate.

When [S] is greater than K_m, the velocity of the reaction is constant and is equal to V_{max}. Now the rate of reaction is independent of substrate concentration and is said to be zero order with respect to substrate concentration is explained in Figure 6.13.

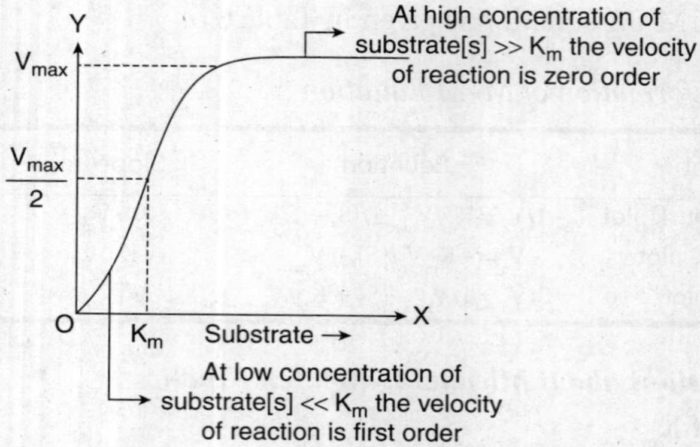

Figure 6.13 *Graphical Representation of Order of Reaction with Related to Substrate Concentration*

EADIE–HOFSTEE TRANSFORMATION

This (Eadie–Hofstee) transformation is used to avoid the bunching of values that—at the lower end of double reciprocal plot as shown in Figure 6.14.

$$V_o = -K_m \frac{V_o}{S} + V_{max}$$

Multiplied $V_o\ V_{max}$

$$\frac{1}{V_o} = \frac{K_m}{V_{max}} \cdot \frac{1}{[S]} + \frac{1}{V_{max}}$$

$$\frac{1}{V_o} V_o V_{max} = \frac{K_m}{V_{max}} \cdot \frac{1}{[S]} V_{max} + \frac{1}{V_{max}} V_{o\,max}$$

$$V_o = -K_m$$

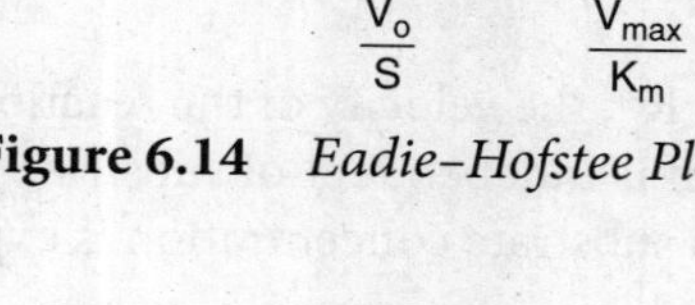

Figure 6.14 *Eadie–Hofstee Plot*

Enzyme Substrate Complex Formation

The prime requisite for enzyme catalysis is that the substrate [S] must combine with the enzymes [E] at the active site to form an enzyme. Substrate complex [ES] ultimately results in the product formation.

$$E + S \rightleftharpoons ES \rightleftharpoons E + P$$

A few theories have been put forth to explain the mechanism of enzyme substrate complex formation:

1. Fischer's template theory
2. Induced fit theory
3. Substrate strain theory

Fischer's template theory (or) lock and key model

This theory was proposed by a German biochemist Emil Fischer in 1898. This is, in fact, the very first model proposed to explain an enzyme-catalysed reaction. According to this model, the substrate or conformation of enzyme is rigid. The substrate fits to the binding site (active site) just as a key fits to the proper lock or a hand into the proper glove as shown in Figure 6.15. Thus, the active site of an enzyme is a rigid and pre-shaped template where only a specific substrate can bind.

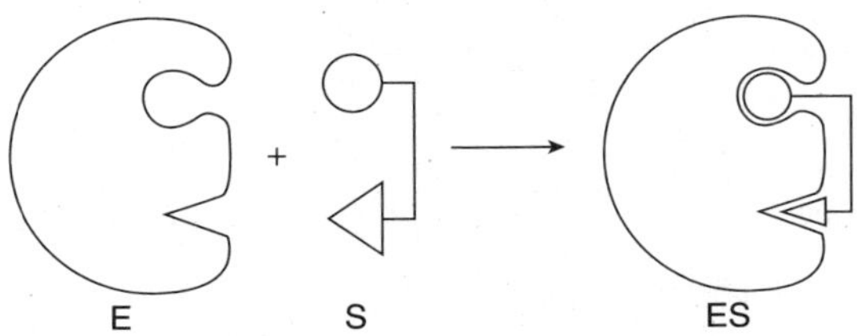

Figure 6.15 *Diagrammatic Representation of Lock and Key Model*

In fact, the enzyme and substrate union depend on molecular substrate of enzyme and substrate. This is also known as concept of intermolecular fit. This enzyme substrate complex is highly unstable and decomposes to produce free enzyme. The ES complex union results in the release of energy. It is this energy, which, in fact, raises the energy level of the substrate molecule, thus, inducing the activated state. In this activated stage, certain bonds of substrate molecule become more susceptible to cleavage.

This model does not give any scope for flexible nature of enzymes. Here, the model totally fails to explain many facts of enzymatic reactions, the most important being the effect of modulators.

Induced fit theory (or) Koshland model

Koshland (1958) proposed a more susceptible and realistic model for enzyme-substrate complex formation. As per this model, the active site is rigid or pre-shaped. The essential features of the substrate binding site are present at the nascent active site. The interaction of the substrate with the enzyme induces fit or a conformation change in enzyme, resulting in the formation of a strong binding site as shown in Figure 6.16. Further, due to induced fit, the appropriate amino acids of the enzyme are reciprocal to form the active site and bring about the catalysis.

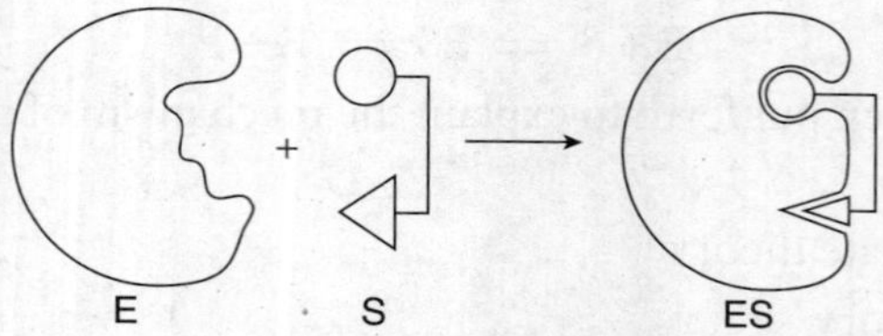

Figure 6.16 *Diagrammatic Representation of Induced Fit Model*

To explain the theory, a hypothetical illustration is given in Figure 6.17. The hydrophobic and charged groups are involved in substrate binding. Phosphoserine and –SH group of cysteine are involved in catalysis. Lysine and methionine involved in neither substrate binding nor catalysis.

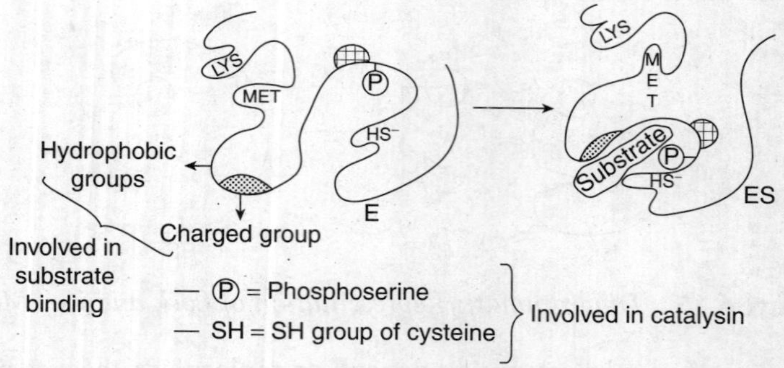

Figure 6.17 *Conformational Changes are Brought About by Induced Fit in an Enzyme Molecule*

In the absence of substrate, the substrate binding and groups are far apart from each other. But the proximity of substrate induces a conformational change in enzyme molecule, aligning the groups for both substrate binding and catalysis. Simultaneously, the spiral orientation of their groups is also changed so that lysine and methionine are now much closer. Koshland also explains the action of allosteric modulators and competitive inhibition of enzymes.

Substrate strain theory

In this model, the substrate is strained due to the induced conformation change in the enzyme. It is also possible that when a substrate binds to the preformed active site, the enzyme induces a strain to the substrate, which leads to the formation of product as given in Figure 6.18. The concept of substrate strain explains the rate of the enzyme, increasing the rate of the reaction. In fact, a combination of induced fit model with the substrate strain is considered to be operative in the enzymatic action.

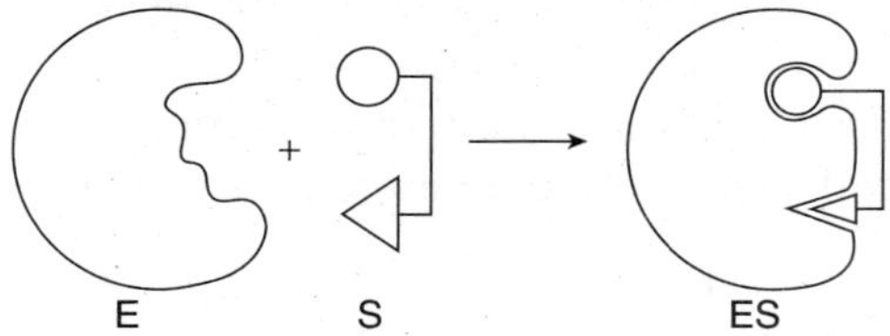

Figure 6.18 *Substrate Strain Model*

ENZYME INHIBITION

Enzyme inhibition is defined as a substance which binds with the enzyme and brings about a decrease in the catalytic activity of that enzyme. The inhibition may be organic or inorganic in nature. There are three broad categories of enzyme inhibition:

1. Reversible inhibition
2. Irreversible inhibition
3. Allosteric inhibition

REVERSIBLE INHIBITION

The inhibitors that bind non-covalently with enzyme are enzyme inhibition, which can be reversed if the inhibitor is removed. The reversible inhibition is further subdivided into the following:

1. Competitive inhibition
2. Non-competitive inhibition
3. Uncompetitive inhibition
   (a) Since enzymes are proteins, any agent which denatures protein will inactivate enzymes. Such agents are known as inhibitors of enzymes and thus may be defined as the chemical substances which reduce the activity of particular enzymes, and this process is called enzyme inhibition.
   (b) Since many inhibitors structurally resemble their enzymes substrate, inhibitor (I) binds at the active site of enzyme [E] and thus results in the failure of attachment of [S] to the active site, thus, decreasing the activity of E on its turnover number.

(c) Inhibitors may be small inorganic ions such as cyanide which inhibits the enzyme called cytochrome oxidase or much more complex inorganic or organic molecules like Di isopropyl flurophosphate.

(d) The poisoning of an E involved in a main metabolic chain will render the whole in operative and will have a profound or even fatal effect upon the organism.

(e) Also, enzyme inhibition is one of the most important ways in which enzyme activity is both artificially and naturally regulated.

Significance

1. In elucidating enzymes catalysis mechanism:
   Inhibitors are commonly used to probe the chemical and conformational nature of a substrate binding site as part of an effect to elucidate the enzymes catalysis mechanism.
2. In elucidating metabolic pathways in cells:
   (a) Enzyme inhibitors are frequently useful as tool to study metabolic pathways. Many of the intermediates in glycolysis and yeast fermentation were discovered largely by the use of inhibitors, blocking the various successive steps and altering the corresponding intermediate to accumulate in sufficient quantities for isolation and identification.

Applications

1. Many enzyme inhibitors are effective chemotherapeutic agent since an unnatural substrate analog can block the action of a specific enzyme in malfunctioning cells.
2. DIPF (inhibitor of acetylcholine esterase) is used as a nerve poison in military.
3. Differences between the reversible and the irreversible inhibitors are given in Table 6.7.

Table 6.7 *Differences between Reversible and Irreversible inhibitors*

Reversible inhibitors	Irreversible inhibitors
The interaction between inhibitor and amino acid of active site is weak.	A covalent bond is formed between the inhibitor and the amino acid of active site. This bond cannot be broken, and so enzyme loses its activity permanently.
The activity returns on merely removing the free inhibitor by dialysis or other means.	The activity does not return on dialysis.
It implies that there is an equilibrium between free I and E. $E + I \leftrightarrow EI$	No such equilibrium between free I and E.

(Continued...)

Reversible inhibitors	Irreversible inhibitors
It is independent of time and depends on the I concentration, provided I is stable.	It increases with time reacting complete inhibition even with very dilute inhibitor, provided I is in excess of the amount of E present.
The effectiveness of I is normally expressed by an equilibrium constant (K_i), which is responsible of the enzyme inhibitor affinity.	The effectiveness of I is expressed not by an equilibrium constant but by a velocity constant, which determines the fraction of the E inhibitor in a given period of time by a certain concentration of inhibitor.
Examples are numerous. Action of cyanide on cytochrome c oxidase. Action of malonate on succinate dehydrogenase	Example: Action of cyanide on xanthine oxidase. Action of nerve gases on cholinesterase.

IRREVERSIBLE INHIBITOR

Diisopropyl Phosphate Fluoridate (DIPF)

1. DIPF irreversible inactivates the enzyme acetyl choline esterase, a serine esterase that catalyses the hydrolysis of acetylcholine.

   NOTE: Acetylcholine is a neurotransmitter. It transmits nerve impulse across the synapses between certain types of nerve cells.

   ↓

   Acetylcholine is degraded by a post synaptic membrane-bound enzyme before the arrival of the next nerve impulse.

   ↓

   The products acetate and choline are transported back into the pre-synaptic terminal for recycling to acetylcholine.

   ↓

   The inactivation of acetylcholine esterase by DIPF prevents the otherwise rapid hydrolysis of the acetylcholine releases by a nerve impulse and thereby interferes with the regular sequence of nerve impulse.

   ↓

   Failure of transmission of nerve impulse propels results in paralysis and thus death by suffocation.

   ↓

   Hence, DIPF was used as nerve gases in military.

2. DIPF combines with the hydroxyl group of an essential serine residue at the active site of acetylcholine esterase to form a catalytically inactive derivative. Once this derivative is formed, the enzyme molecules can no longer function.

$$CH_3\text{—}COO\text{—}CH_2\text{—}CH_2\text{—}N^+(CH_3)_3 + H_2O \xrightarrow{\text{Acetylcholine esterase}} CH_3\text{—}COO + HO\text{—}CH_2\text{—}CH_2\text{—}N^+(CH_3)_3 + H^+$$

Acetylcholine → Acetate + Choline

3. DIFP has been found to inhibit a whole class of enzymes; many of them are capable of catalysing hydrolysis of peptide on ester linkages. They include not only acetylcholine esterase but also trypsin chymotrypsin, elastase, phosphoglucose esterase, and cocoonase (an enzyme secreted by silkworm larvae to hydrolyse the silk fibres of the cocoon, allowing the larva to escape). All the DIFP-inhibited enzymes have an essential serine residue in their active site, which participates in their catalytic activity.
4. There is a useful side of DIFP. It has led to the development of malathion and other insecticides that are relatively non-toxic for people and animals. Malathion is inactive by itself and degraded by higher animals into product that are harmless to them, but it is converted by enzymes of insects into an active inhibitor of their acetylcholine esterase.

Iodoacetamide

1. Another irreversible inhibitor of some enzymes is iodoacetamide, which can react with sulfhydryl groups of essential cysteine residues or the imidazole group of essential histidine residues.
2. With the help of an inhibitor, the OH group of serine and the SH group of cysteine and imidazole are participating in the catalytic activity of different classes of enzymes.

Competitive Inhibition (Reversible)

A compound that competes with the substrate for enzymatic binding site is known as competitive inhibitor.

The inhibitor usually resembles the substrate structure such that it specifically binds to the active site but differs from the substrate so as to be unreactive.

Salient features

1. It is a reversible inhibition.
2. In this type of inhibition, there is structurally similarity between the inhibitor and the substrate, and hence there is competition between I and S for the active site of the enzyme.
3. In this type, both ES and EI complexes are formed during the reaction.

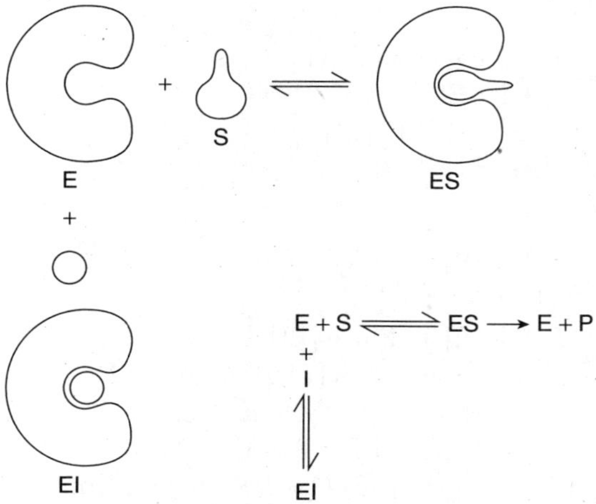

$$\text{Inhibitor constant } K_I = \frac{[E][I]}{[EI]}$$

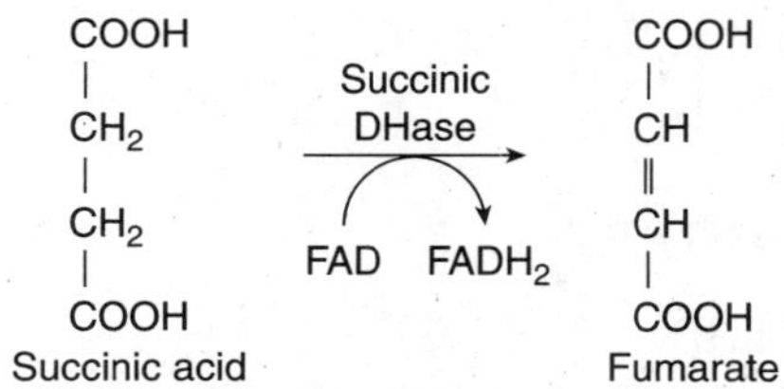

Inhibitors for this enzymes are Figure 6.19:

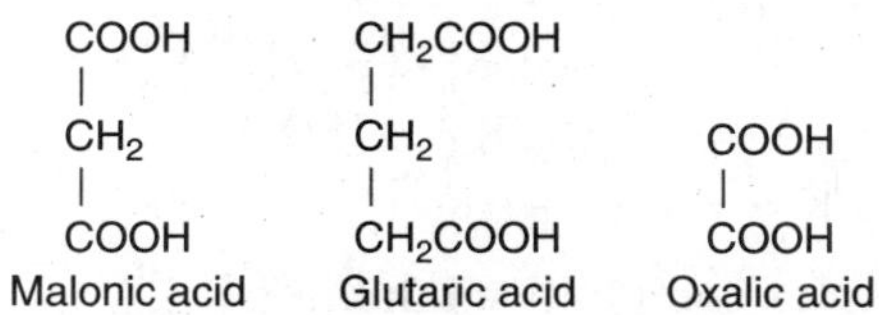

Figure 6.19 *Competitive Inhibitors for Succinic Dehydrogenase*

However, the actual amount of ES and EI will depend on the following:.

1. Affinity between enzymes and substrate inhibitor.
2. Actual concentration of substrate and inhibitor present.

Thus, if I is present in sufficiently high concentration, it can replace the substrate entirely and thus block the reaction or lower the rate of the reaction.

Thus, K_m is high, but V_{max} is the same in competitive inhibitor.

However, when the concentration of [S] is increased, the effect of I can be reversed, forcing it out from EI complex; thus, [I] and [S] are mutually exclusive.

Kinetics

$$E + S \rightarrow ES \rightarrow E + P$$
$$E + I = EI$$
$$K_i = [E]\,[I]/[EI] \quad (1)$$
$$[EI] = [E][I]/K_i \quad (2)$$

In the equal state,

$$K_m = [E][S]/[ES] \quad (3)$$
$$[E] = K_m[ES]/[S] \quad (4)$$
$$[ET] = [E] + [ES] + [EI]$$
$$= [E] + [ES] + [E][I]/K_I \text{ from (2)}$$
$$= [E] + [E][I]/K_i + [ES]$$
$$= [E](1 + I/K_i) + [ES]$$
$$= K_m[ES]/[S]\,(1 + I/K_i) + [ES] \text{ from (4)}$$
$$E_T = [ES](K_m(1 + I/K_i)/[S] + 1)$$
$$[ET]/[ES] = K_m(1 + I\,/K_I)/[S] + 1 \quad (5)$$

For enzyme-catalysed reactions,

$$V = K_3[ES] \quad (6)$$
$$V_{max} = K_3[E_t] \quad (7)$$

Divide (7) by (6).

$$V_{max/}V = ET/[ES] \quad (8)$$

Substituting (8) in (5), we get,

$$V_{max/}V = K_m(1 + I/K_i)/[S] + 1$$
$$V_{max/}V = K_m(1 + I/K_i) + [S]/[S]$$
$$V_{max}[S] = V(K_m(1 + I/K_i) + [S]$$
$$V = V_{max}\,[S]/K_m(1 + I/K_i) + [S] \quad (9)$$
$$V = V_{max}[S]/K_m^{app} + [S]) \quad (10)$$

Where $K_m^{app} = K_m(1 + I/K_I)$

This is the M–M equation for competitive inhibitor.
From the equation, it is clear that when an enzyme is inhibited competitively,

1. V_{max} does not change at all
2. K_m value increases as the concentration of I increases.

I ↑es
↓
$K_m{}^{app}$ ↑ es
↓
K_m ↑ es
↓
Affinity towards [S] ↓ es

Diagnostic plot

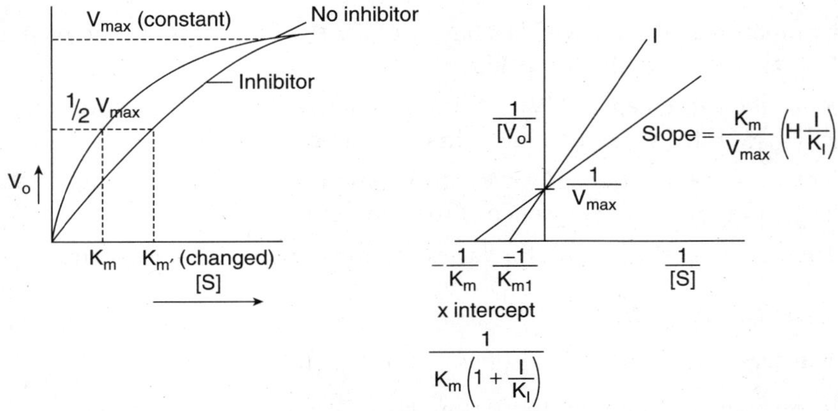

Figure 6.20 *Competitive Inhibitor*

The graph Figure 6.20 shows that

1. As the concentration of inhibitor ↑ es
   → *the affinity for* [*s*]↓ es
2. But here V_{max} remains constant.

Note that as [S] → ∞, $V_O \rightarrow V_{max}$, for any value of inhibitor concentration.
(However, the greater the concentration of I, the greater the [S] must be to approach V_{max}. Thus, I does not affect the turnover number of the enzyme. Rather, the presence of I has the effect of making [S] appear more dilute than it actually is, or attentively, making K_m appear larger than it really is.)
From these two points, we can make out that I is competitive inhibitors.
NOTE: We can determine K_m value from the hyperbolic curve but not K_i, so we have to find out with LB plot.

LB plot

$$V = V_{max}[S]/K_m(1 + I/K_i) + [S]$$
$$1/V = K_m(1 + I/K_i) + [S]/V_{max}[S]$$
$$1/V = K_m(1 + I/K_i)/V_{max}[S] + [S]/V_{max}[S]$$
$$1/V = K_m(1 + I/K_i)/V_{max}.1/[S] + 1/V_{max} \text{ (refer Table 6.8)}$$

Table 6.8 *LB Equation for Competitive Inhibition*

Slope (m)	$K_m(1 + I/K_i)$
Y intercept	$1/V_{max}$
X intercept	$-1/K_m(1 + I/K_i)$

1. It is an LB plot (double-reciprocal plot) obtained in the absence of I and with two different concentration of a competitive inhibitor.
2. Competitive inhibitor yields a family of lines with a common intercept on the $1/V_o$ axis but with different slopes.
3. Since the *y* intercept is $1/V_{max}$, this states that an infinity high concentration of [S] (1/[S] = 0). V_i is the same as in the absence of inhibitor.
4. However the *x* intercept varies with inhibitor concentration and becomes a larger number in the presence of the competitive Inhibitor.
5. Thus, a competitive inhibitor varies the apparent K_m for the substrate.

Non-Competitive Inhibition

Salient features

1. In this type of inhibitor, there is no structural similarity between I and S and hence no mutual competition between them for the active site of the enzyme.
2. The sites of attachment of I and S are different. Inhibitor does not bind with the active site of E but binds with a site other than the active site.
3. The non-competitive inhibitor binds with both free enzyme and ES complex. Thus, both EI and ESI complexes are formed as shown in Figure 6.21.
4. When I binds to E, it brings about three-dimensional structural changes in E, thus, inactivating it completely.
5. In this case, the extent of inhibition is independent of score because the percent of inhibitor is not reduced even by increasing the concentration of S. The combination with either S or I does not affect the affinity for the other.
6. Here V_{max} is lowered, but K_m is kept constant.
7. Chymotrypsin, where the catalytic site includes a proton acceptor, may be inhibited by increasing hydrogen conc.

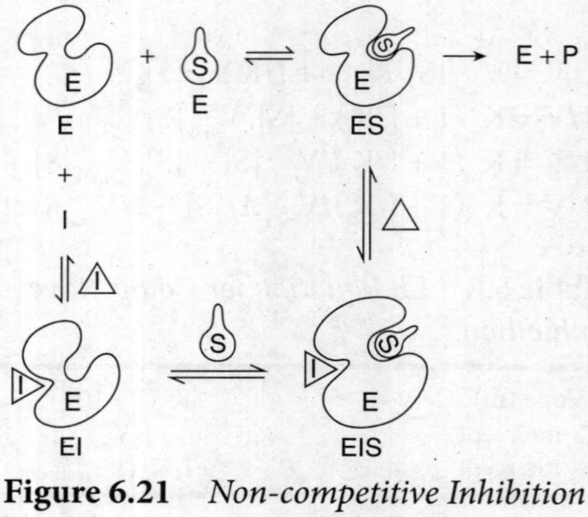

Figure 6.21 *Non-competitive Inhibition*

Kinetics

$$E + S \rightarrow ES \rightarrow E + P$$

$$+ +$$

$$I\ I$$

$$I\ I$$

$$I\ I$$

$$EI\ ESI$$

$$K_I = [E][I]/[EI]$$

$$[EI] = [E][I]/K_I \qquad (1)$$

$$K_i = [ES]/[ESI]$$

$$[ESI] = [ES][I]/K_I \qquad (2)$$

$$K_m = [E][S]/[ES] \qquad (3)$$

$$[E] = K_m[ES]/[S] \qquad (4)$$

$$ET = E + ES + ESI + EI$$

$$= [E] + [ES] + [ES][I]/K_I + [E][I]/K_I$$

$$= [E]\ (1 + I/K_i) + [ES](1 + I/K_i)$$

$$= K_m\ [ES]/[S] + [ES](1 + I/K_I)$$

$$= [ES](K_m/[S] + 1)\ (1 + I/K_I)$$

$$E_{T/}[ES] = (K_m/[S] + 1)\ (1 + I/K_I) \qquad (5)$$

For enzyme-catalysed reaction:

$$V = K_3[ES] \qquad (6)$$

$$V_{max} = K_3[ET] \qquad (7)$$

$$V_{max}/V = E_T/[ES] \qquad (8)$$

Substituting (8) in (5), we get

$$V_{max}/V = (K_m/[S] +)(\ 1 + I/K_I)$$

$$V_{max}/V = (K_m + [S]/[S])(1 + I/K_I)$$

$$V/V_{max} = [S]/K_m + [S]) \times 1/1 + I/K_I)$$

$$V = V_{max}[S]/(K_m + [S]) \times 1/1 + I/K_I)$$

Dividing by $(1 + I/K_I)$ in numerator and denominator,

$$V = V_{max/}1 + I/K_I[S]/K_m + [S] \qquad (9)$$

$$V = V_{max}{}^{app}\ [S]/K_m + [S] \qquad (10)$$

Where $\quad V_{max}{}^{app} = V_{max/}1 + I/K_i$

This is the M–M equation for non-competitive inhibitor refer Table 6.9.

From equation (10), it is clear that when an enzyme is inhibited non-competitive, V_{max} decreases and K_m value does not change; that is, the combination with either [S] or [I] does not affect the affinity for the other as shown in Figure 6.22.

Table 6.9 *M–M equation (Slope, X-intercept, Y-intercept)*

Slope(M)	$K_m(1 + I/K_I)/V_{max}$
Y intercept	$1 + I/K_{I/}V_{max}$
X intercept	$-1/K_m$

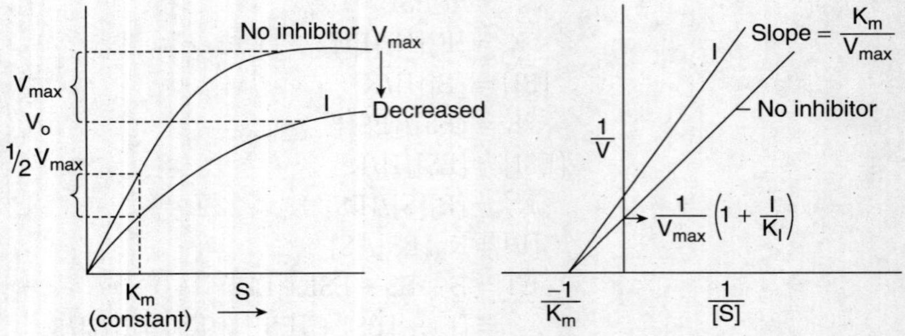

Figure 6.22 *Effect of Non-competitive Inhibitor*

LB plot

$$V = V_{max/}1 + I/K_I\,[S]/K_m + [S]$$
$$1/V = K_m + [S]/V_{max/}1 + I/K_I[S]$$
$$1/V = K_m/V_{max/}1 + I/K_I[S] + [S]/V_{max/}1 + I/K_I\,[S]$$
$$1/V = K_m(1 + I/K_I)/V_{max} \cdot 1/[S] + 1 + I/K_I/V_{max}$$

1. Non-competitive inhibitor yields a family of lines having a common intercept on [S] axis indirectly that K_m for the substrate is not altered by a non-competitive inhibitor but V_{max} decreases.
2. As K_m is unchanged, the addition of high concentration of [S] does not relieve the inhibition.

Uncompetitive Inhibition

1. Reversible
2. In this case, I recognises ES complex and not the free enzyme.
3. Thus the complex formed is ESL as shown in Figure 6.23.

$$
\begin{array}{c}
E + S \rightarrow ES \rightarrow E + P \\
+ \\
I \\
ESI \rightarrow P
\end{array}
$$

Since [I] binds slowly with ES, it produces proportional changes in K_m and V_{max} as binding of I to ES will decreases the rate of breakdown of the ES complex to [E] and [S], and hence, K_S or K_m is decreased.

4. Also binding of I with ES will decrease the rate of breakdown of ES to product [P] and hence decrease V.
5. There need not be any structural similarity between S and I.

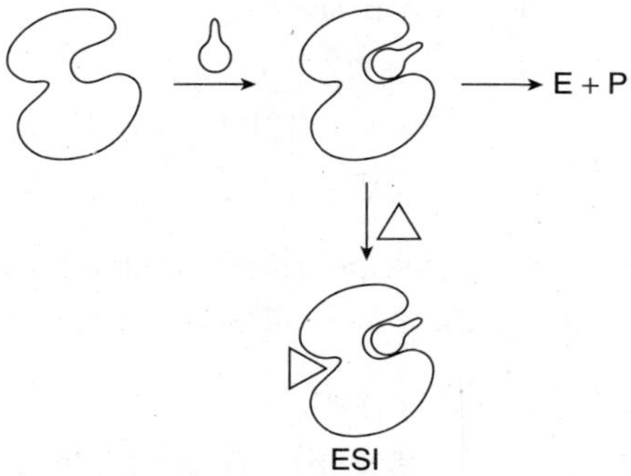

Figure 6.23 *Uncompetitive Inhibition*

6. When [S] binds to [E], some changes take place which change the inhibitor binding site in such a manner that I can bind at that site.
7. Uncompetitive inhibition is extremely rare in single substrate reaction, although a case has been reported for the inhibition of rare intestinal alkaline phosphatase by L-phenylalanine.
8. Uncompetitive Inhibitor requires that I affect the catalytic function of E but not its substrate binding.

Kinetics

$$\begin{array}{c} E + S \rightleftharpoons ES \rightarrow E + P \\ + \\ I \\ \upharpoonleft\downharpoonright \\ ESI \end{array}$$

$$K_I = [ES][S]/[ESI] \quad (1)$$
$$[ESI] = [ES][I]/K_I \quad (2)$$
$$K_m = [E][I]/[ES] \quad (3)$$
$$[E] = K_m[ES]/[S] \quad (4)$$

M–M equation

$$[E_T] = [E]+[ES]+[ESI]$$
$$= K_m[ES]/[S] + [ES] + [ES][I]/K_I \text{ from (2) and (4)}$$
$$= [ES]\ (K_m/[S] + 1 + I/K_I)$$
$$= [ES]\ (K_m + [S]\ (1 + I/K_I)/[S]$$

$$E_{T/}[ES] = K_m + [S](1 + I/K_I)/[S] \quad (5)$$
$$V_o = K_3[ES] \quad (6)$$
$$V_{max} = K_3[E_T] \quad (8)$$
$$V_{max/}V = E_T/ES \quad (9)$$

Substitute (8) in (5).

$$V_{max/}V = K_m + [S](1 + I/K_I)$$
$$V = V_{max}[S]/K_m + [S](1 + I/K_I)$$

This is the M–M equation for an uncompetitive inhibited E catalysed reaction. Dividing the right side of the equation by $1 = I/K_I$ on numerator and denominator,

$$V = (V_{max/}1 + I/K_I)\ [S]/K_m/1 + I/K_I + [S] \quad (10)$$
$$V = V_{max}{}^{app}[S]/K_m{}^{app} + [S] \quad \text{(refer Table 6.10)} \quad (11)$$

Here both V_{max} and K_m change with [I] concentration of M–M plot as shown in Figure 6.24.

LB plot

$$1/V = K_m{}^{app} + [S]/V_{max}{}^{app}[S]$$
$$1/V = K_m{}^{app/}V_{max}{}^{app}[S] + 1/V_{max}{}^{app}$$

Table 6.10 *Characteristics of M–M Equation (slope, X-intercept and Y-intercept)*

Slope (M)	$K_m/(1 + I/K_I)/V_{max}/1 + I/K_I$ (or) K_m/V_{max}
Y intercept	$1 + I/K_I/V_{max}$
X intercept	$1 + I/K_I/K_m$

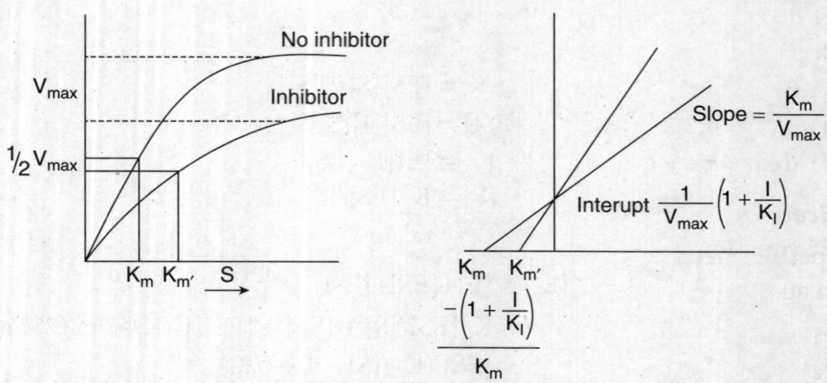

Figure 6.24 *M–M and LB Plot of Uncompetitive Inhibition*

EXAMPLES OF INHIBITION

COMPETITIVE INHIBITION

1. The classical example for competitive inhibition is that of succinate dehydrogenase by malonate anion.
2. Succinate dehydrogenase catalyses the conversion of succinate to fumarate with FAD as cofactor.

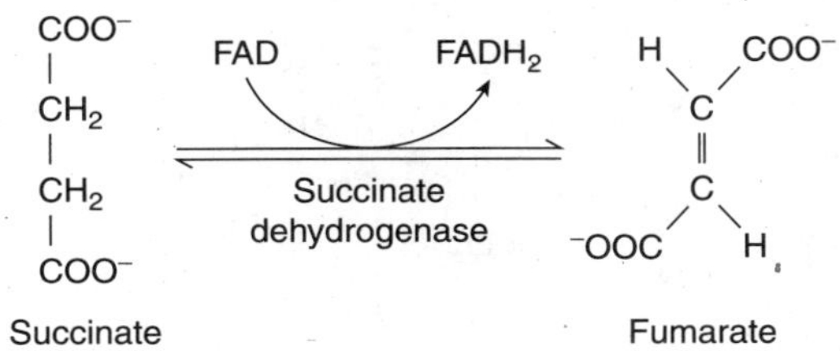

3. The enzyme catalyses the removal of two hydrogen atoms from succinate: one from each of the two methylene (—CH_2) groups and FAD saves as hydride ion acceptor.
4. This enzyme is competitively inhibited by malonate, which resembles succinate in having two ionised carbonyl groups at pH 7.0 but differs in having only there carbon atoms.
5. However, malonate is not dehydrogenated by succinate dehydrogenase; it simply occupies the active site, keeping it away from acting on its normal substrate.
6. The catalytic site of succinate dehydrogenase has two approximately spaced positively charged groups capable of attracting the two negatively charged carboxylate group of succinate anion.
7. The effectiveness of malonate in competitively inhibiting succinate dehydrogenase is due to the complementarity between the active site of enzyme and its structure.
8. However, increasing the succinate concentration will reduce the extent of inhibitions by a given concentration of malonate (reversible).
9. Among other competitive inhibitors of succinate dehydrogenase having distactly placed anionic groups is oxaloacetate.

Applications of Competitive Inhibitors

Competitive inhibitors are used in clinical and biological situations. They are frequently called antagonists or antimetabolites of the substrates with which they compete.

1. *Treatment of Gout*
   (a) Xanthine oxidase is an enzyme of purine catabolism, which catalyses the oxidation of hypoxanthine to xanthine and xanthine to uric acid.

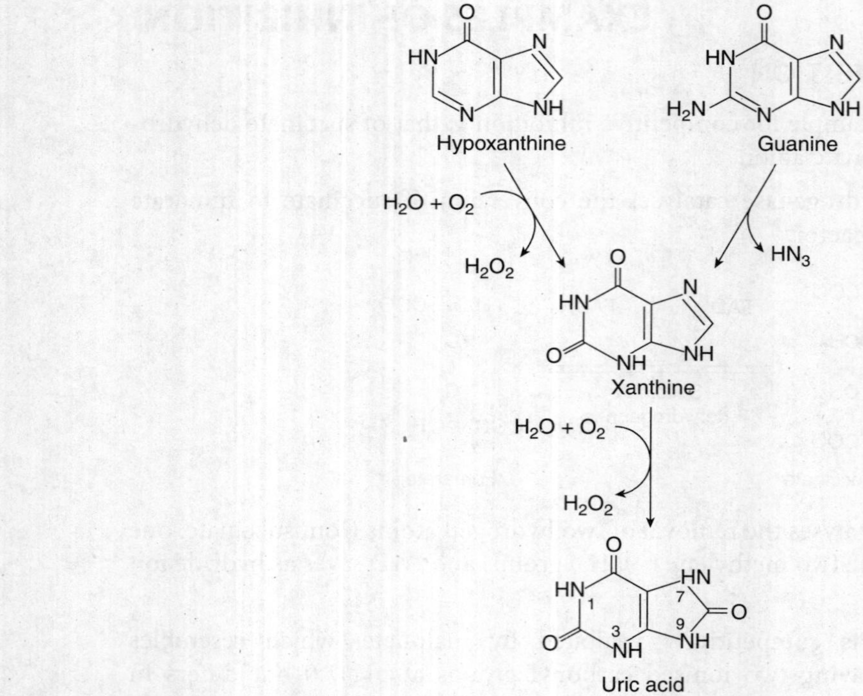

(b) Xanthine oxidase is a dimeric proteins of identical 130kD subunits, each of which contains a few election transfer agents – an FAD, an Mo complex that cycles between its Mo(VI) and Mo(IV) oxidation state and two different Fe-S clusters.

(c) Allopurinol, an analog of hypoxanthine, in which the position of N_7 and C_8 are interchanged competitively, inhibits xanthine oxidase.

Mechanism of Action of Allopurinol

(a) Allopurinol first acts as a substrate and as an inhibitor of the enzyme xanthine oxidase.

(b) The enzyme hydroxylate allopurinol to alloxanthine, which then remains tightly bound to the active site of the enzyme.

(c) The molybdenum atom of the enzyme is kept tightly bound with the alloxanthine in the +4 oxidation state, and thus the normal +6 oxidation state of molybdenum, which is necessary for the catalytic activity of the enzyme, is not restored. Thus, the product uric acid is not formed.

(This mode of action of allopurinol is an example of suicide inhibition, in which an enzyme converts a compound into a potent inhibitor that immediately inactivates the enzyme.)

(a) Gout is a disease characterised by gout pain and swelling in joints due to excessive deportation of sodium water, crystals in joints, subcutaneous tissues, and so on.

(b) Allopurinol alternates the symptoms of gout by decreasing the rate of uric acid production while increasing the levels of the more solute hypoxanthine and xanthine.

2. *Anticancer Agents/Chemotherapy*

(a) dTMP is synthesised from dUMP by thymidylate synthase with N^5, N^{10} methylenetetrahydrofolate as the methyl donor.

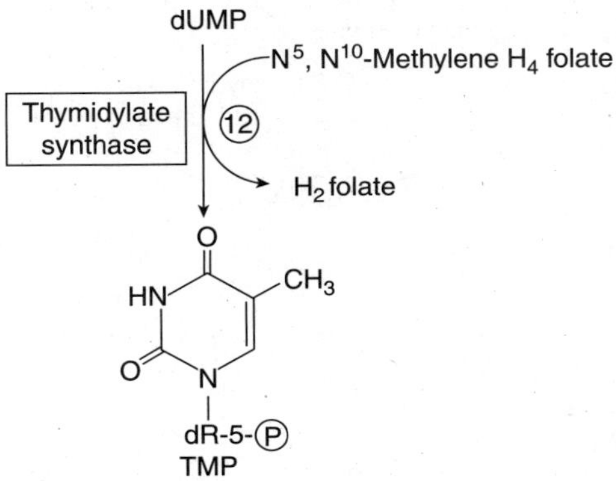

(b) Dihydrofolate reductase catalysed the reduction of DHF to THF by NADPH.

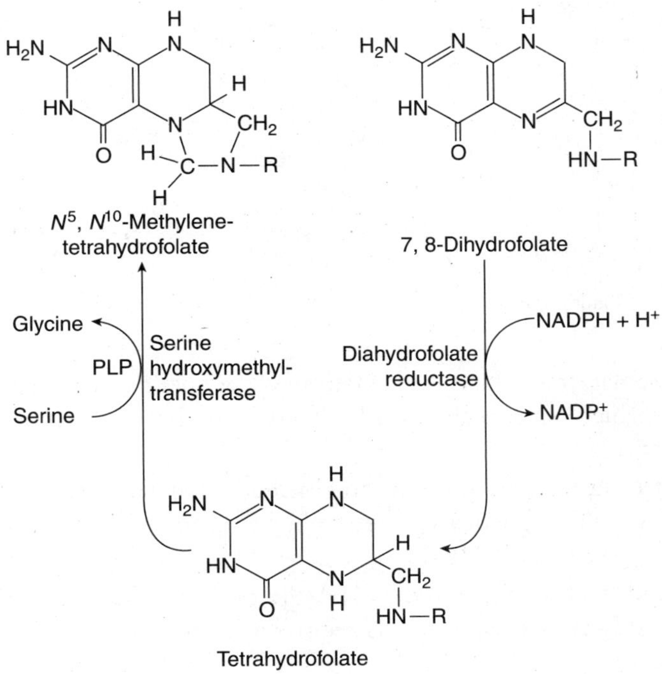

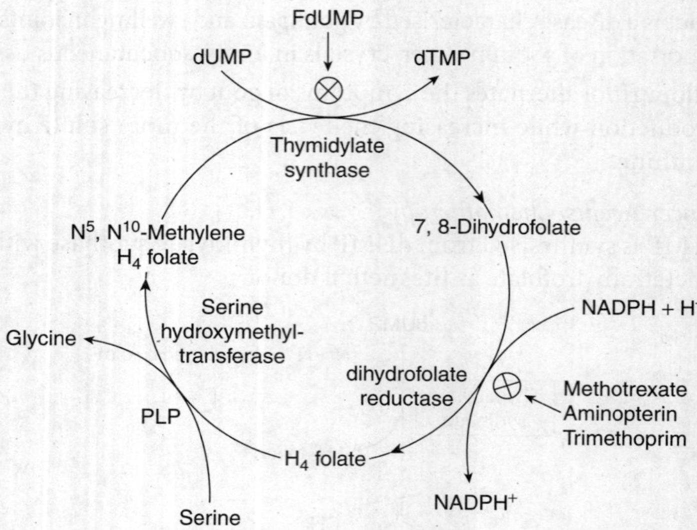

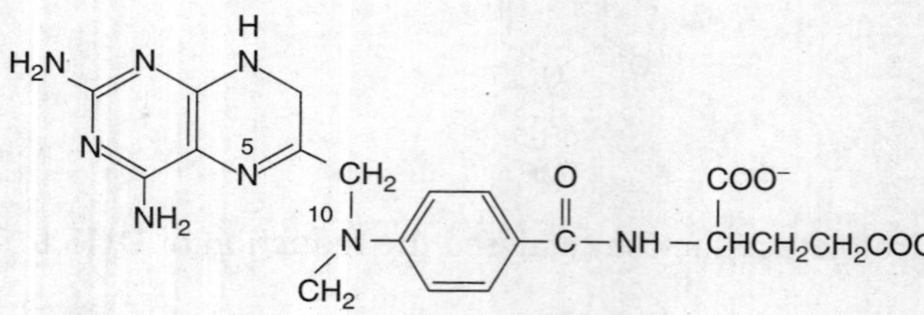

Methotrexate

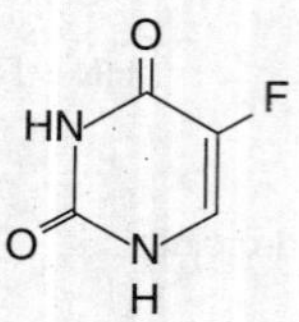

Fluorouracil

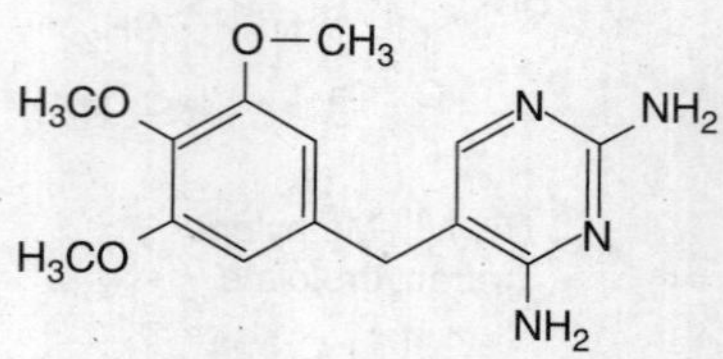

Trimethoprim

(c) Dihydrofolate reductase is very strongly competitively inhibited by methotrexate, which is an analog of FH_2 (DHF); another is an important folic acid antagonist.

(d) Methotrexate, amino protein, and trimethoprim are all DHF analogs that completes, although nearly irreversibly bind to reductase, with a 1000-fold greater affinity than DHF does.

(e) These antifolates which reduce DHFR prevents dTMP synthesis. Rapidly proliferating cells as cancer cells require a steady supply of dTMP in order to survive. Thus, methotrexate is used as an effective anticancer agent, particularly against childhood leukaemia.

3. *Antibiotics*
Folic acid is a coenzyme used in certain biosynthetic reaction although it is not synthesised in man. Some bacteria synthesise it using para aminobenzoic acid (PABA).

$$\underset{\text{(Sulphonamide)}}{\text{PABA} \rightarrow \text{Folic acid}}$$

Sulphonamide are chemically and physically similar to PABA and can competitively inhibit the synthesis of folic acid (an essential vitamin for microbes), thus, leading to the inhibition of the growth of susceptible organism.

Sulphonamide can be toxic to micro-organism and virtually harmless to the host as the host obtains its folic acid sequence from the diet and not from PABA.

4. *Treatment of Ethylene Glycol Poisoning*
   (a) About fifty deaths occur annually from the ingestion of ethylene glycol, a constituent of permanent-type automobile anti-freeze.
   (b) Ethylene glycol itself is not lethally toxic; rather, the harm is done by oxalic acid, an oxidation product of ethylene glycol.
   (c) The first step is the conversion as the oxidation of ethylene glycol by alcohol dehydrogenase.
   (d) This reaction can be effectively inhibited by the administration of a nearly intoxicating dose of ethanol, which acts as a competitive inhibitor of alcohol dehydrogenase.
   (e) The ethylene glycol is then excreted harmlessly.

   (The same rationale is used for the therapy of methanol poisoning.)

5. *Treatment of Hypotensive Shock*
   (a) Competitive inhibition explains the action of several other drugs. For example, ephedrine and amaphetamine prolong the effect of the hormones like adrenaline/noradrenaline, and 5-hydroxytryptamine by competitively inhibiting the enzyme monoamine oxidase (MAO), which brings about the oxidative deamination of the above-mentioned hormones.
   (b) A similar explanation has been put forward for the action of cocaine and antidepressive drugs like proniazid and tranylcypromine.

IRREVERSIBLE INHIBITION

The inhibitors bind covalently with the enzymes and indicate those which are irreversible. Formation of covalent link between irreversible inhibitor and an enzyme is common. Irreversible inhibitors are another useful tool for studying reaction mechanisms.

For example, iodoacetate is an irreversible inhibitor to enzymes like papain and glyceraldehyde-3-phosphate dehydrogenase. It combines with sulfhydryl (SH) groups at active site of three enzymes and makes them inactive.

Diisopropyl fluorophosphates (DFP) is a nerve gas developed by Germans during the Second World War. DFP irreversibly bind with enzymes containing serine at the active site of acetylcholine esterase (essential for nerve condition) and results in paralysis of vital body function.

SUICIDE INHIBITION

It is a special type of irreversible inhibition of enzyme activity. It is also known as mechanism based on inactivation. The substrate link analogue (inhibitors) initially binds with the enzyme, and the first few steps of the pathway are catalysed. This new product irreversibly binds to the enzyme and inhibits the further reaction.

Example: Difluoro methyl ornithine (DFMO) is a suicide inhibitor for treating African sleeping sickness Trypanosomiasis and Indian kala azar disease, which is induced by parasite trypanosoma. Ornithine decarboxylase (ODC) catalyses the conversion of ornithine to putrescine, which is necessary for polyamine synthesis as shown in Figure 6.25.

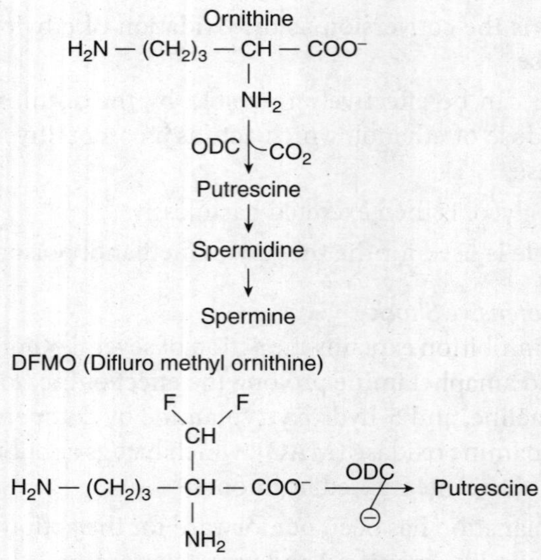

Figure 6.25 *Synthesis of Polyamines*

The vulnerable point in the metabolism of trypanosome (parasite) or the pathway of polyamine synthesis, which is required for the DNA packaging, is required in large amount in rapidly dividing cells. In parasites, the half life of ODC is many hours. This half life of ODC in man is only five minutes.

In man, as enzyme molecules are constantly synthesised, the drug will not affect human. But DFMO inhibits polyamine synthesis. So parasites cannot divide, and the immune system of host can kill them. Yet another example of suicide inhibition is the effect of purine and pyrimidine analogues in cancer chemotherapy as depicted in Figure 6.26. They get converted to nucleotide derivatives.

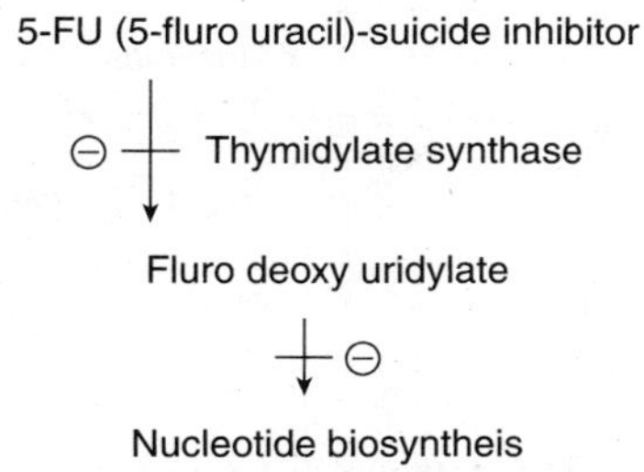

Figure 6.26 *Sucide Inhibition (5FU–5 Fluro Uracil) in Cancer Chemotherapy*

ALLOSTERIC INHIBITION

Some enzymes possess additional sites known as allosteric sites (*allo* - other, *steric* - solid or shape) besides the active site. Such enzymes are known as allosteric enzymes. Allosteric enzymes are usually multiple subunit enzymes, and allosteric sites may be located on different subunits. The binding of regulatory molecules and modulators or modifiers or effectors can either enhance the activity of enzyme (allosteric activation) refer Figure 6.27 – the effector is called positive effector – or inhibit enzyme activity (allosteric inhibition) refer Figure 6.27 – the effector is called negative effector.

Action of Allosteric Enzymes

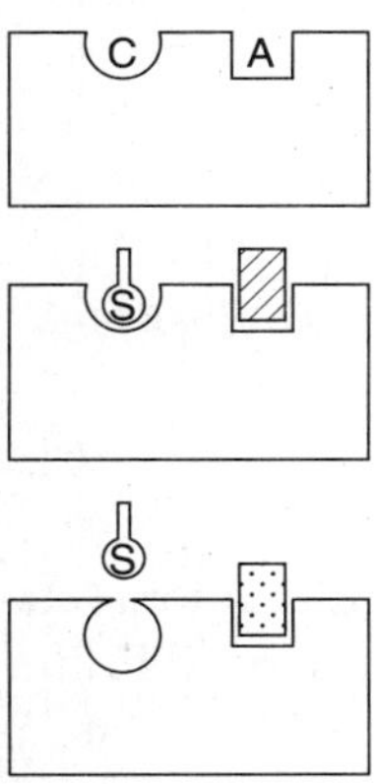

Allosteric enzyme with C = Catalytic or active site
A = Allosteric site.

Allosteric activation – When modulator is bound to allosteric site it changes the confirmation of active site, so that substrate can bind. In enhance the enzyme activity and the effect in called +ve co-operativity.

Allosteric inhibition – When –ve modular is bound to allosteric site it changes the active site so that substrate is not bound. If inhibit the enzyme activity and the effect is called –ve co-operativity.

Figure 6.27 *Action of Allosteric Enzymes*

Homotropic Effectors

When the substrate itself serves as an effector, the effector is said to be homotropic. Most often, an allosteric functions is a positive effector. This enzymes show a sigmoidal curve (Figure 6.28) when reaction velocity plots against substrate concentration.

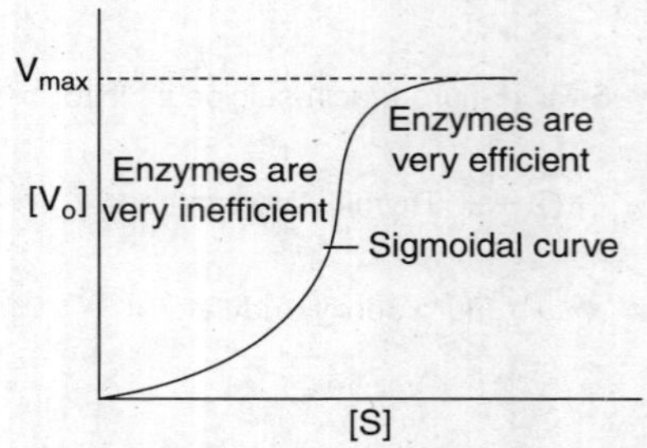

Figure 6.28 *Homotropic Effectors–sigmoidal Curve*

The concept of cooperative binding is analogous to the binding of oxygen to hemoglobin – substrate is comparable to oxygen.

Heterotrophic Effectors

When the effector may be is different from substrate, in which case, the effector is said to be heterotrophic (Figure 6.29).

Example: Feedback inhibitors. Aspartate transcarbamoylase.

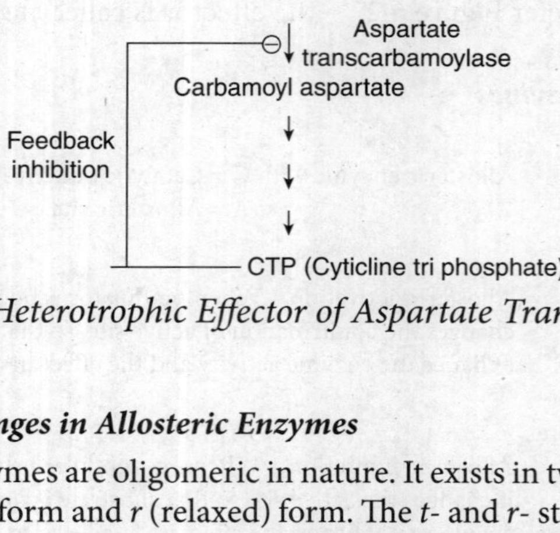

Figure 6.29 *Heterotrophic Effector of Aspartate Transcarbamoylare*

Conformational Changes in Allosteric Enzymes

Most of allosteric enzymes are oligomeric in nature. It exists in two conformational statuses: *t* (tense or taut) form and *r* (relaxed) form. The *t*- and *r*- states are in equilibrium. Allosteric inhibitors favour *t*-state, whereas activators and substrate favour *r*- state as shown in Figure 6.30.

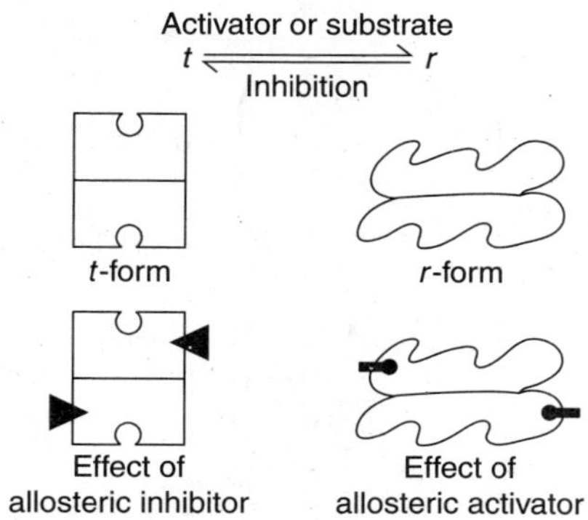

Figure 6.30 *Allosteric Enzyme with t and r-state with Respect to the Activator and Inhibitor*

Kinetics of Allosteric Enzymes

Allosteric enzymes show relationships between V_0 and [S] that differ from Michaelis-Menten kinetics. They do exhibit saturation with the substrate [S] is sufficiently high, but for some allosteric enzymes, when V_0 is plotted against [S], a sigmoid saturation curve occurs.

Archibald Hill, in1910, first analysed an equation for the allosteric enzyme kinetics. For a protein with N binding sites, the equilibrium equation is as follows:

$$P + L \rightleftharpoons PL$$

$$P + nL \rightleftharpoons PL_n$$

$$Kd = \frac{[P][L]^n}{[PL_n]}$$

$$\theta = \frac{\text{binding sites occupied}}{\text{total binding site}}$$

$$\theta = \frac{[L]^n}{[L]^n + Kd}$$

Rearranging, then taking the log of both sides, yields.

$$\log\left(\frac{\theta}{1-\theta}\right) = n\log[L] - \log Kd$$

$$\theta = V_o$$

$$I = V_{max}$$

$$L = \text{Substrate}$$

or

Simply as
$$\log\left(\frac{V_0}{V_{max}-V_0}\right) = n\log[S] - \log K'$$

K' = complex constant

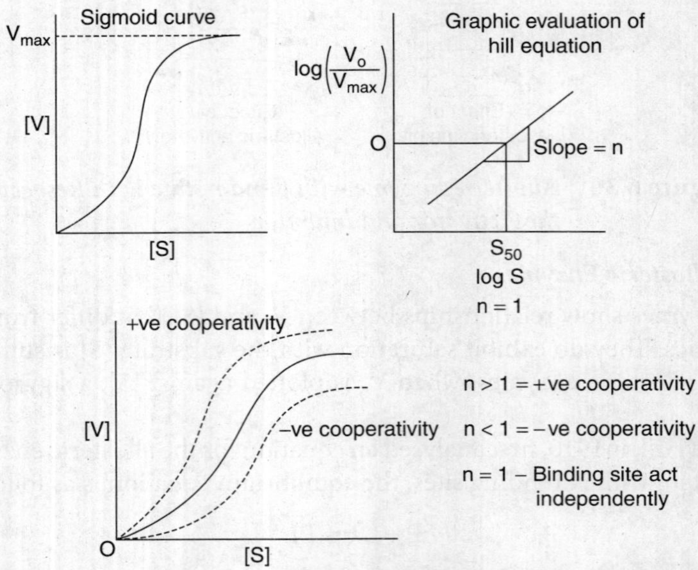

Figure 6.31 *Types of Cooperativity Based on the Binding Site of Substrate*

The equation states that when [s] is low, compared to k', the reaction velocity increases as the nth power of [s]. Where n-value depend on the number of substrate binding sites and the type of interactions between these binding sites.

∴ n > l, the sites are cooperative and greater the value of n, the stronger is the cooperativity. n < l, the sites are said to inhibit negative cooperativity as shown in Figure 6.31.

S_{50} – the concentration of substrate that produces half maximal velocity drops a perpendicular line to *x*-axis from the point where log $(V_0/V_{max} - V_0) = 0$.

REGULATION OF ENZYME ACTIVITY

The various factors that influence enzyme activity in vitro are temperature and pH, which are quite constant in living cell. In biological system, regulation of enzyme activities occurs at different stages in one or more of the following ways to achieve cellular economy:

1. Allosteric inhibition
2. Activation of proenzymes or zymogens

3. Control of enzyme synthesis
4. Enzyme degradation

ALLOSTERIC INHIBITION: FEEDBACK INHIBITION

The process of inhibiting the first step by the final product, in a series of enzyme-catalysed reactions of a metabolic pathway, is referred to as feedback inhibition or regulation.

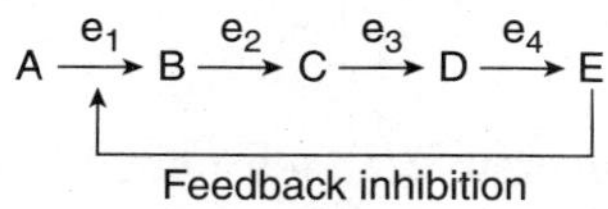

A is the initial substrate, B, C, and D are the intermediates, and E is the end product in a pathway catalysed by four different enzymes (e_1, e_2, e_3, and e_4). The very first step (A→B) by e1 is the most effective for regulating the pathway, but the final end product of this type is often called negative feedback inhibition. This is real cellular economy to serve all from the wasteful expenditure of synthesising a compound which is already within the cell. Feedback inhibition or end product inhibition is a specialised type of allosteric inhibition necessary to control metabolic pathways for efficient cellular functions.

Example:

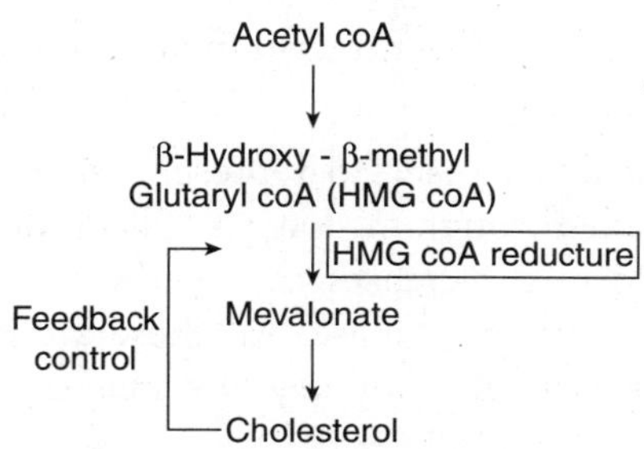

ACTIVATION OF ZYMOGENS AND PROENZYMES

Zymogen Activation

Some enzymes are synthesised in an inactive precursor from, which is activated at a physiologically appropriate time and place. Digestive enzymes exemplify this kind of control, which is called *proteolytic activation or zymogen activation.*

Example: Trypsinogen is synthesised in pancreas and activated in small intestine to form active enzyme trypsin.

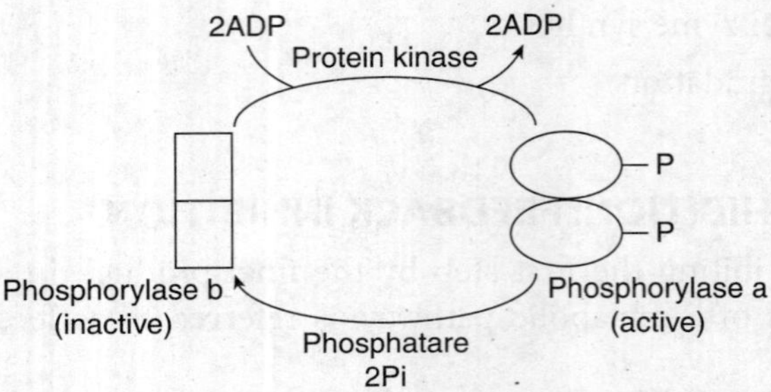

Figure 6.32 *Covalent modification of Phosphorylase*

Covalent Modification

Enzymes are controlled by reversible attachment of phosphoryl groups to specific serine and threonine residues.

Example: The activities of enzymes that synthesise and degrade glycogen are coordinate in this way as shown in Figure 6.32.

CONTROL OF ENZYME SYNTHESIS

Most of the enzymes, particularly the rate-limiting ones, are present in very low concentration. Many rate-limiting enzymes have short, half lives. This helps in efficient regulation of enzyme levels.

There are two types of enzymes:

1. Constitutive enzymes: Housekeeping enzymes. The levels of these enzymes are not controlled and remain fairly constant.
2. Adaptive enzymes: Their concentration increases or decreases as per the body needs and are well regulated.

Induction: The term *induction* is used to represent increased synthesis of enzymes which are under gene control through mediation of hormones or other substances.

Example: Hormone insulin increases the synthesis of synthetase, glucokinase, phosphofructokinase, and pyruvate kinase. All these are involved in the utilisation of glucose repress the synthesis of many key enzymes involved in glucose synthesis (gluconeogenesis).

Repression: The term *repression* is used to represent decreased synthesis.

Example: In many instances, substrate can repress the synthesis of enzyme. Pyruvate carboxylase is the key enzyme in the synthesis of glucose from non-carbohydrate sources like pyruvate and amino acids. If there is sufficient glucose available, there is no necessity for its synthesis. This is achieved through repression of pyruvate carboxylase by glucose.

ENZYME DEGRADATION

Enzymes have a lot of variability in the half lives of individual enzymes.

Example: LDH 4 = 5–6 days; LDH 1 = 8–12 hours; amylase = 3–5 hours. In general, the key and regulatory enzymes are rapidly degraded. If not needed, they immediately disappear, and as and when required, they are quickly synthesised.

NON-PROTEIN ENZYMES

Ribozymes, RNA molecules, present in the enzyme can function as enzymes. Altman and coworker in 1983 found that ribonuclease P (Figure 6.33), an enzyme-cleaved precursor of tRNA, was functional due to RNA component present in the enzyme and not protein part of the enzyme. This RNA ribozymes also obey Michaelis-Menten kinetics.

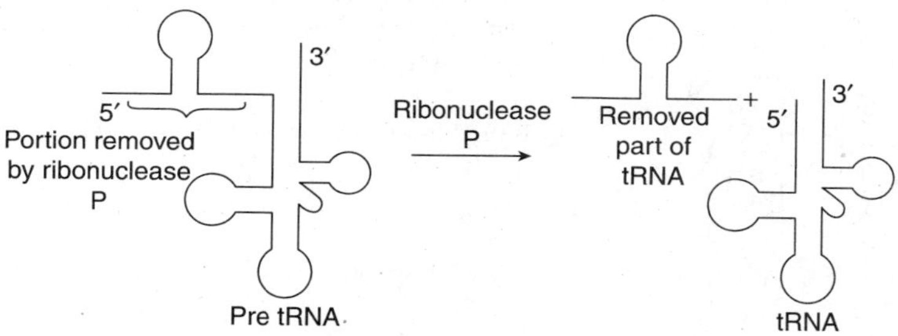

Figure 6.33 *Ribozyme–Ribonuclease P*

Isoenzymes

The multiple forms of an enzyme catalysing the same reaction are isoenzymes or isozymes; they, however, differ in their physical and chemical properties, which include structure, electrophoretic and immunological properties, K_m and V_{max} values, optimum pH, relative susceptibility to inhibitors, and degree of denaturation.

Example: Lactate dehydrogenase LDH inter-converts the lactate and pyruvate (Figure 6.34).

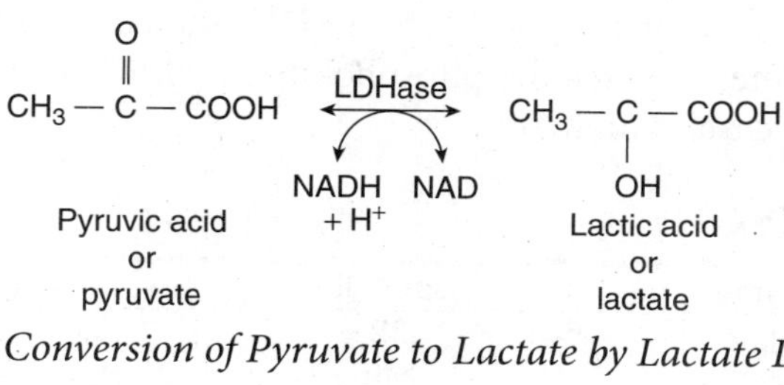

Figure 6.34 *Conversion of Pyruvate to Lactate by Lactate Dehydrogenase*

LDHase have five distinct isoenzymes. It is an oligomeric enzyme made up of four polypeptide subunits. Two types of subunits namely M (muscle) and H (heart) are produced by different genes. The five distinct isoenzymes LDH_1, LDH_2, LDH_3, LDH_4, and LDH_5, can be separated by electrophoresis. LDH_1 – more negative charge and LDH_5 – more positive charge refer Figure 6.35.

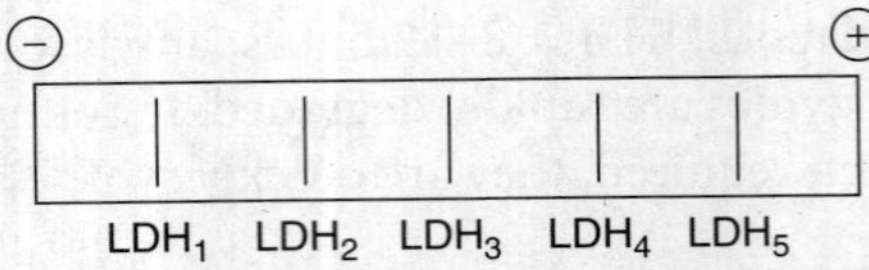

Figure 6.35 *Electrophoritic Movement of LDH*

 – H subunit   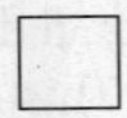 – M subunit

LDH_1 (H_4)  is predominantly found in heart muscle and is inhibited by pyruvate. Hence pyruvate is not converted to acetyl COA which enters TCA.

LDH_2 (H_3M) 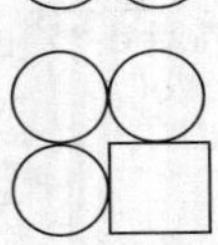 is predominantly found in Heart and RBC.

LDH_3 (H_2M_2) 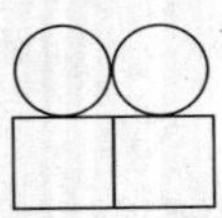 is predominantly found in Brain and Kidney.

LDH_4 (HM_3) 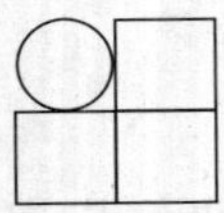 is predominantly found in Liver and skeletal muscle.

LDH_5 (M_4) 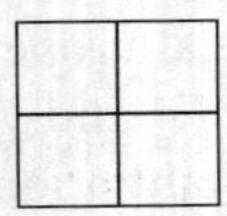 is found in skeletal muscle and Liver. The inhibition of this enzyme by pyruvate is minimal and therefore pyruvate is converted to lactate.

Diagnostic importance of LDH

Isoenzymes of LDH have immense value in diagnosis of heart and liver disorders. LDH_2 increases in myocardial infraction and LDH_5 increases in liver diseases. Other examples of isoenzymes are creatine phosphokinase (CPK) refer table 6.11, which is dimer M and B (muscle and brain)

Table 6.11

Isozyme	subunit	Tissue of origin
CPK_1	BB	Brain
CPK_2	MB	Heart
CPK_3	Mm^{-}	Skeletal muscle

Immobilised enzymes

Enzymes can be used as catalytic agent in industrial and medical applications. Such enzymes are immobilised by binding them to a solid, insoluble matrix, which will not affect the enzyme stability or its catalytic activity. Beaded gels and cyanogens bromide-activated sepharose are commonly used for immobilisation of enzymes. The bound enzymes can be preserved for a long period without loss of activity.

Example: Glucose oxidase and peroxidase are immobilised and coated on a strip of paper and are used in the clinical laboratory for the detection of glucose in urine. The intensity of blue colour depends on the concentration of glucose. Hence the strip method is useful for semi-quantitative estimation of glucose in urine.

$$\text{Glucose} \xrightarrow{\text{Oxidase}} \text{Gluconic acid} + H_2O_2$$

$$H_2O_2 \xrightarrow{\text{Peroxidase}} H_2O$$

$$\underset{\text{(colourless)}}{\text{O-Toluidine}} \xrightarrow{O} \underset{\text{(blue colour)}}{\text{Oxidised toludine}}$$

MODIFIERS OF ENZYME ACTIVITY

The catalytic activity of certain enzymes is reversibly altered by certain inorganic and organic molecules called modifiers or modulators or effectors. Those molecules which increase the enzyme activity are called positive modifiers or activators. And those which decrease the enzyme activity are called negative modifiers or inhibitors. Many metals act both as positive and negative modifiers, whereas certain organic molecules retard enzyme activity, thus, acting as negative modifiers.

ENZYME ACTIVATORS

Some of the enzymes require certain inorganic metallic cations like Mg^{2+}, Mn^{2+}, Zn^{2+}, Ca^{2+}, Co^{2+}, Cu^{2+}, Na^{+}, K^{+}, and so on, for their optimum activity. Rarely anions are also needed for enzyme activity (e.g. chloride ion Cl^{-} for amylase).

Two categories of enzymes requiring metals for their activity are distinguished:

1. Metal-activated enzymes: The metal is not tightly held by the enzyme and can be exchanged easily with other ions.

   Example: ATPase – (Mg^{2+} and Ca^{2+})
   Glucokinase – (Mg^{2+})

2. Metalloenzymes: The enzymes hold the metals rather tightly which are not readily exchanged.

   Example: Cytochrome oxidase (Fe^{2+}, Cu2^{+}) and Pyruvate oxidase (Mn^{2+})

METAL-ACTIVATED ENZYMES AND METALLOENZYMES

Almost ¼ of enzymes require metal atoms for activity. Metal atoms usually exist as cations and often have more than one oxidation state. Examples: Ferrous (Fe^{2+}) and Ferric (Fe^{3+}).

Metal ion can bind a particular number of groups (ligands) by accepting free electron pair to form coordinate bonds in specific orientation.

CATALYSIS

Metal ions can be involved in enzyme catalysis in a variety of ways.

1. They may accept or donate electrons to activate electrophiles or nucleophiles even in neutral solution.
2. They themselves act as neutrophiles.
3. They may mask nucleophiles to prevent unwanted side reaction.
4. They may bring together enzyme and substrate by means of coordinate bonds.
5. They may hold reacting groups in the required three-dimensional orientations.
6. They may catalytically active conformation of the enzyme.

TYPES OF METAL-ACTIVATED ENZYMES

Metalloenzymes: Metal is tightly bound and retained by the enzyme for purification.

Metal-activated enzyme: Binding is less tight, and purified enzymes may have to be activated by addition of metal ions.

Activation by Alkali Metal Cations (Na^+ and K^+)

Alkali metal cations bind only weakly to form complexes with enzymes, but K^+, the most abundant intracellular cation, is known to activate a great many enzymes, particularly those catalysing phosphoryl transfer or elimination reaction. It appears that K^- is largely bound to be a negatively charged group on an inactive form of the enzyme and thus causes a change in conformation to a more active form.

Example: Muscle pyruvate kinase – tetrameric enzyme

$$\underset{\text{PEP}}{\text{Phosphoenol pyruvate}} \xrightarrow[\text{Mn}^{2+}]{\text{Pyruvate kinase}} \text{Pyruvate}$$

This reaction is required for alkali metal cations and Mn^{2+} (or Mg^{2+}), all of which bind in the region of the active site. The carboxyl group of PEP binds to the enzyme bound K^+, whereupon a conformational change takes place which facilitates the progress of the reaction via an E-Mn^{2+}-PEP complex.

Activation by Alkaline Earth Metal Cations (Ca^{2+} and Mg^{2+})

Oxygen atoms are often involved in the bonds of both alkali metal and alkaline earth metal cations, bonds of the latter being relatively stronger. Mg^{2+} is accumulated by cells in exchange for transport of Ca^{2+} in the opposite direction. So Ca^{++} activation enzymes are mostly extracellular ones. Example: Salivary and pancreatic α-amylases.

Intracellular enzymes require Mg^{2+} for activity, and in most cases, this requirement can be replaced in vitro by one for Mn^{2+}. Example: Arginase –M^{2+}.

For Mg^{2+}. Example: Muscle creatine kinase.

$$\text{Creatine} + \text{MgATP} \rightleftharpoons \text{MgADP} + \text{Phosphocreatine} + \text{H}^+$$

The true substrate is MgATP, and the reaction proceeds via the formation of the complex creatine E—ATP—Mg. Cohen and colleagues showed that divalent cation binds to α and β phosphate of nucleotide, but not in the terminal γ-phosphate, which is transferred to creative.

In contrast, the reaction catalysed by pyruvate kinase involves a cyclic metal bridge complex.

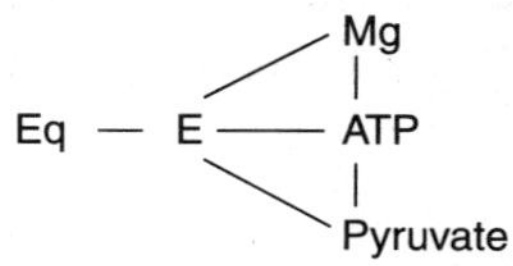

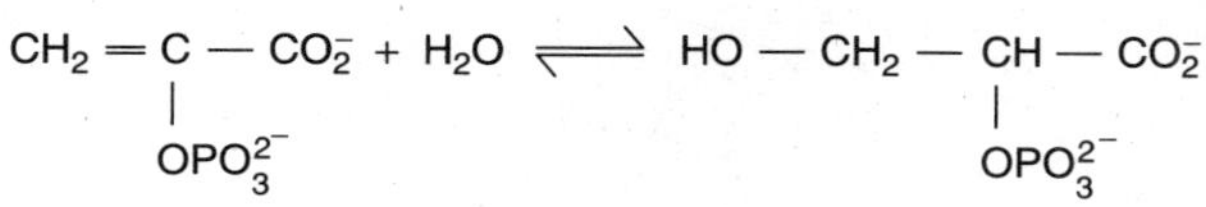

Phosphoenol pyruvate 2-Phosphoglycerate

Enolase is a dimeric enzyme, having two Mg^{2+} ions that are needed to stabilise the active dimer. In addition, two more Mg^{2+} ions are required if each of two sites binds to a substrate.

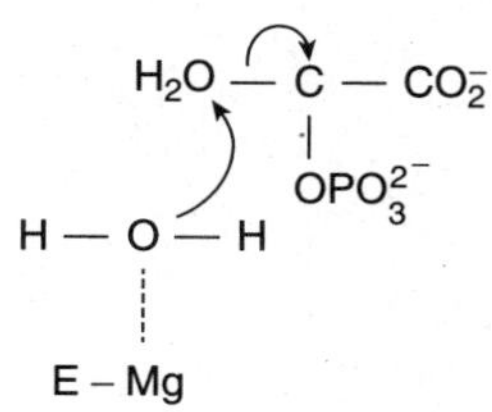

Activation by Transition Metal Cations

Cu, Zn, Mo, Fe, and Co cations. Transition metal ions bind to enzymes much more strongly than the metal ions and form metalloenzymes. Mo and Fe are found in nitric-oxide reductase; Fe is a component of Hb, the O_2^- carrying haemoprotein of erythrocytes. Co is found in vitamin B12. Superoxide dismutase is a copper-metalloenzyme, which catalyses the removal of highly reactive O_2-produced.

$$2O_2^- + 2H^+ \rightarrow H_2O_2 + O_2$$

Bovine erythrocyte superoxide dismutase is a dimeric protein containing two Cu^{2+} ions and two Zn^{2+} ions. The Zn^{2+} ions appear to have a structural rather than a catalytic role, while Cu^{2+} ions are involved in reaction sequence.

$$E-Cu^{2+} + O_2^- \rightarrow E-Cu^+ + O_2$$

$$E-Cu^+ + O_2^- \xrightarrow{+2H^+} E-Cu^{2+} + H_2O_2$$

In contrast to above, Zn^{2+} has a catalytic role in the reaction catalysed by carboxypeptidase A, where the C-terminal amino acid of a polypeptides is removed, provided it has a non-polar side chain.

Carboxypeptidase A – Zn^{2+} C-terminal acid of polypeptide is removed. Bovine pancreas carboxypeptidase A is a monomeric enzyme which contains one atom of Zn. The active site contains the coordinated Zn^{2+} ions bound to histidine-69, glumate-72, histidine-196, and H_2O as well as a groove for the polypeptide substrate and hydrophobic pocket for binding the side chain of the C-terminal amino acid. The terminal carboxyl group of the substrate forms an electrostatic interaction with arginine-145 (Figure 6.36).

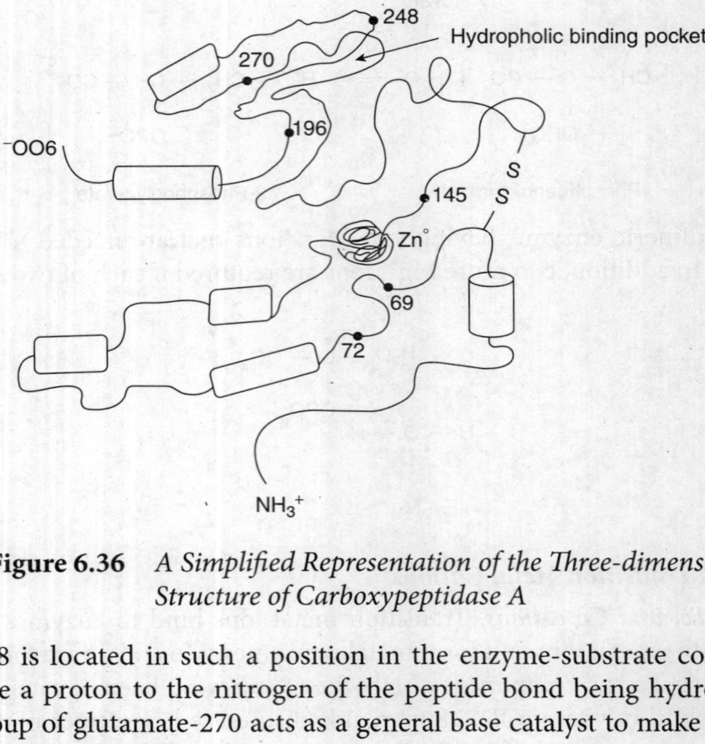

Figure 6.36 *A Simplified Representation of the Three-dimensional Structure of Carboxypeptidase A*

Tyrosine-248 is located in such a position in the enzyme-substrate complex that it could donate a proton to the nitrogen of the peptide bond being hydrolysed, and a carboxyl group of glutamate-270 acts as a general base catalyst to make the attacking water molecule nucleophilic.

The amino acids Arginine in position 145, glutamate in 270, and tyrosine-248 are moving to the substrate and forcing water molecules out of active site, thereby creating a hydrophobic environment for the substrate to bind.

COENZYMES

The protein part of the enzyme on its own is not always adequate to bring about the catalytic activity; many enzymes require certain non-protein small additional factors, collectively referred to as cofactors. *The non-protein, organic, low molecular weight and dialysable substance associated with enzyme function are known as coenzyme.*

Coenzymes are second substrates
Coenzymes are often regarded as the second substrates or cosubstrates, since they have affinity with the enzymes comparable with that of the substrates. Coenzymes undergo alterations during the enzymatic reactions which are later regenerated. This is constant to the substrate which is converted to the product.

Coenzymes participate in various reactions
Coenzymes participate in transfer of atoms or groups like hydrogen, aldehyde, keto, amino, acyl, methyl, carbon dioxide, and so on.

Coenzymes from B-complex vitamins
Most of the coenzymes are the derivatives of water-soluble B-complex vitamins. Refer Table 6.12.

Table 6.12 *Coenzymes of B-complex Vitamins*

Coenzymes	Derived from vitamin	Atom or group transferred	Example
Thiamine pyrophosphate (TTP)	Thiamine	Aldo to keto	Transketolase
Flavin mononucleotide (FMN)	Riboflavin	Hydrogen and electron	α-amino acid oxidase
Flavin dinucleotide (FAD)	Riboflavin	Hydrogen and electron	D-amino acid oxidase
Nicotinamide adenine dinucleotide (NAD)	Niacin	Hydrogen and electron	Lactate dehydrogenase
Nicotinamide adenine dinucleotide phosphate ($NADP^+$)	Niacin	Hydrogen and electron	Pyruvate dehydrogenase
Coenzyme A (CoA)	Pantothenic acid	acyl	Thiokinase
Tetrahydrofolate (FH4)	Folic acid	One carbon formyl methenyl	Formyl transferase
Biotin coenzyme	Biotin	Co_2	Pyruvate carboxylase

Non-vitamin coenzymes
Not all coenzymes are vitamin derivatives; there are some other substances, which have no relation with vitamins but function as coenzymes. They may be considered as no vitamin coenzymes. Examples: ATP, CTP, and UTP. Refer Table 6.13.

Nucleotide coenzymes
Some of the coenzymes possess nitrogenous bases, sugars, and phosphates. Such coenzymes are, therefore, referred to as nucleotide coenzymes.

Examples: NAD^+, $NADP^+$, FMN, CoA, UDPG, and so on.

Coenzymes do not decide enzyme specificity

A particular coenzyme may participate in catalytic reaction along with different enzymes.

Example: NAD^+ acts as coenzymes for lactate dehydrogenase and alcohol dehydrogenase. The specificity of the enzyme is mostly dependent on the apoenzyme and not the coenzyme.

Table 6.13 *Coenzymes Not Related to B-complex Vitamins*

Coenzymes	Abbreviation	Biochemical function
Adenosine triphosphate	ATP	Donate phosphate, adenosine, and AMP moieties
Cytidine diphosphate	CDP	Required in phospholipids synthesis as carrier of choline and ethanolamine
Uridine diphosphate	UDP	Carrier of mononucleotides (glu and gal) required for glycogen synthesis
S-adenosyl methionine	SAM	Donate methyl group in biochemical reaction
Phosphoadenosine phospho sulphate	PAPS	Donate sulphate for the synthesis of mucopolysaccharide

NOMENCLATURE AND CLASSIFICATION

The International Union of Biochemistry (IUB) appointed an enzyme commission in 1961. This committee made a thorough study of the existing enzyme and devised some basic principles for the classification and nomenclature of enzyme.

SUBSTRATE ACTED UPON BY THE ENZYME

The substrate upon which an enzyme acts is called the substrate. Dudaux (1883) named the enzymes by adding the suffix, *ase*. Example: Enzymes acting upon carbohydrates are carbohydrase, proteins – protease, lipids – lipase. These are known as trivial names of enzymes.

TYPES OF REACTION CATALYSED

The enzymes are highly specific as to the reaction they catalyse. Example: Hydrolase (hydrolysis); isomerase (isomerisation); dehydrogenase (dehydrogenation).

SUBSTRATE ACTED UPON AND THE TYPE OF REACTION CATALYSED

The name of some enzymes gives due to both the substrate utilised and the type reaction catalysed. Example: Succinate dehydrogenase (catalyses the dehydrogenation of the substrate succinic acid).

SUBSTANCE THAT IS SYNTHESISED

Few enzymes have been named based on the name of the substance synthesised. Example: Rhodoase.

CHEMICAL COMPOSITION OF AN ENZYME

Based on their chemical composition, the enzymes have been classified into three categories:

1. Enzyme molecule containing protein only. Example: Pepsin, trypsin, urease, and papain.
2. Enzyme molecule containing a protein and a cation. Example: Carbonic anhydrase (Zn^{2+} as cation) and arginase (Zn^{2+} as cation).
3. Enzyme molecule containing a protein and a non-protein and organic compound known as prosthetic group. Example: Iron phorphyrin enzyme – catalase, cytochrome c peroxidase, flavoprotein enzyme – glycine oxidase, and coenzymes – phosphorylase.

SUBSTANCE HYDROLASED AND THE GROUP INVOLVED

Carbohydrate – hydrolysing enzymes. Glycosidase – cellulose, amylase, and sucrase.

Protein – hydrolysing enzymes. Peptide bonds – pepsin and rennin.

Non-peptide bonds—C—N linkages and urease.

Lipid – hydrolysing enzymes – lipases, esterase, and lecithinase.

OVERALL CHEMICAL REACTION TAKEN INTO CONSIDERATION

In 1964, the IUB of enzyme classified into six major groups. Each class on its own represents the general type of reaction brought about by the enzymes of that class.

Oxidoreductase

Enzymes involved in oxidation–reduction reactions.

$$AH_2 + B \rightarrow A + BH_2$$

Transferase

Enzymes that catalyse the transfer of functional group.

$$A - X + B \rightarrow A + B - X$$

Hydrolases

Enzymes that bring about hydrolysis of various compound.

$$A - B + H_2O \rightarrow AH - B - OH$$

Lyases

Enzymes specialised in the addition or removal of water, ammonia, CO_2, and so on.

$$A - B + X - Y \rightarrow AX - BY$$

Addition $\rightarrow$ Elimination

Isomerases

Enzymes involved in all the isomerisation reactions.

Inter-conversion of isomers:

$$A \rightarrow A'$$

Ligases

Enzymes catalysing the synthetic reactions, where two molecules are joined together and ATP is used.

$$A + B \rightarrow A - B$$

OTHLIL

It is the first letter of each class. Each class in turn is subdivided into many classes, which are further divided. A four-digit enzyme commission (E.C) number is assigned to each enzyme representing class (first digit shows which of six main classes in OTHLIL), sub-class (second digit), sub-sub class (third digit), and individual enzyme (fourth digit).

Class (First Digit)

It shows to which of six main classes in OTHLIL.

Subclass (Second Digit)

Oxidoreductase indicates the type of group in the donors which undergo oxidation (e.g. CH—OH, aldehyde, or keto group).

Transferases indicates the nature of group which is transferred.

Hydrolases shows the type of bond hydrolysed.

Lyases shows the type of link which is broken.

Isomerase shows the type of isomerisation involved.

Ligases indicates the type of bond formed.

Sub-sub Class (Third Digit)

Oxidoreductase shows for each type of donor the type of acceptor involved (NAD and NADP).

Transferases shows the type of groups transferred.

Hydrolase shows, more precisely, the type of bond hydrolysed.

Lyase shows the nature of group removed.

Isomerase indicates, more precisely, the nature of transformation.

Ligases shows the nature of the substance formed.

Fourth digit – enzyme of its sub-sub class.

Examples:

1. *Oxidoreductase*

Alcohol dehydrogenase (alcohol NAD^+ oxidoreductase E.C.1.1.1.1)

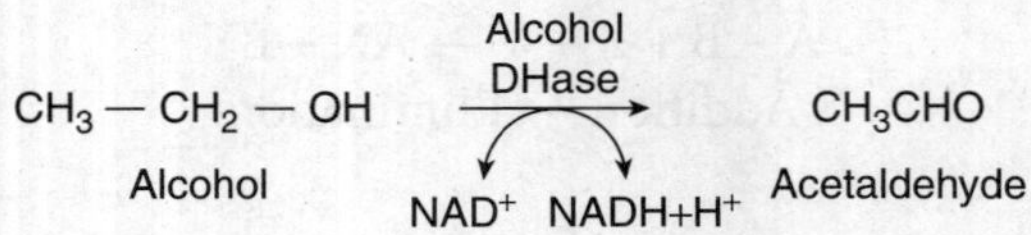

E.C.1.1.1.1
1 – Oxidoreductase
1 – Acting on CH—group of donors, which undergoes oxidation
1 – with NAD^+ or $NADP^+$ as acceptor
1 – Alcohol dehydrogenase

2. *Transferases*

Hexokinase (ATP: D-hexose-6-phosphotranferase E.C.2.7.1.1)

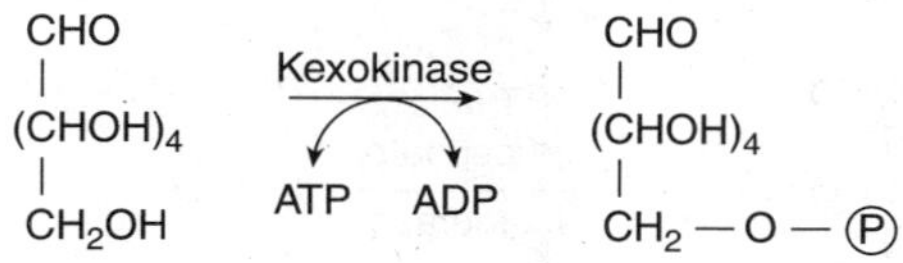

E.C.2.7.1.1
2 – Trasferase
7 – Transferring phosphate containing group
1 – Phosphotransferase with a alcohol group as acceptor
1 – Hexokinase

3. *Hydrolases*

Lipase (triacylglycerol acylhydrolase E.C 3.1.1.3)

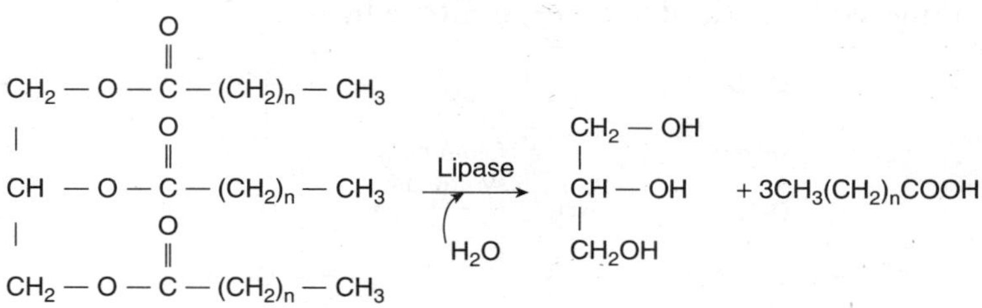

E.C. 3.1.1.3
3 – Hydrolases
1 – Acting an ester bonds
1 – Carboxylic ester hydrolases
3 – lipases

4. *Lyases*

Aldolase (ketose1-phosphate aldehyde lyase E.C 4.1.2.7)

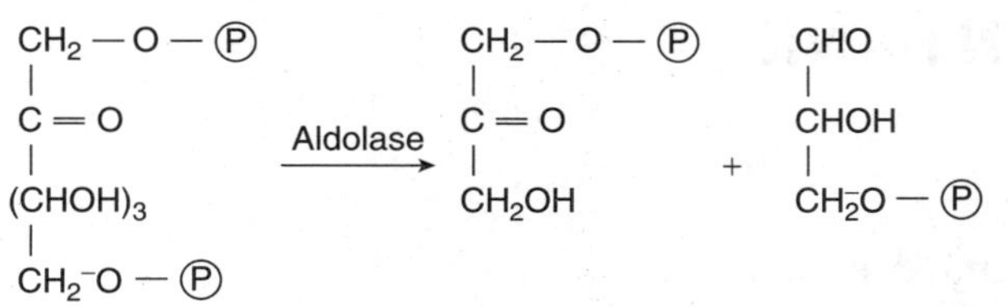

Fructose 1, 6 bisphosphate — Dihydroxy acetone phosphate — Glyceraldehyde 3 — Ⓟ

E.C. 4.1.2.7
4 – Lyases
1 – Carbon-carbon lyases
2 – Aldehyde lyases
7 – Aldolase

5. *Isomerases*
Triose phosphate isomerase (D-glyceraldehyde 3-phosphate ketoisomerase E.C 5.3.1.1)

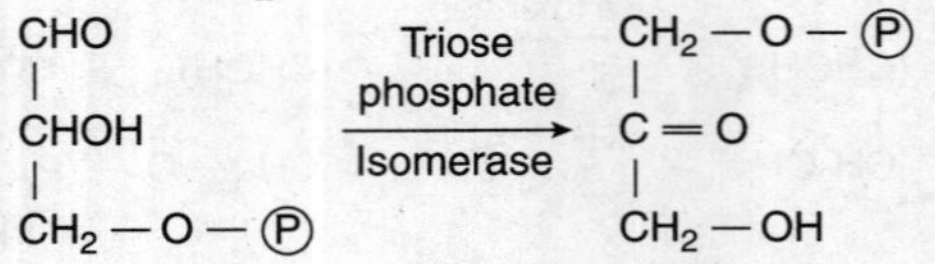

E.C.5.3.1.1
5 – Isomerases
3 – Intramolecular oxidoreductase
1 – Inter-converting aldoses and ketoses
1 – Triose phosphate isomerase

6. *Ligases*
Glutamine synthase (α-glutamate ammonia ligase E.C 6.3.1.2)

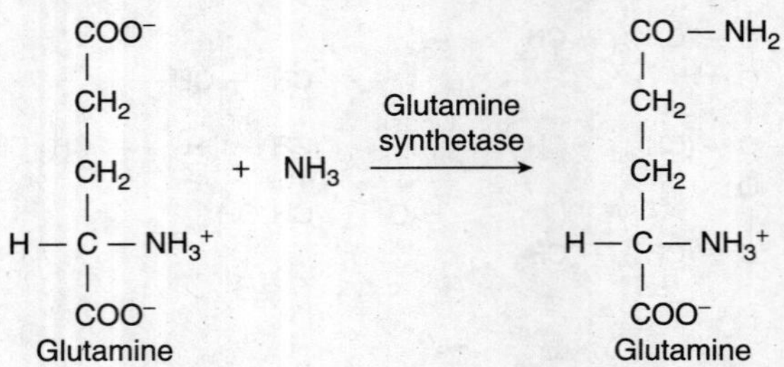

E.C.6.3.1.2
6 – Ligases
3 – Forming nitrogen carbon bonds
1 – Acid ammonia ligases
2 – Glutamine synthase

QUESTION BANK

SECTION A

1. Define turnover number.
2. What is an active site?

3. Explain lock and key theory.
4. What is competitive inhibition?
5. Give the importance of LB plot.
6. Give the role of biotin.
7. What are coenzymes?
8. What are abzymes?
9. Mention any two applications of immobilised enzymes.
10. Name the major classes of enzymes. Give an example for each class.
11. Define the terms activity and specific activity of enzymes.
12. Describe the effect of pH on enzyme activity.
13. Define active site.
14. Define katal.
15. Give the differences between catalytic site and regulatory site of an allosteric enzyme with example.
16. Define holoenzymes.
17. What is the effect of temperature on enzyme activity?
18. Define absolute specificity and relative specificity.
19. Define allosteric enzymes with an example.
20. Define transferase with an example.
21. What is oxidoreductase?
22. Give an example for hydrolase.
23. Define substrate specificity.
24. Define rate of reactions.
25. Explain order of reactions.
26. Define zero order reactions with example.
27. What are optimum pH and optimum temperature?
28. Define non-competitive inhibitors with an example.
29. Define uncompetitive inhibitors with an example.
30. Define competitive inhibitors with an example.
31. Define K_m.
32. What is substrate specificity?
33. Give the equation of LB plot.
34. What is induced fit theory?

35. Define activation energy.
36. What is lock and key hypothesis?
37. Give the structure and functions of NAD.
38. Write the structure and functions of FAD.
39. What are the structure and functions of COA?
40. What are the structure and functions of TPP?
41. What are the structure and functions of PLP?
42. What are the structure and functions of THF?
43. What is the significance of K_m?
44. Define multienzyme complex.

SECTION B

1. How is the active site of an enzyme determined?
2. Briefly explain the mode of action of chymotrypsin.
3. Write a note on metallo enzymes.
4. Explain multienzymes complex with an example.
5. What is uncompetitive inhibition?
6. Write a brief note on enzyme specificity?
7. Write short notes on allosteric enzyme with an example.
8. State the Michaelis–Menten equation. How it is modified to obtain a linear graph (LB plot)?
9. Explain non-competitive inhibition and show how it differs from uncompetitive inhibition.
10. Explain rate and order of reactions.
11. Describe the role of activation energy in enzyme action.
12. Write notes on lock and key hypothesis.
13. Explain induced fit theory for the binding of a substrate to its enzyme action.
14. Discuss lock and key hypothesis and Edie–Hofstee equation.
15. Classify enzymes with examples.

SECTION C

1. Detail the classification of enzymes
2. Describe in detail IUB nomenclature of enzymes with an example.

3. Outline the characteristic features of active site.
4. Derive the Michaelis–Menten equation.
5. Describe the factors affecting enzyme activity.
6. Discuss the competitive, non-competitive, and uncompetitive inhibitions of enzyme action.
7. Describe the mechanism of enzyme action.
8. Express the equation connecting [S] and the velocity of an enzyme-catalysed reaction.

CHAPTER 7

HIGH-ENERGY COMPOUNDS

Concept Outline

- ❑ Bioenergetics
- ❑ Phosphoryl Group Transfer Through ATP

BIOENERGETICS

Bioenergetics or biochemical thermodynamics deals with the study of energy changes (transfer and utilisation) in biochemical reactions. The reactions are broadly classified as exergonic (energy releasing) and endergonic (energy consuming).

FREE ENERGY ΔG

The energy actually available to do work during a reaction at constant temperature and pressure is called free energy, ΔG. When a reaction proceeds with the release of free energy, the free energy change, ΔG, is negative, and the reaction is said to be *exergonic*. In *endergonic* reaction, the system gains free energy, and ΔG is positive.

ENTHALPY ΔH

Enthalpy refers to the heat content of the reducing system. It indicates the number and kinds of chemical bonds in the reactants and products. A chemical reaction that releases heat is known as an exothermic reaction. The heat content of the product is less than that of reactants, and ΔH is negative. When a chemical reaction takes up heat from surroundings, it is said to be *endothermic* and has positive ΔH value.

ENTROPY ΔS

It represents a change in the randomness or disorder of the reactants and products. Entropy attains maximum as the reaction approaches equilibrium. The relation connecting the changes of free energy, enthalpy and entropy is expressed as

$$\Delta G = \Delta H - T\,\Delta S$$

T = Absolute temperature in Kelvin (K = 273 + °C)

STANDARD FREE ENERGY $\Delta G°$

If a reacting system is not in equilibrium, the force used to attain the equilibrium is defined as free energy change, ΔG. *Standard free energy* ΔG is defined as the force required to attain equilibrium at standard condition (298K = 25°C) and when the reactants and products are present at 1M concentration at 1 atm.

At constant temperature and pressure, ΔG is dependent on actual concentration of the reactants and products.

Example

Consider the reaction

$$A \rightarrow B$$

A mathematical relation for the free energy change can be derived.

$$\Delta G = \Delta G° + RT \ln \frac{[B]}{[A]}$$

Where $\Delta G°$ = Standard free energy change
R = Gas constant
T = Absolute temperature
ln = Natural logarithm
B = Concentration of product
A = Concentration of reactant

$\Delta G°$ is related to equilibrium constant (K_{eq})

When a reaction $A \leftrightarrow B$ is at equilibrium (eq), the free energy change is zero. The above equation may be written as,

$$\Delta G = 0 = \Delta G^{\circ} + RT \text{ In} \frac{[B]_{eq}}{[A]_{eq}}$$

Hence $\Delta G^{\circ} = -RT \ln K_{eq}$.
When ΔG° is negative (the products contain less free energy than the reactants), the reaction will proceed to form the products under standard conditions.

When ΔG° is positive (the products of the reaction contain more free energy than the reactants), the reaction will tend to go in the reverse direction.

Relationship among K'_{eq} and ΔG° and direction of the chemical reaction:

When K'_{eq} is	ΔG° is	Starting with 1*M* concentration of substrate
> 1.0	negative	proceeds forward
1	zero	remains at equilibrium
< 1.0	positive	proceeds backward

PHOSPHORYL GROUP TRANSFER THROUGH ATP

Adenosine Triphosphate (ATP) is the universal currency of free energy in biological systems; living objects require a continuous supply of free energy mainly for the following four purposes:

1. To synthesize macromolecules from simpler and smaller precursors.
2. To transport of molecules and ions across membranes against gradients.
3. To perform mechanical work, as in muscle contraction.
4. To ensure fidelity of information transfer.

The free energy in these processes is derived from the environment. Autotrophs obtain this energy by trapping light energy from the sun. On the other hand, heterotrophs obtain it by oxidation of foodstuffs in the form of organic material present in other organisms, living or dead, before it is used for biosynthesis, transport, motion, and fidelity.

ATP plays a major role in the transfer of free energy from exergonic (energy yielding) to endergonic (energy requiring) reaction in almost all the cells. Karl Lohmann discovered ATP in the extracts of skeletal muscles. In 1941, Fritz Lipmann postulated that ATP is the primary and universal carrier of chemical energy in cells. He also postulated the

ATP cycle. ATP and its successive hydrolysis products ADP and AMP are nucleotides consisting of adenine, ribose, and 3, 2, 1 phosphate groups, respectively. ATP occurs not only in cell cytosol but also in mitochondria and nucleus as shown in Figure 7.1.

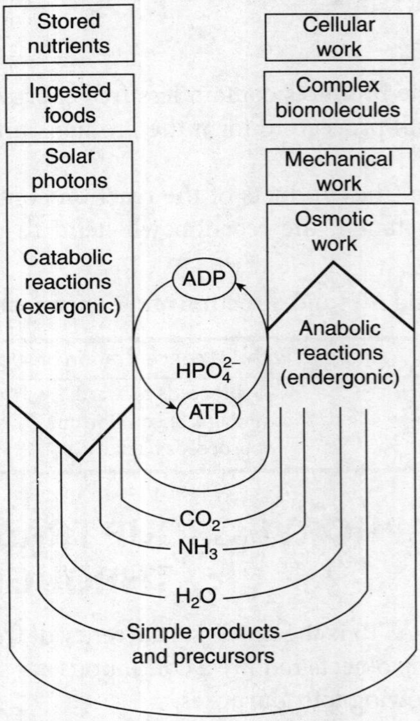

Figure 7.1 *The ATP Cycle*

At pH 7.0, the phosphate groups of ATP, ADP, and AMP are almost fully ionised so that they occur as multiple charged anions ATP^{4-}, ADP^{3-}, and ADP^{2-}. But as the cell fluid contains high concentrations of Mg^{2+}, both ATP and ADP exist largely as Mg ATP^{2-} and Mg ADP^{-} complexes.

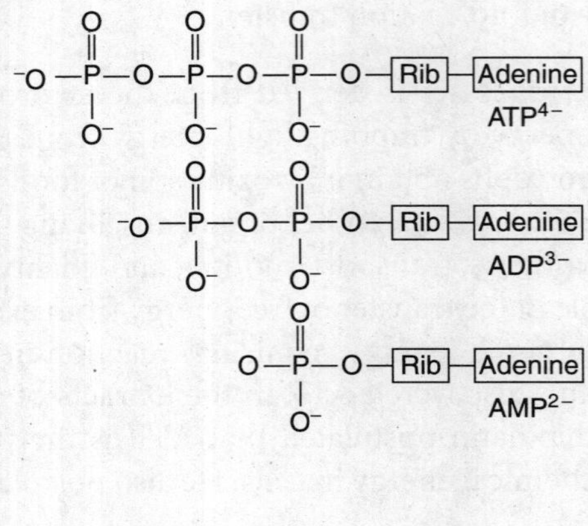

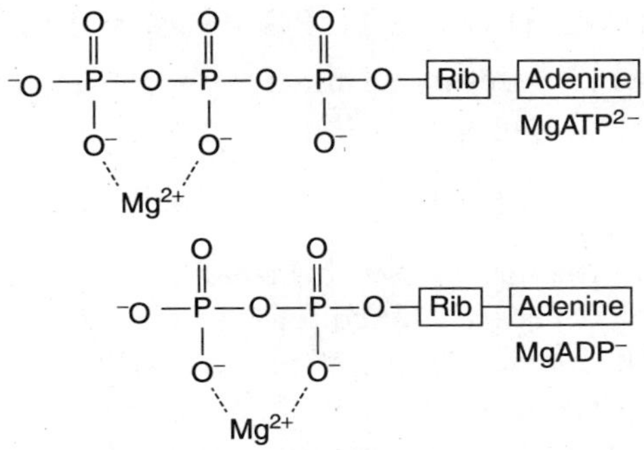

ATP plays a major role in the transfer of free energy in the systems rather than in the storage of energy. In a cell, an ATP molecule is consumed within a minute of its production. The turnover of ATP is usually high. For example, a human body consumes 40 kg/day. During strenuous exercise, the rate of consumption is 0.5 kg/minute. The free energy change for ATP is large and negative.

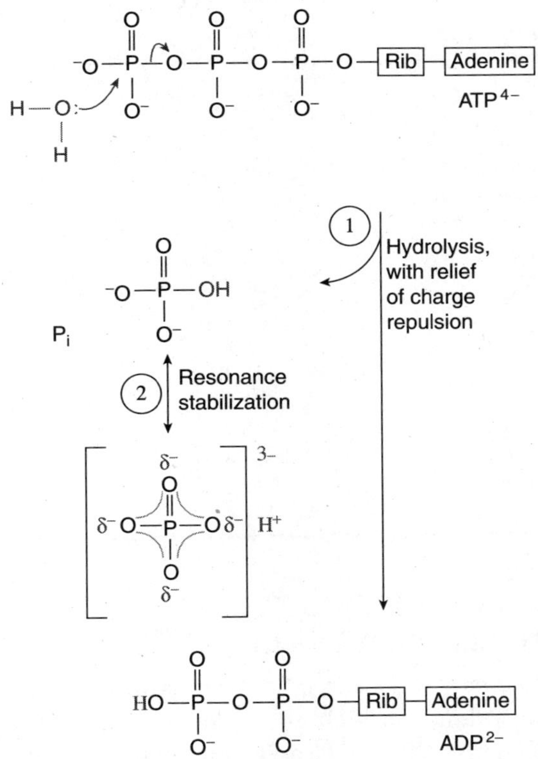

$$ATP^{4-} + H_2O \longrightarrow ADP^{3-} + P_i^{2-} + H^-$$
$$\Delta G'^{\circ} = -30.5 \text{ k/mol}$$

HYDROLYSIS OF ATP AND OTHER ORGANOPHOSPHATES

When ATP is hydrolysed, it loses its terminal γ-phosphate group to form ADP and orthophosphate or inorganic phosphate (P_i).

$$ATP + H_2O \longrightarrow ADP + P_i$$

The standard free energy change $\Delta G°$ for this reaction is –7.3 Kcal/mol. Standard free energy changes have also been determined for the hydrolysis of other phosphorylated compounds or organophosphates. Under standard conditions, some phosphates yield more and some yield less free energy than ATP on hydrolysis. This intermediate position makes ATP to function as a high-energy compound. Two classes of organophosphates have been noticed in metabolism—enolphosphates (e.g., phosphoenolpyruvate) and phosphoguanidines (e.g., creatine phosphate, arginine phosphate).

Standard free energy of hydrolysis of some phosphorylated compounds is given in Table 7.1.

Table 7.1 *Standard Free Energy of Hydrolysis of Some Phosphorylated Compounds*

	DG′°	
	(kJ/mol)	(kcal/mol)
Phosphoenolpyruvate	–61.9	–14.8
1,3-bisphosphoglycerate (→ 3-phosphoglycerate + Pi)	–49.3	–11.8
Phosphocreatine	–43.0	–10.3
ADP (→ AMP + Pi)	–32.8	–7.8
ATP (→ ADP + Pi)	30.8	–7.3
ATP (→ AMP + Pi)	45.6	–10.9
AMP (→ adenosine+ Pi)	–14.2	–3.4
PPi (→ 2Pi)	–19.2	–4.0
Glucose 1-phosphate	–20.9	–5.0
Fructose 6-phosphate	–15.9	–3.8
Glucose 6-phosphate	–13.8	–3.3
Glycerol 1-phosphate	–9.2	–2.2
Acetyle-CoA	–31.4	–7.5

ROLE OF HIGH-ENERGY PHOSPHATES AS UNIVERSAL CURRENCY OF THE CELL

ATP has a position midway down the list of standard free energies of organophosphates. So ATP acts as a donor of high-energy phosphates to the compounds which have free energy less than the ATP. ADP can accept high-energy phosphate in the presence of enzyme to form ATP from the compounds whose free energy is greater than ATP. Hence, ATP/ADP cycle connects these two types of processes, that is, those which generate ~P and those which utilize ~P.

Process Which Generates ~P

The processes that feed ~P into ATP/ADP cycle fall mainly within four groups:

1. *Oxidative phosphorylation*: The greatest quantitative source of ~P is from reactions catalysed by the enzyme ATP synthase, which, in fact, reverses the hydrolysis of ATP. The free energy to drive this process is obtained from the respiratory chain within mitochondria, and it is called oxidative phosphorylation.
2. *Catabolism of glucose to lactic acid*: Energy is captured during catabolism of glucose to lactic acid in a sequence of reactions commonly called glycolysis. In this pathway, there is a net formation of two high-energy phosphate groups, which produce two ATP molecules from two ADP molecules.
3. *Oxidation of pyruvic acid*: The energy capture occurs at the succinyl thiokinase step of the Krebs cycle, wherein succinyl CoA, one of the intermediates of the cycle, transforms to succinate. Here again, one ATP molecule is formed from one ADP molecule.
4. *The muscle compounds*: Another group of compounds acting as storage forms of high-energy phosphate within muscles include creatine, occurring in muscle.

Process Which Utilises ~P

When ATP donates a phosphate group, it is converted to a group with low energy, forming phosphoric esters of alcohols. This reaction is catalysed by kinases. Among them is hexokinase, which catalyses the transfer of phosphate group from ATP to D-glucose. This reaction is proceeded by an orthophosphate cleavage of ATP, in which, ATP loses a single orthophosphate group as shown in Figure 7.2.

$$\text{D-glucose} + \text{Adenosine} \sim P \sim P \sim P \longrightarrow \text{D-glucose 6-phosphate} + \text{Adenosine} \sim P \sim P$$

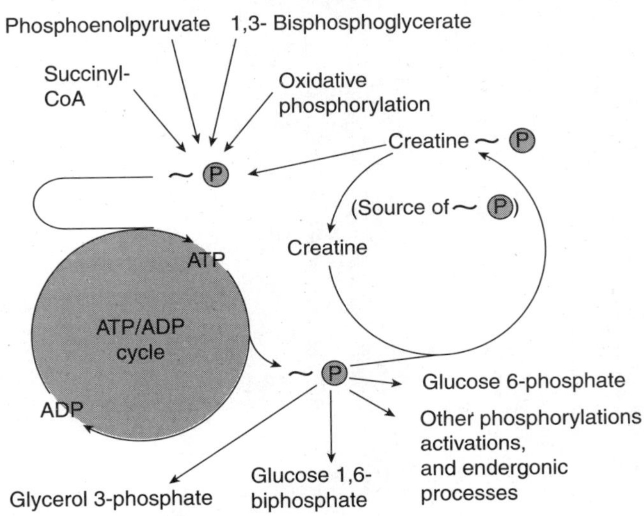

Figure 7.2 *Role of ATP/ADP Cycle in the Transfer of High-energy Phosphate*

An important feature is that all super high-energy phosphate compounds must pass their phosphate groups via ATP to various acceptor molecules to form their low-energy phosphate derivatives.

QUESTION BANK

SECTION A

1. What is exergonic reaction? Explain with specific example.
2. How much energy is released during the hydrolysis of ATP?
3. Define standard free energy.
4. What is known as enthalpy?
5. Explain entropy.

SECTION B

1. Why is ATP accepted as universal currency of energy?
2. Explain the free energy produced during the hydrolysis of ATP.

SECTION C

1. Describe the ATP/ADP cycle.

CHAPTER

8

INTRODUCTION TO METABOLISM

Concept Outline

- Introduction
- Outline of Catabolism and Anabolism
- Lipid Metabolism
- Amino Acid Metabolism
- Metabolic Pathways – Levels of Organisation
- Glycolysis
- Tricarboxylic Acid Cycle
- Glycogenesis
- Electron Transport Chain
- ATP Synthase
- Shuttle Systems: Mitochondrial Oxidation of Cytosolic NADH
- Hexose Monophosphate Pathway
- Gluconeogenesis

INTRODUCTION

The term metabolism is defined as the entire of spectrum of chemical reactions that occur in the living system. A metabolic pathway consists of a series of enzymatic reactions to produce desired products. The term *metabolite* is applied to a substrate or an intermediate or a product in metabolic reactions. Metabolism is broadly divide into two categories: catabolism and anabolism (Figure 8.1).

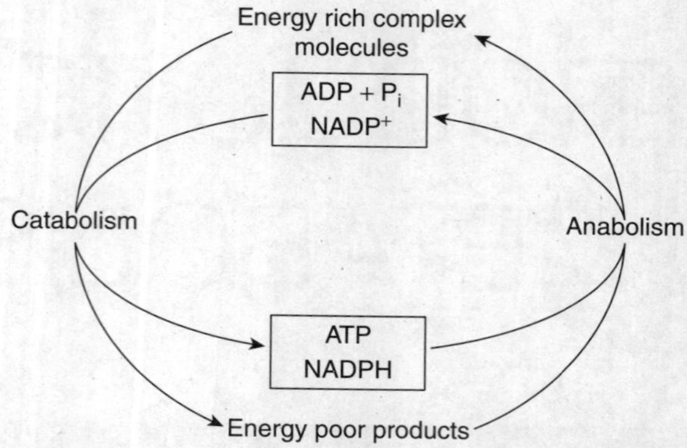

Figure 8.1 *Outline of Catabolism and Anabolism*

OUTLINE OF CATABOLISM AND ANABOLISM

CATABOLISM

Catabolism is defined as the degradative process concerned with the breakdown of complex molecules to simpler ones, with a release of energy. The very purpose of catabolism is to trap the energy of the biomolecules in the form of ATP and generate the substances (precursors) required for the synthesis of complex molecules. Catabolism occurs in three stages as shown in Figure 8.2.

Stage I – Conversion of complex into their building blocks

Polysaccharides are broken down to monosaccharide, lipids to free fatty acids and glycerol, and proteins to amino acids.

Stage II – Formation of simple intermediates

The building blocks produced in stage I are degraded to simple intermediates such as pyruvate and acetyl-CoA. These intermediates are not readily identifiable as carbohydrates, lipids, or proteins. A small quantity of energy (as ATP) is captured in stage II.

Stage III – final oxidation of acetyl-CoA

Acetyl-CoA is completely oxidised to CO_2, liberating NADH and $FADH_2$ that finally get oxidised to release large quantity of energy (as ATP). Krebs's cycle of citric acid cycle is the common metabolic pathway involved in the final oxidation of all energy-rich molecules. This pathway accepts the carbon compounds (pyruvate, succinate, and so on) derived from carbohydrates, lipids, or proteins.

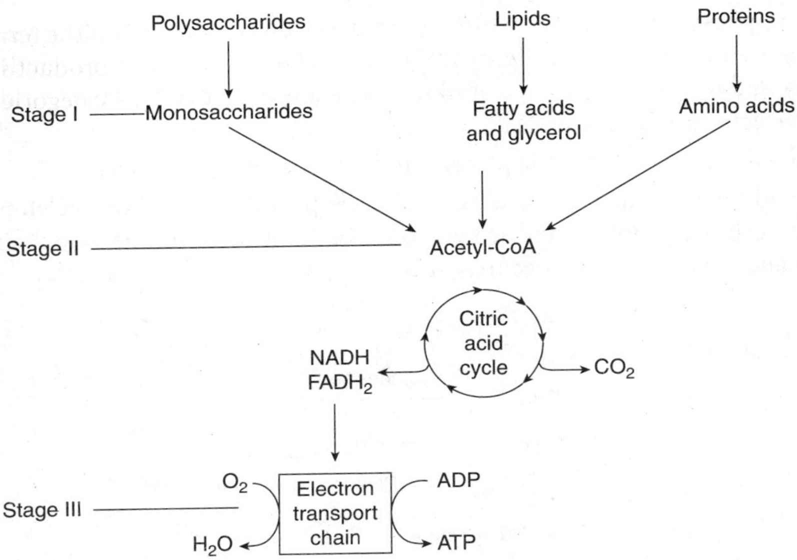

Figure 8.2 *Outline Pathways for the Catabolism of Dietary Carbohydrates, Protein and Fat*

ANABOLISM

It is the biosynthetic reaction involving the formation of complex molecules from simpler precursors. For the synthesis of a large variety of complex molecules, the starting materials are relatively few. These include pyruvate, acetyl-CoA, and the intermediates of citric acid cycle. Besides the availability of precursors, the anabolic reactions depend on the supply of energy (as ATP or GTP) and reducing equivalents as (NADPH + H^+).

CARBOHYDRATE METABOLISM IS CONCERNED WITH THE FATE OF GLUCOSE

Glucose is metabolised to pyruvate and lactate in all mammalian cells by the pathway of glycolysis. Phosphorylation is necessary for glucose to enter this pathway. Glycolysis can occur in the absence of oxygen (anaerobic) when the end product is lactate only. Tissues that can utilise oxygen (aerobic) are able to metabolise pyruvate to acetyl-CoA, which can enter the citric acid cycle for complete oxidation to CO_2 and H_2O, with the liberation of much free energy as ATP in the process of oxidative phosphorylation. Thus, glucose is a major fuel of many tissues (Figure 8.3). But it also takes part in other processes as follows:

1. Conversion to its storage polymer, glycogen, particularly in skeletal muscle and liver.

2. The pentose phosphate pathway, which arises from intermediates of glycolysis. It is a source of reducing equivalents (2H) for biosynthesis. Example includes fatty acids, and it is also the source of ribose, which is important for nucleotide and nucleic acid formation.
3. Triose phosphate gives rise to the glycerol moiety of acylglycerol (fat).
4. Pyruvate and intermediates of the citric acid cycle provide the carbon skeletons for the synthesis of amino acid and acetyl-CoA is building block for the long chain fatty acids and cholesterol, the precursor of all steroids synthesised in the body.

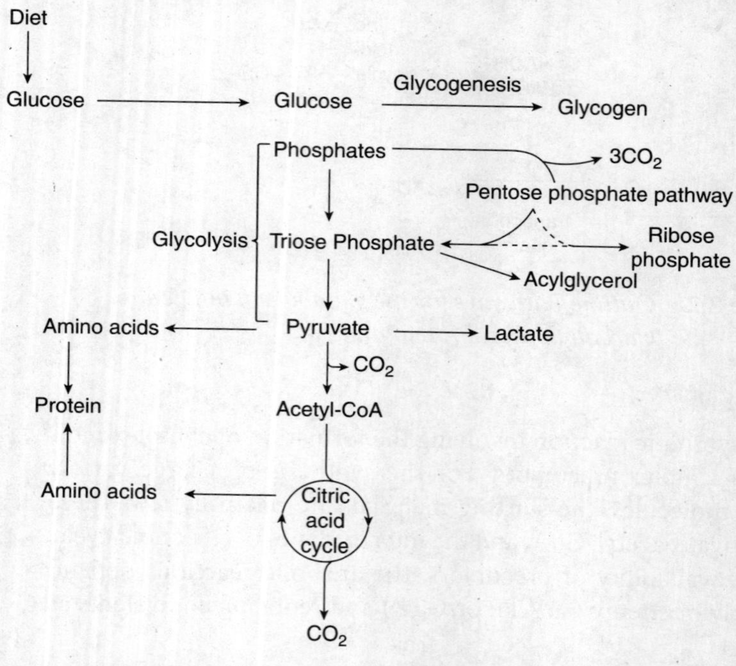

Figure 8.3 *Overview of Carbohydrate Metabolism*

LIPID METABOLISM

The source for the fatty acid synthesis is acetyl-CoA. Fatty acid is produced either by de novo synthesis from acetyl-CoA derived from carbohydrate or from dietary lipid. In the tissues, fatty acids are oxidised to acetyl-CoA (β-oxidation) or esterified to acylglycerol, whereas in triacylglycerol (fat), they constitute the body's main caloric reserve.

Acetyl-CoA formed by β-oxidation has several important facts:

1. As in the case of acetyl-CoA derived from carbohydrate, it is oxidised completely to CO_2 and H_2O via the citric acid cycle. Fatty acids yield considerable energy both in β-oxidation and in the citric acid cycle and are, therefore, very effective of tissue fuels.

2. It is a source of the carbon atoms in cholesterol and other steroids.
3. In the liver, it forms acetoacetate, the parent ketone body. Ketone bodies are alternative water-soluble tissue fuels, which become important sources of energy under certain conditions (Example: starvation) (Refer Figure 8.4).

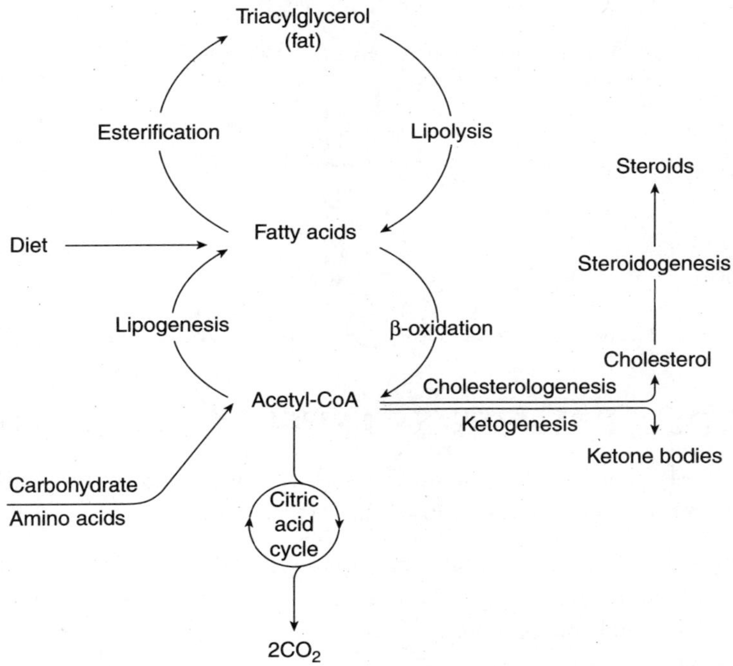

Figure 8.4 *Overview of Fatty Acid Metabolism*

AMINO ACID METABOLISM

Amino acids are necessary for protein synthesis. Some must be supplied specifically in the diet (the essential amino acids) since the tissues are unable to synthesis them. The reminder or non-essential amino acids are also supplied in the diet, but they can be formed from intermediates by transamination using the amino nitrogen from other surplus amino acids. After deamination, excess amino nitrogen is removed as urea, and the carbon skeletons that remain after transamination (Figure 8.5).

1. are oxidised to CO_2 via citric acid cycle,
2. form glucose (gluconeogenesis) and
3. form ketone bodies.

In addition to their requirement for protein synthesis, the amino acids are also the precursors of many other important compound. *Examples:* Purine, pyrimidine, and hormones such as epinephrine and thyroxine.

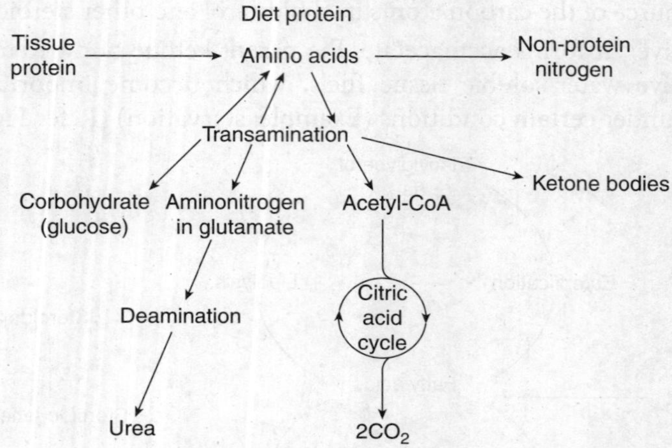

Figure 8.5 *Overview of Amino Acid Metabolism*

METABOLIC PATHWAYS – LEVELS OF ORGANISATION

The location and integration of metabolic pathways are released by studies at lower levels of organisation, namely the following:

1. At the subcellular level, each cell organelle (mitochondria) or compartment (cytosol) carries out specific biochemical roles that from part of a subcellular pattern of metabolic pathway.
2. At the tissue and organ level, the nature of the substrate entering and metabolites are leaving tissues and organs.

At the tissue and organ level, the blood circulation integrates metabolism. Amino acids resulting from digestion of dietary protein and glucose resulting from the digestion of carbohydrate share a common route of absorption via the hepatic portal vein. This ensures that both of these metabolites and other water-soluble products of digestion are initially directed to liver.

Liver has the primary metabolic function of regulating the blood concentration of most metabolites, particularly glucose and amino acids. In the case of glucose, this is achieved by taking up excess glucose and converting it to glycogen (glycogenesis) or fat (lipogenesis). Between meals, it can draw upon its glycogen.

GLYCOLYSIS

INTRODUCTION

Glycolysis was derived from the Greek words *glykys*, which means sweet, and *lysis*, which means splitting. In the process of glycolysis a molecule is catabolised in a series of reactions to yield two molecules of three-carbon compound called pyruvate. In the glycolysis pathway, a series of reaction takes place during which some of the energy is released from glucose and conserved in the form of ATP and NADH. Glycolysis is the pathway that was first to be elucidated and is probably the best understood (See Figure 8.6).

BIOMEDICAL IMPORTANCE

1. Glycolysis is an almost universal central pathway of glucose catabolism, the pathway with the largest flux of carbon in most cells.
2. The glycolytic breakdown of glucose is the sole source of metabolic energy in some mammalian tissues and cell types (erythrocytes, renal medulla, brain, and sperm, for example).
3. Many anaerobic microorganisms are entirely dependent on glycolysis.

OUTLINE OF GLYCOLYSIS

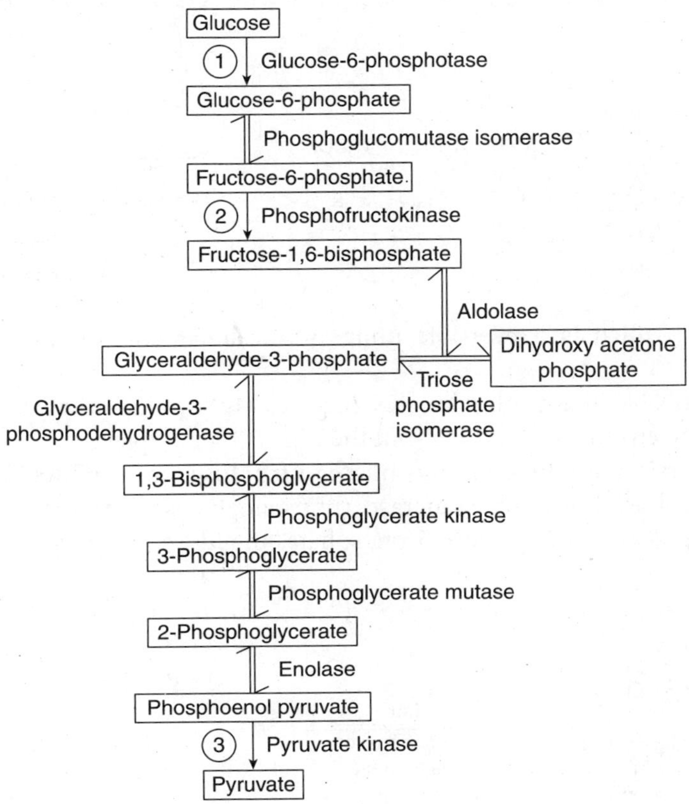

Figure 8.6 *Outline of Glycolysis*

Location

Glycolysis occurs in all cells, and in some cells, it is the sole source of energy like brain cells, erythrocytes.

Organelle: It occurs in cytoplasm of the cells.

Steps

Glycolysis is also called *Embden Meyerhof pathway.* In glycolysis, the breakdown of the six-carbon glucose into two molecules of the three-carbon pyruvate occurs in ten steps,

Glycolysis is divided in to two phases:

1. Preparatory phase
2. Pay-off phase

1. *Preparatory phase*

Step 1: Glucose to glucose 6-phosphate (Phosphorylation of glucose)

In the first step of glycolysis, glucose is activated for subsequent reactions by its phophorylation at C-6 to yield glucose 6-phosphate with ATP as the phosphoryl donor.

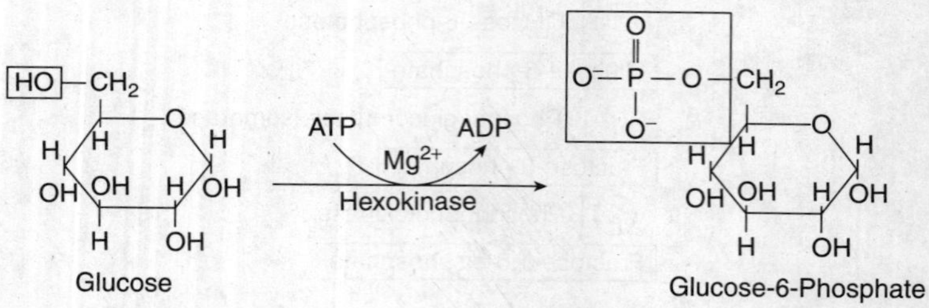

This reaction, which is irreversible under intracellular conditions, is catalysed by hexokinase.

Hexokinase, like many other kinases, requires Mg^{2+} for its activity because the true substrate of the enzyme is not ATP^{4-} but the Mg ATP^{2-} complex.

Step 2: Conversion of glucose 6-phosphate to fructose 6-phosphate

The enzyme phosphohexose isomerase (phosphoglucose isomerase) catalyses the reversible isomerisation of glucose 6-phosphate, an aldose, to fructose 6-phosphate, a ketose.

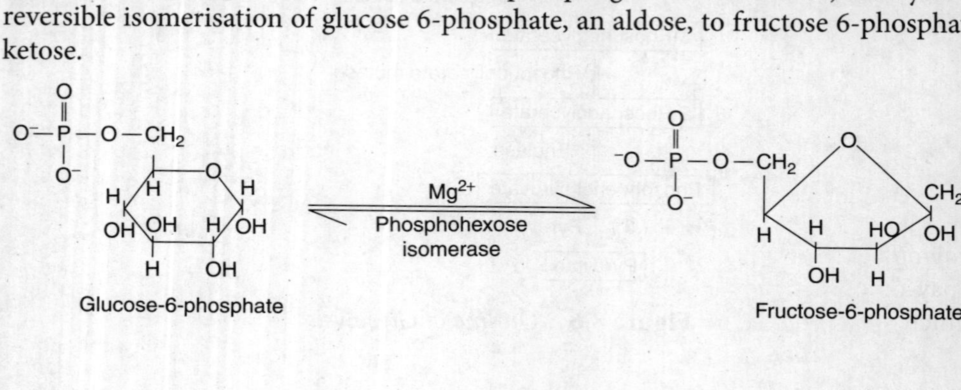

Step 3: Phosphorylation of fructose 6-phosphate to fructose 1,6-bisphosphate
In the second of the two priming reactions of glycolysis, phosphofructokinase-1 (PFK-1) catalyses the transfer of a phosphoryl group from ATP to fructose 6-phosphate to yield fructose 1,6-bisphosphate.

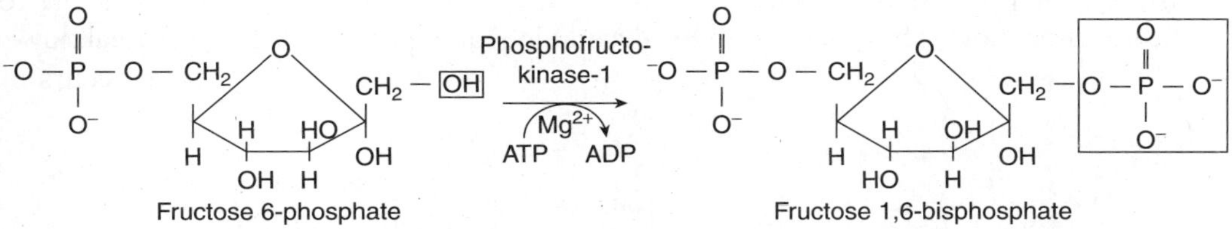

The PFK-1 reaction is essentially irreversible under cellular conditions. It s the major point of regulation in glycolysis.

Step 4: Cleavage of fructose 1, 6-bisphosphate
The enzyme fructose 1,6-bisphosphate aldolase, often called simply aldolase, catalyses a reversible aldol condensation. Fructose 1,6-bisphosphate is cleaved to yield two different triose phosphates, glyceraldehyde 3-phosphate, an aldose, and dihydroxyacetone phosphate, a ketose.

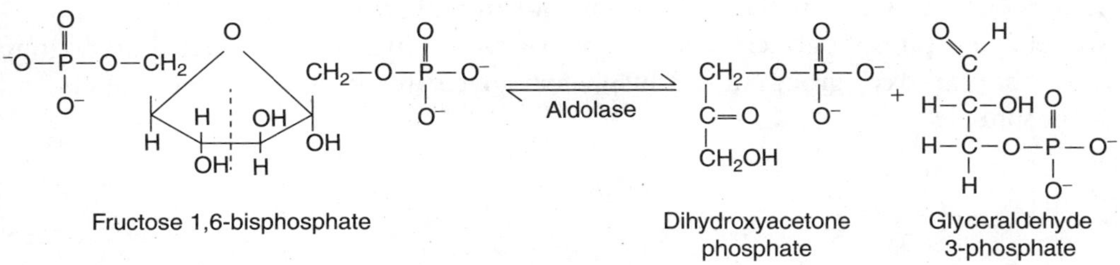

Step 5: Inter-conversion of the triose phosphates
Only one of the two triose phosphates formed by aldolase, glyceraldehydes 3-phosphate, can be directly degraded in the subsequent steps of glycolysis.

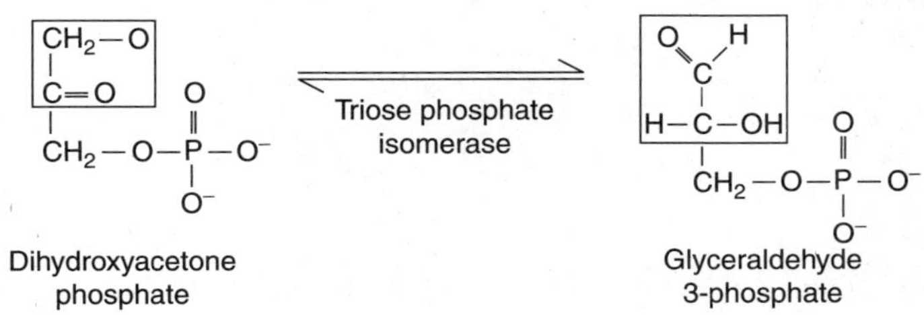

2. *Pay-off phase*
The pay-off phase of glycolysis includes the energy-conserving phosphorylation steps in which some of the free energy of the glucose molecule is conserved in the form

of ATP. Remember that one molecule of glucose yields two molecules of glyceraldehyde 3-phosphate; both halves of the glucose molecule follow the same pathway in the second phase of glycolysis.

Step 6: Oxidation of glyceraldehyde 3-phosphate to 1,3-bisphosphoglycerate

The first step in the pay-off phase is the oxidation of glyceraldehyde 3-phosphate to 1,3-bisphosphoglycerate, catalysed by glyceraldehyde 3-phosphate dehydrogenase.

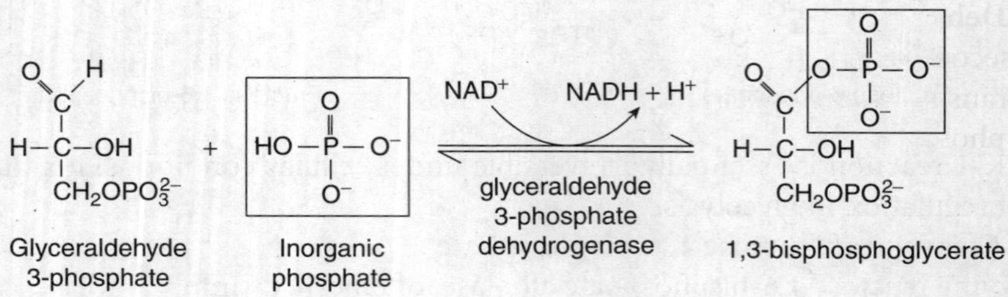

This is the first of the two energy-conserving reactions of glycolysis that eventually lead to the formation of ATP.

Step 7: Phosphoryl transfer from 1,3-bisphosphoglycerate to ADP

The enzyme phosphoglycerate kinase transfers the high-energy phosphoryl group from the carboxyl group of 1,3-bisphosphoglycerate to ADP, forming ATP and 3-phosphoglycerate.

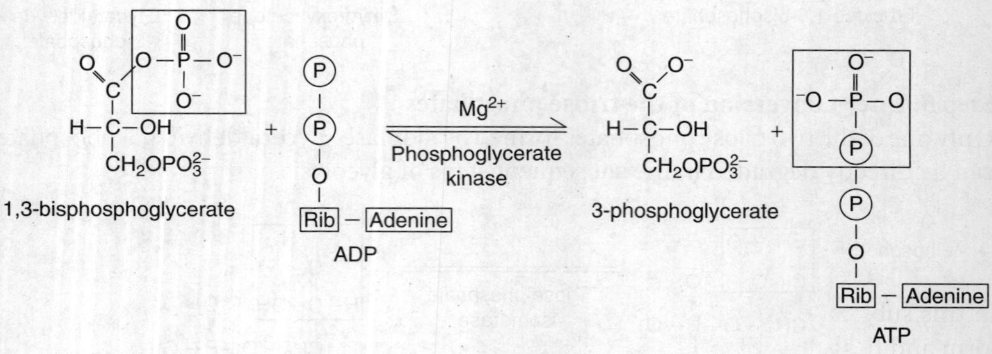

The formation of ATP by phosphoryl group transfer from a substrate such as 1,3-bisphosphoglycerate is referred to as a substrate-level phosphorylation.

Step 8: Conversion of 3-phosphoglycerate to 2-phosphoglycerate

The enzyme phosphoglycerate mutase catalyses a reversible shift of the phosphoryl group between C-2 and C-3 of glycerate; Mg^{2+} is essential for this reaction.

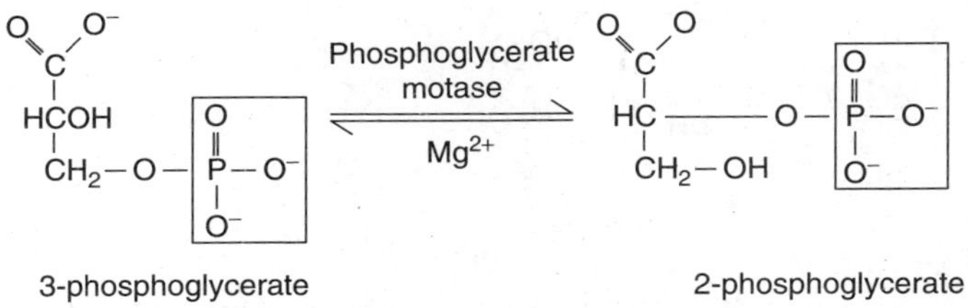

Step 9: Dehydration of 2-phosphoglycerate to phosphoenolpyruvate
In the second glycolytic reaction that generates a compound with high phosphoryl group transfer potential, enolase promotes reversible removal of a molecule of water from 2-phosphoglycerate to yield phosphoenolpyruvate (PEP).

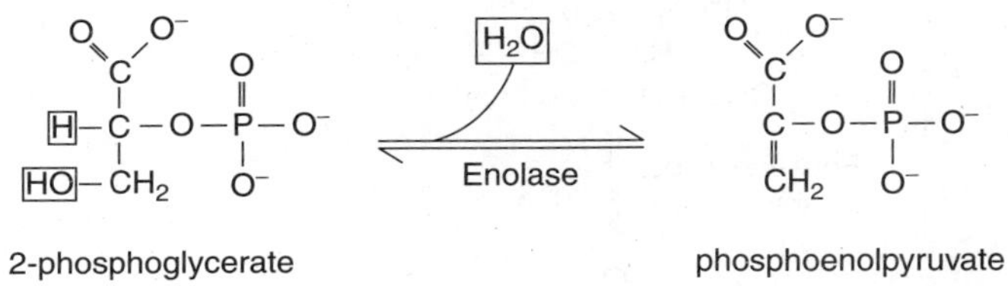

Step 10: Transfer of the phosphoryl group from phosphoenolpyruvate to ADP
The last step in glycolysis is the transfer of the phosphoryl group from phosphoenol-pyruvate to ADP, catalysed by pyruvate kinase, which requires K^+ and either Mg^{2+} or Mn^{2+}.

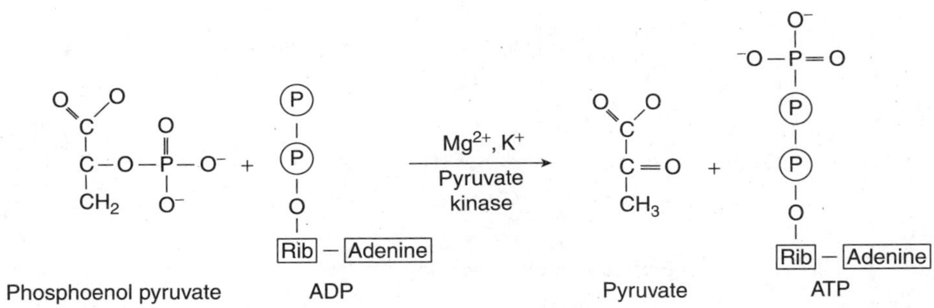

In this substrate-level phosphorylation, the product pyruvate first appears in its enol form and then tautomerises rapidly and non-enzymatically to its keto form.

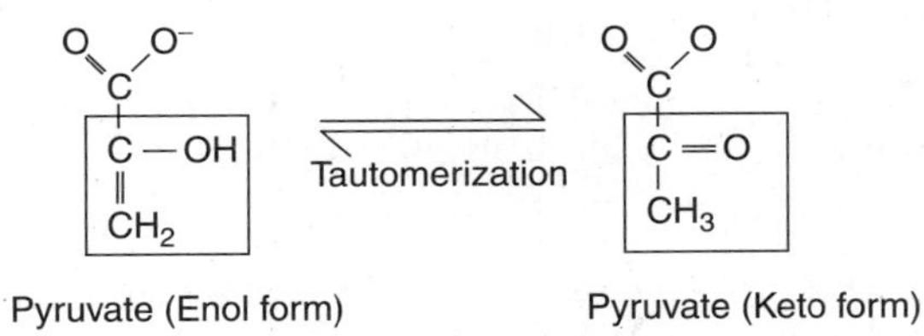

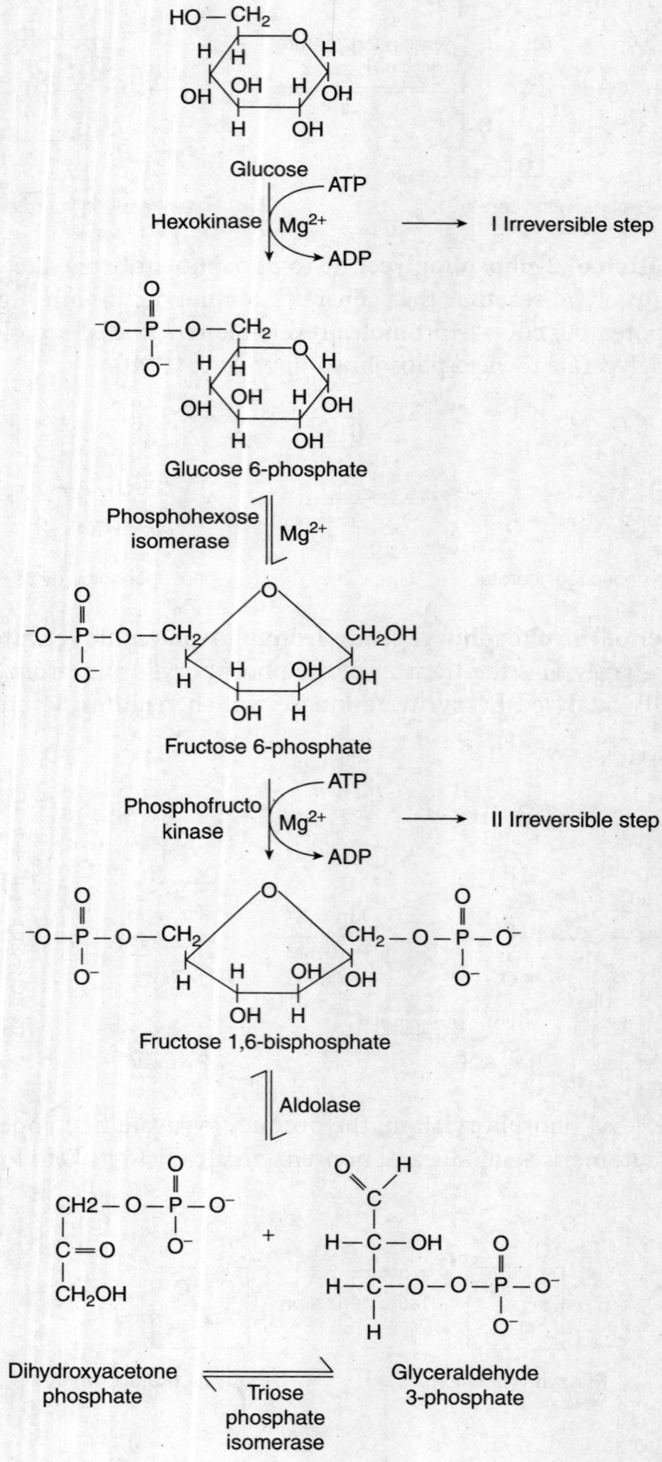
Glucose
ATP
Hexokinase
Mg2+
ADP
I Irreversible step
Glucose 6-phosphate
Phosphohexose isomerase
Mg2+
Fructose 6-phosphate
ATP
Phosphofructo kinase
Mg2+
ADP
II Irreversible step
Fructose 1,6-bisphosphate
Aldolase
Dihydroxyacetone phosphate
Glyceraldehyde 3-phosphate
Triose phosphate isomerase

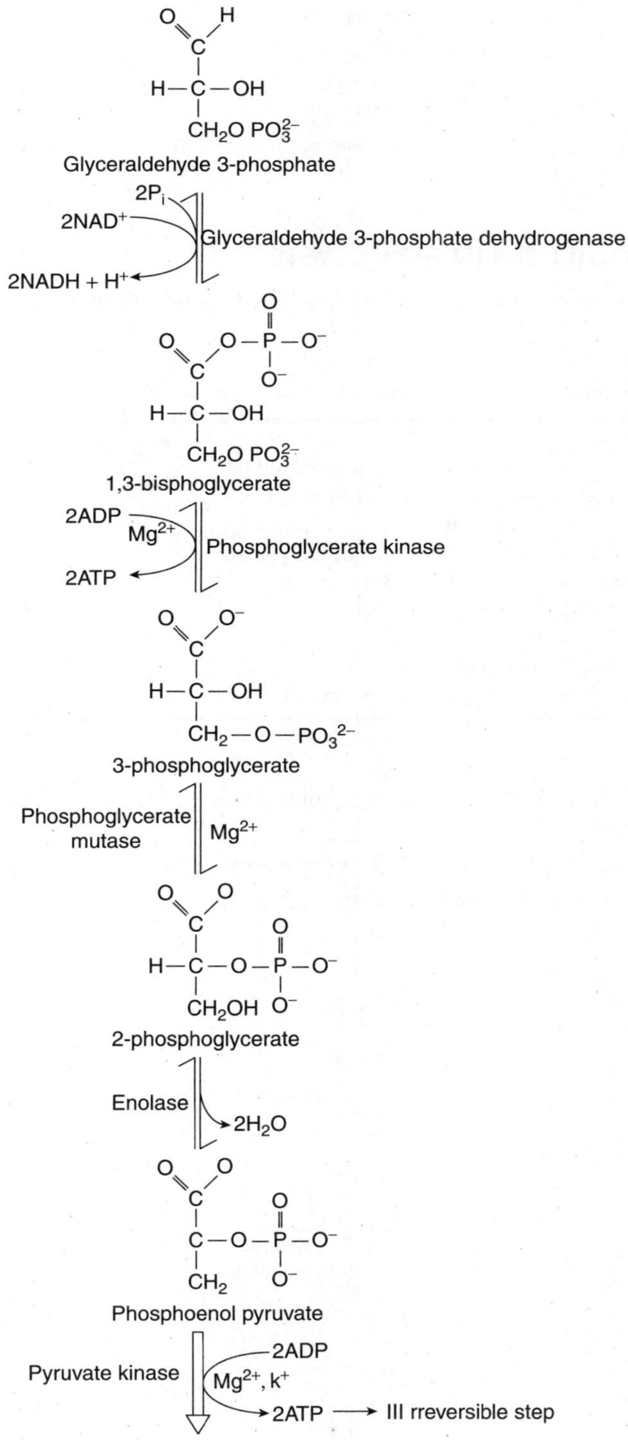
Glyceraldehyde 3-phosphate
2Pi
2NAD+
Glyceraldehyde 3-phosphate dehydrogenase
2NADH + H+
1,3-bisphoglycerate
2ADP
Mg2+
Phosphoglycerate kinase
2ATP
3-phosphoglycerate
Phosphoglycerate mutase
Mg2+
2-phosphoglycerate
Enolase
2H2O
Phosphoenol pyruvate
Pyruvate kinase
2ADP
Mg2+, k+
2ATP
III rreversible step

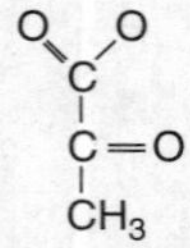
Pyruvate

ENERGY PRODUCTION IN GLYCOLYSIS

During glycolysis, ATP is formed and used in the following reactions (Refer Table 8.1):

Table 8.1 *Production of High-energy Phosphate in the Catabolism of Glucose*

S.No	Reactions	ATP used	ATP formed
1	Glucose + ATP → Glucose 6-phosphate + ADP	1	
2	Fructose 6-phosphate + ATP → Fructose 1,6-bisphosphate + ADP	1	
3	Reoxidation of (2NADH$^+$H$^+$) formed in reaction catalysed by glyceraldehydes 3-phosphate dehyrogenase, in respiratory chain.		6
4	2(1,3-bisphosphoglycerate) → 2(3-phosphoglycerate)		2
5	2(phosphoenol pyruvate) → 2(pyruvate)		2

The net gain of ATP molecules during glycolysis is eight as seen from the table (10 – 2 = 8).

During anaerobic condition, NAD$^+$ is regenerated from NADH by the reduction of pyruvate to lactate catalysed by lactate dehydrogenase.

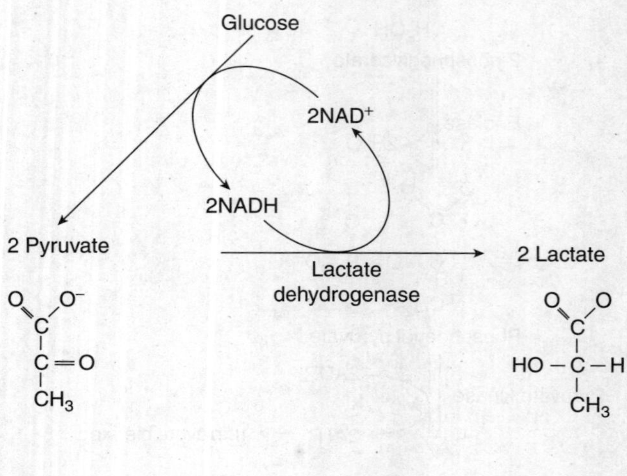

REGULATION OF GLYCOLYSIS

Carbohydrate metabolism provides the precursors for biosynthetic process and also produces ATP. It is very important for a cell to maintain the concentration of ATP at a nearly constant level. The regulation of glycolysis is carried by an enzyme-linked rate-limiting step. The rate-limiting step for this pathway is an irreversible reaction, and hence, the rate of the reaction is also slow.

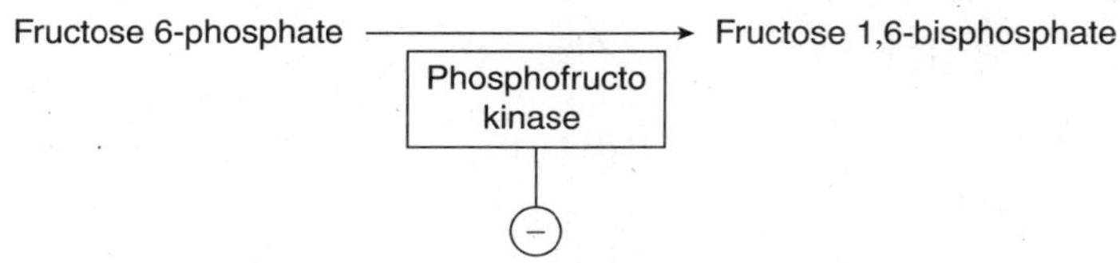

Phosphofructokinase-1 Is under Complex Allosteric Regulation

PFK-1 catalysed step is the irreversible step that commits a cell to convert glucose to glycolysis. Apart from the substrate-binding sites, this enzyme contains additional sites (regulatory sites) where allosteric activators or inhibitors bind. ATP not only serves as a substrate for PFK-1 but it also becomes the end product of the glycolytic pathway. When the concentration of ATP is more in the cells, it inhibits PFK-1 by binding to an allosteric site and lowering the affinity of the enzyme for fructose 6-phosphate. ADP and AMP concentration increases and they act allosterically to relieve this inhibition of ATP production. Thus a high level of ADP and AMP activates the enzyme, whereas increase in ATP level lowers the activity of PFK-1.

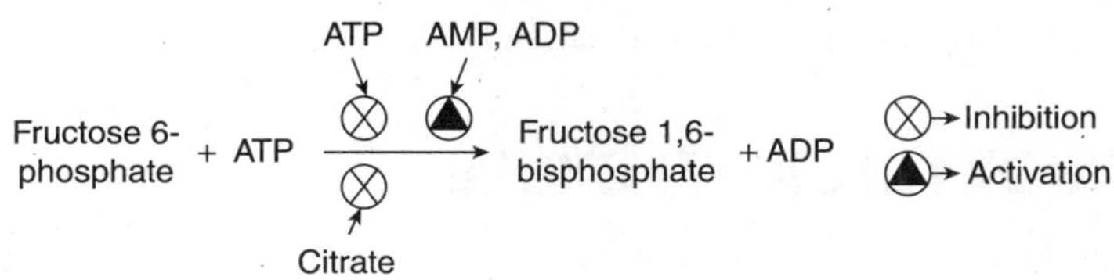

Citrate also acts as an allosteric regulator of PFK-1, which is produced as the key intermediate in the aerobic oxidation of pyruvate, fatty acids, and amino acids. When the concentration of citrate increases, the ATP concentration decreases; as a result, it inhibits PFK-1 and thereby further reduces the flow of glucose through the glycolysis pathway. Allosteric inhibition of PFK-1 by ATP is shown by substrate activity curve as shown in Figure 8.7.

At low ATP, the K0.5 for fructose 6-phosphate is relatively low, enabling the enzyme to function at a high rate at relatively low concentrations of fructose 6-phosphate.

INHIBITION OF GLYCOLYSIS

Inhibition can occur during two reactions in glycolysis.

Glyceraldehydes 3-phosphate + H_3Po_4 → 1,3-diphosphoglyceric acid + NADH + H^+

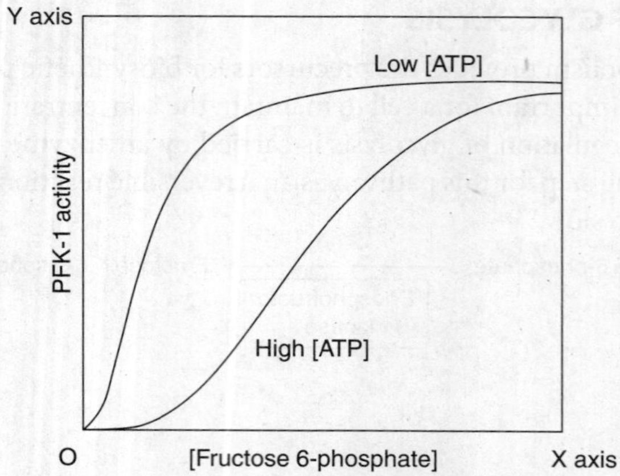

Figure 8.7 *Graphical Representation of Allosteric Inhibition of Phosphofructokinase with Respect to ATP Concentration*

This reaction may be inhibited by iodoacetate; arsenate can also interfere with this reaction, not by inhibition, but by competing with phosphate in the first carbon atom to form 1-areseno 3-phosphoglycerate, which hydrolyses spontaneously to give 3-phosphoglyceric acid without generating ATP in the next step. This is an example of uncoupling of oxidation from phosphorylation.

$$\text{2-phosphoglyceric acid} \rightarrow \text{Phosphoenol pyruvate} + H_2O$$

Enolase, which catalyses this reaction, is inhibited by fluoride. Addition of fluoride to blood prior to glucose estimation prevents glycolysis.

TRICARBOXYLIC ACID CYCLE

INTRODUCTION

The citric acid cycle, Krebs cycle, or tricarboxylic acid cycle is the pathway that occurs in mitochondria that oxidises acetyl-CoA and reduces the coenzymes $NADH^+$ and $FADH_2$ reoxidised through the electron transport chain. The third phase of metabolism is the citric acid cycle. The carbohydrate, lipid and protein finally meet in this pathway, because glucose, fatty acids and most amino acids are catabolised to acetyl-CoA or any of the intermediates of this pathway.

SITE OF SYNTHESIS

Tricarboxylic acid (TCA) cycle takes place in most of the tissues, but liver is the only tissue where it occurs to a significant extent.

The enzymes of the citric acid cycle are located in the mitochondrial matrix, either free or attached to the inner mitochondrial membrane and the crista membrane, where the enzymes of the respiratory chain are also found.

REACTIONS OF TCA CYCLE

TCA cycle is called amphibolic pathway as explained in Figure 8.8 because both catabolic and anabolic processes are involved in this pathway.

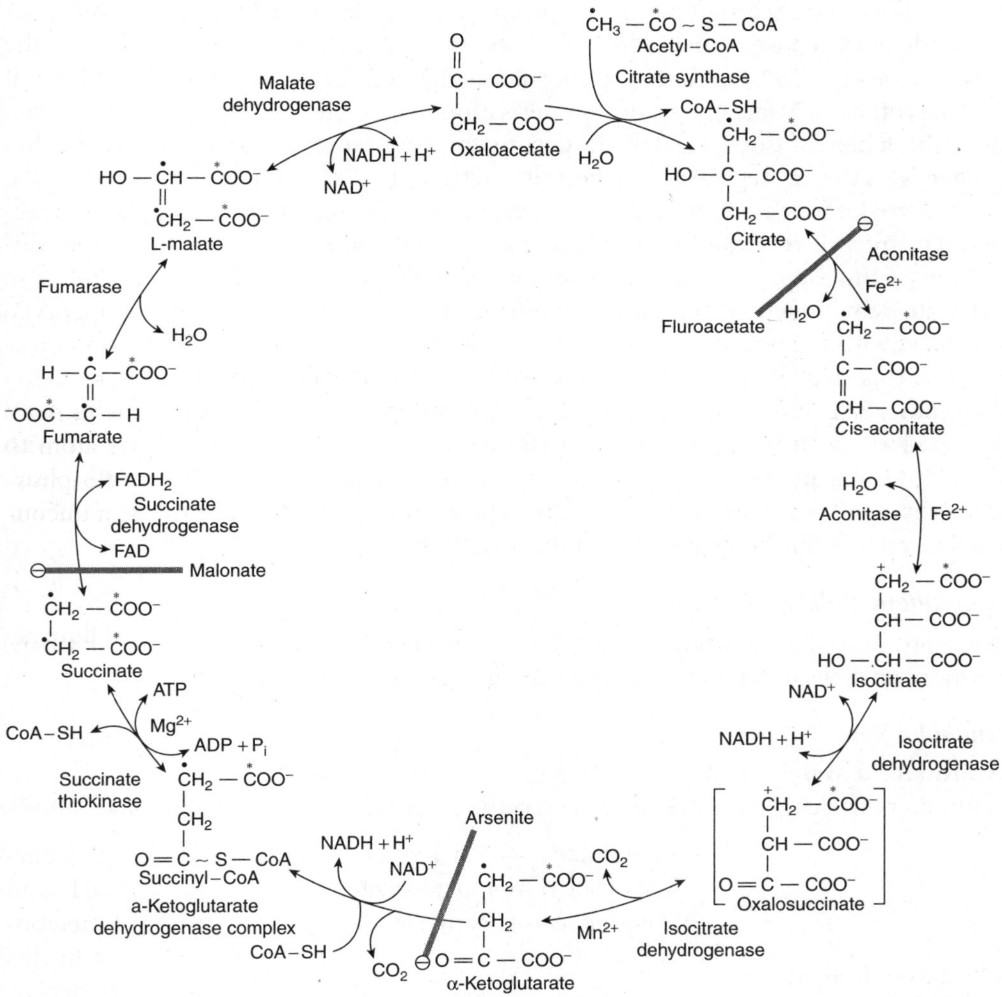

Figure 8.8 *Tricarboxylic Acid Cycle or Citric Acid Cycle*

Catabolic process: The cycle helps in the degradation of acetyl residues, which are derived from carbohydrates, fats, proteins, and so on.

Anabolic process: The intermediates of TCA cycle are used as precursors in the biosynthesis of many compounds.

Overall reaction of TCA cycle:

$$\text{Pyruvate} + \text{GDP} + P_i + 3\text{NAD}^+ + \text{FAD} \longrightarrow 3CO_2 + \text{GTP} + 3\text{NADH} + \text{FADH}_2 + 3H^+$$

$$\text{Acetyl-CoA} + \text{GDP} + \text{P}_i + 2\text{NAD}^+ + \text{FAD} \longrightarrow 2\text{CO}_2 + \text{GTP} + 2\text{NADH} + \text{FADH}_2 + 2\text{H}^+$$

Most of the enzymes required for TCA occur in the mitochondrial matrix, except succinate dehydrogenase, which is found in the inner mitochondrial membrane. In the first step, acetyl-CoA combines with oxaloacetate and forms citrate with the help of citrate synthase. Citrate undergoes isomerisation in an oxidative decarboxylation reaction with a help of isocitrate dehydrogenase, forming carbon dioxide. Isocitrate dehydrogenase is the key regulatory enzyme; it is activated by ADP and inhibited by NADH.

α-ketoglutarate is converted to succinyl-CoA in the second decarboxylation reaction. During this reaction CO_2 is released and $NADH^+$ and H^+ are produced. The conversion of this step requires α ketoglutarate dehydrogenase, which is very similar to pyruvate dehydrogenase complex. The next step is the conversion of succinyl-CoA to succinate with the help of succinate thiokinase. Substrate level phosphorylation takes place in this step. Succinate is oxidised to fumarate and catalysed by succinate dehydrogenase; FAD is the coenzyme involved in this step. FAD yields 2 ATPs. Fumarate is hydrated to malate with the help of fumarase. Malate is converted to oxaloacetate with the help of malate dehydrogenase and NAD^+ is the coenzyme involved in this reaction. Three ATPs are produced as electrons pass through the electron transport chain. Oxaloacetate is finally regenerated in this reaction.

Amphibolic Role of TCA Cycle

The word 'amphi' means both—the anabolic and catabolic reactions occur in the same pathway or cycle and hence TCA cycle is regarded as 'amphibolic'.

Anabolic Role

Amino Acid Metabolism

A number of glycogenic amino acids enters the TCA cycle via transamination reactions.

$$\text{Glutamate} \longrightarrow \alpha\text{-Ketoglutarate}$$
$$\text{Aspartate} \longrightarrow \text{Oxaloacetate}$$
$$\text{Alanine} \longrightarrow \text{Pyruvate}$$

Lipid Metabolism

Acetyl-CoA and glycerol are needed for the synthesis of steroids and fatty acids. For porphyrin synthesis, succinyl-CoA and glycerol are required as precursors. TCA cycle is thus involved in synthesis of haemoglobin, cytochromes and haemoproteins.

Nitrogen Metabolism

Aspartate, which is formed from oxaloacetate through transamination reaction, is used for the synthesis of argininosuccinate and purines. The final product of this process is fumarate, which enters the TCA cycle.

Catabolic Role

The oxidation of acetyl-CoA of TCA cycle produces energy, CO_2, and water.

REGULATION OF TCA CYCLE

The cellular demands of ATP are crucial in controlling the rate of citric acid cycle. The regulation is brought about either by enzymes or the levels of ADP. Three enzymes, namely citrate synthase, isocitrate dehydrogenase, and α-ketoglutarate dehydrogenase, regulate citric acid cycle.

Regulatory enzymes of the TCA cycle include citrate synthase, isocitrate dehydrogenase, and α-ketoglutarate dehydrogenase complex. High accumulation of ATP: ADP, so high concentration of acetyl-CoA and NADH : NAD^+ ratio inhibit the TCA cycle at the respective steps.

Citrate synthase is inhibited by ATP, NADH, acyl-CoA, and succinyl-CoA.

Isocitrate dehydrogenase is activated by ADP and inhibited by ATP and NADH.

α-Ketoglutarate dehydrogenase is inhibited by succinyl-CoA and NADH.

Availability of ADP is very important for the citric acid cycle to proceed. This is due to the fact that unless sufficient levels of ADP are available, oxidation (coupled with phosphorylation of ADP to ATP of NADH and FADH2) through electron transport chain stops. The accumulation of the enzymes also limits the supply of NAD+ and FAD, which are essential for TCA cycle to proceed.

GLYCOGENESIS

INTRODUCTION

Glycogen is the major storage form of carbohydrate in animals and starch in plants. The main stores of glycogen in the body are found in skeletal muscle and liver, although most other cells may store minute amounts. The function of muscle glycogen is to serve as a fuel reserve for the synthesis of ATP during muscle contraction. That of liver glycogen is to maintain the blood glucose concentration.

FUNCTIONS OF LIVER AND MUSCLE GLYCOGEN

Periods fasting (a few days) and is only moderately decreased in prolonged fasting (few weeks) as shown in Figure 8.9.

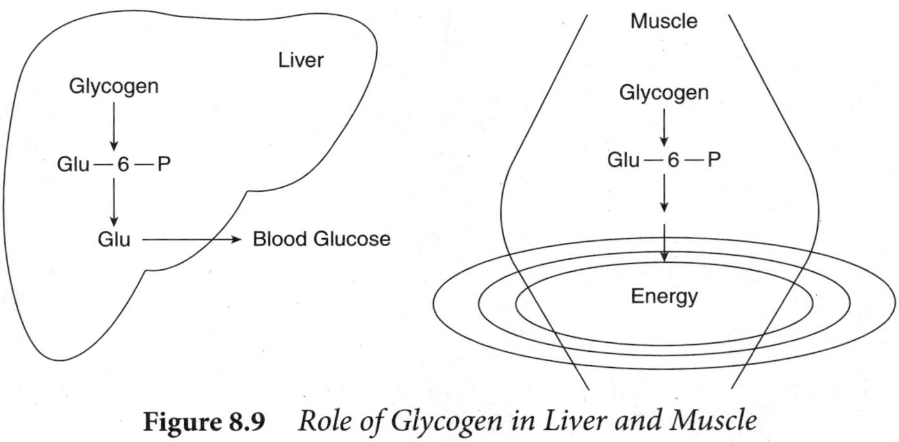

Figure 8.9 *Role of Glycogen in Liver and Muscle*

AMOUNTS OF LIVER AND MUSCLE GLYCOGEN

Liver stores 6%–8% of glycogen and muscle 1%–2%. Due to more muscle mass, the quantity of glycogen in muscle is about three times higher than that of liver. Liver glycogen stores increase during well-fed state and are depleted during fast. Muscle glycogen is not affected by short period's fasting (a few days) and is only moderately decreased in prolonged fasting (a few weeks).

STRUCTURE OF GLYCOGEN

The molecular weight of a single glycogen is 108 daltons. Glycogen is present in cytoplasmic granules. The granules contain most of the enzymes necessary for glycogen synthesis and degradation. The structure of glycogen is basically a branched-chain of homopolysaccharides, which consist of α-D-glucose units made of up to 100,000 residues. The glucose units are joined together by α 1-4 linkage with 8–10 residues of glucose. Branching occurs through α 1-6 linkage as shown in Figure 8.10.

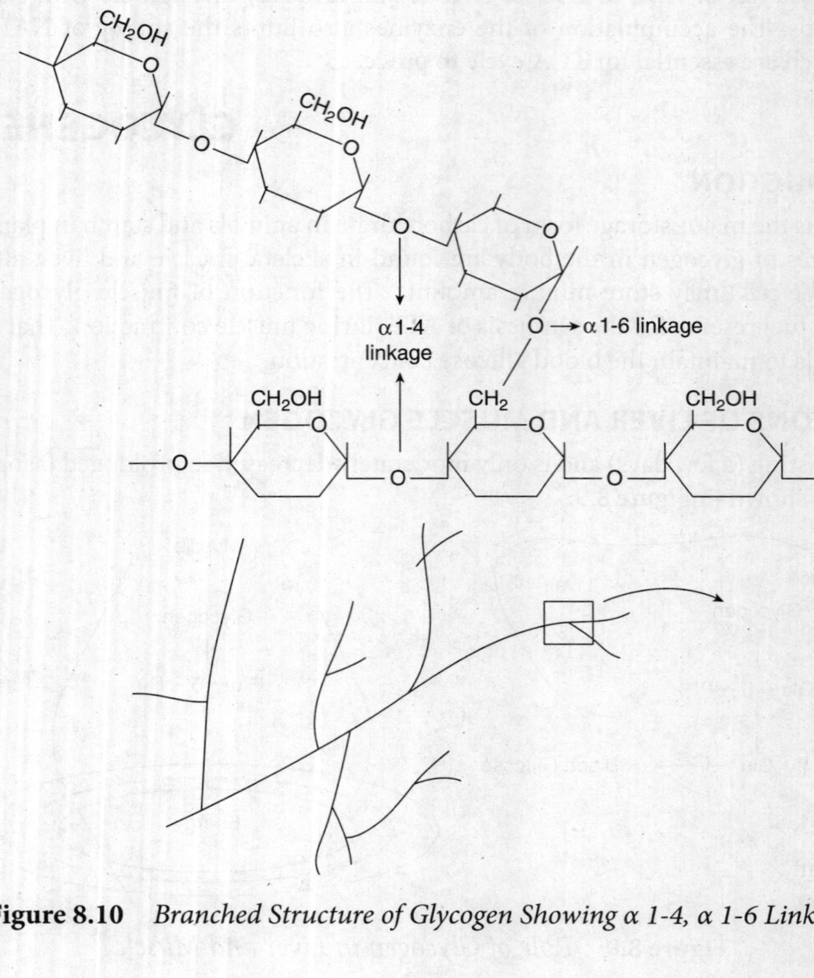

Figure 8.10 *Branched Structure of Glycogen Showing α 1-4, α 1-6 Linkage*

SYNTHESIS OF GLYCOGEN

The synthesis of glycogen from glucose is termed as glycogenesis. Glycogenesis takes place in the cytosol and requires ATP and UTP besides glucose refer Figure 8.11.

Overall reaction of glycogen synthesis is

$$(\text{Glucose})_n + \text{glucose} + 2\text{ATP} \rightarrow (\text{glucose})_{n+1} + 2\ \text{ADP} + \text{Pi}$$

Of this two ATP utilised, one is required for the phophorylation of glucose while the other is needed for the conversion of UDP to UTP.

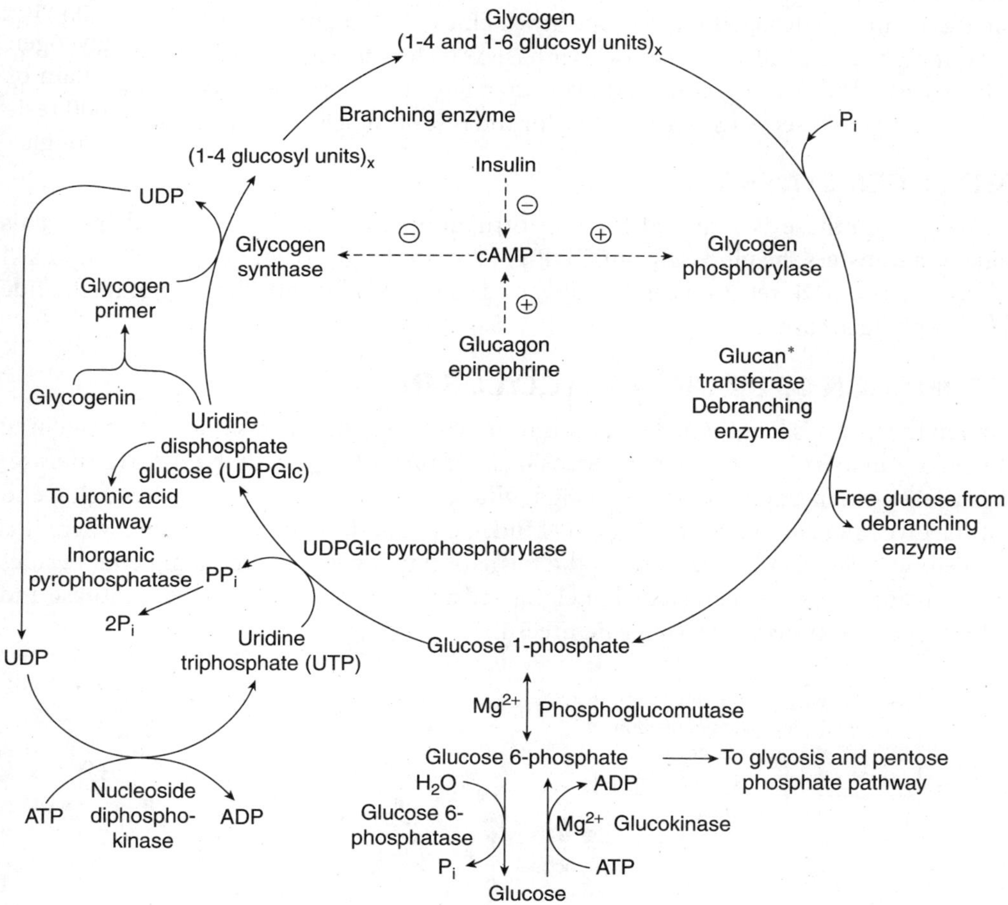

Figure 8.11 *Mechanism of Branching Step in Glycogenesis*

SYNTHESIS OF UDG–GLUCOSE

The enzymes hexokinase in muscle and hexokinase in liver convert glucose to glucose 6-phosphate. Phosphoglucomutase catalyses the conversion of glucose 6-phosphate to glucose 1-phosphate. Uridine diphosphate glucose (UDPG) is synthesised from glucose 1-phosphate and UTP by UDP-glucose phosphorylase. Pyrophosphate (PPi) produced in this reaction is hydrolysed to inorganic phosphate, which will ensure the optimal synthesis of UDPG.

REQUIREMENT OF PRIMER TO INITIATE GLYCOGENESIS

A small fragment of pre-existing glycogen must act as a primer to initiate glycogen synthesis. It is recently fount that in the absence of glycogen primer, a specific protein namely *glycogenin* can accept glucose from UDPG. The hydroxyl group of amino acid tyrosine of glycogenin is the site at which the initial glucose unit is attached. The enzyme glycogen initiator glucose synthase transfers the first molecule of glucose to glycogenin. Then glycogenin itself takes up a few glucose residues to form a fragment of primer, which serves as an acceptor for the rest of the glucose molecules.

GLYCOGEN SYNTHASE

Glycogen synthase is required for the formation of α 1-4 glycosidic linkages. This enzyme transfers the glucose from UDP-glucose to the glycogen primer to form α 1-4 linkages. The UDP released can be converted again to UTP with the help of nucleotide diphosphate kinase.

FORMATION OF BRANCHES IN GLYCOGEN

When the chain is lengthened to a minimum of eleven glucose residues, the branching enzyme amylo (α 1-4, α 1-6) transglucosidase (Figure 8.12) (glucosyl α 4, 6-transferase) transfers a small fragment of five to eight glucose residues from the non-reducing end of the glycogen chain by breaking α 1-4 linkage to another glucose residue, where it is linked by α 1-6 bond. This leads to the formation of a new non-reducing end, besides the existing one. Glycogen is further elongated and branched by glycogen synthase and glucosyl α-4, 6 transferase. (refer Figure 8.13)

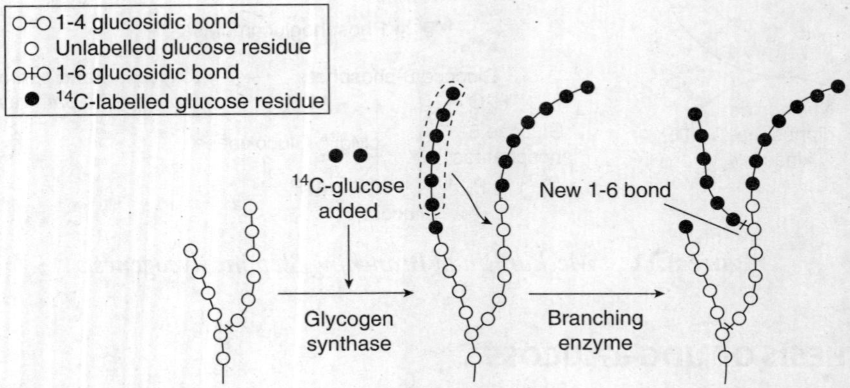

Figure 8.12 *Mechanism of Branching Step in Glycogenesis*

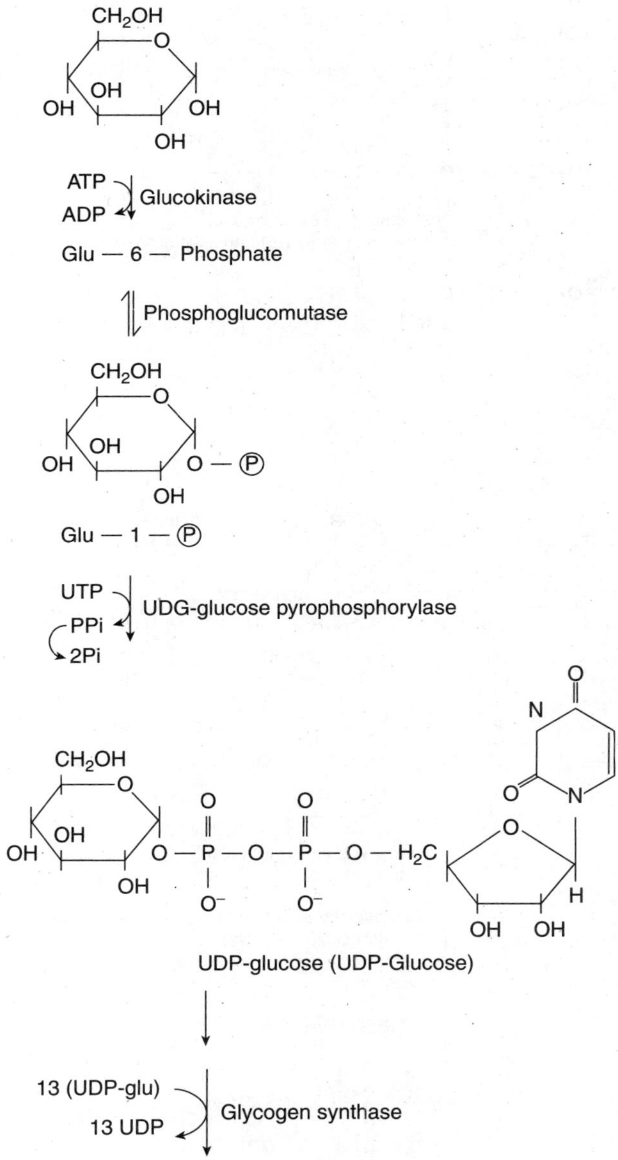

Figure 8.13 *Biosynthesis of Glycogen*

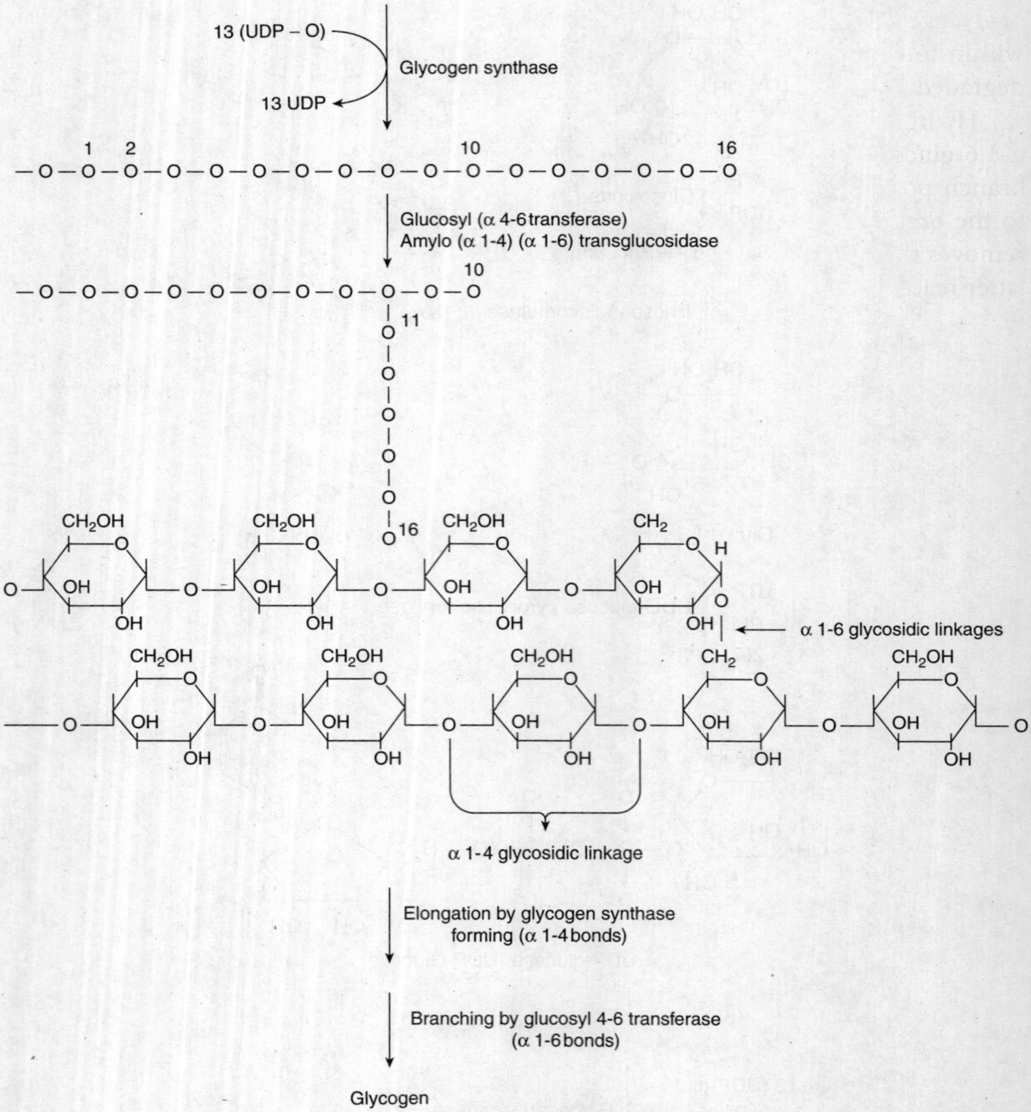

Figure 8.13 *(Continued)*

GLYCOGENOLYSIS

Glycogen degradation requires the following two reactions:

Removal of glucose forms the non-reducing ends of glycogen. Using inorganic phosphate (Pi), glycogen phosphorylase cleaves the (1,4) linkages on the outer branches of glycogen to yield glucose-1-phosphate. Glycogen phosphorylase stops when it comes within four glucose residues of a branch point. A glycogen molecule that has been degraded to its branch points is called a limit dextrin.

Hydrolysis of the α-1,6- glycosidic bonds at the branch points of glycogen. Amylo α-1,6-glucosidase, also called debranching enzyme, begins the removal of α(1,6) branch points by transferring the outer three of the four glucose residues attached to the branch point to a nearby non-reducing end as shown in Figure 8.14. It then removes the single glucose residue attached at each branch point. The product of this latter reaction is free glucose.

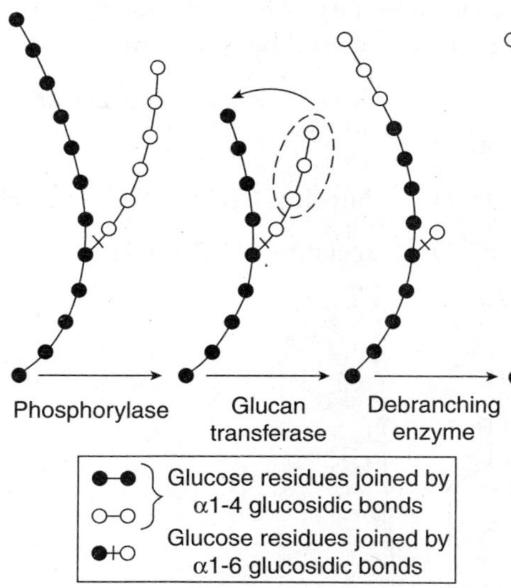

Figure 8.14 *Steps Involved in Glycogenolysis*

ELECTRON TRANSPORT CHAIN

ELECTRON CARRIERS

Electron carriers in the electron transport chain are the following:

1. NAD^+ and $NADP^+$
2. FAD and FMN
3. Q, CoQ, or ubiquinone
4. Cytochromes – *b*, *c*, *a*

1. *NAD^+ and $NADP^+$*

Nicotinamide nucleotide linked dehydrogenase catalyse reversible reactions of the following general types:

$$\text{Reduced substrate} + NAD^+ \rightleftharpoons \text{Oxidised substrate} + NADH + H^+$$

$$\text{Reduced substrate} + NADP^+ \rightleftharpoons \text{Oxidised substrate} + NADP + H^+$$

NAD–linked dehydrogenase dehydrogenates two hydrogen atoms from their substrates. One of the atoms is transferred in the form of a hydride ion (H^- is the equivalent of a proton and two electrons) to NAD^+ and $NADP^+$. The second hydrogen atom is released as H^+ in the medium. The reduced form of NAD^+ or $NADP^+$ converts the nicotinamide benzene ring (with a fixed charge of the ring nitrogen) to the quinonoid form (with no charge on the nitrogen ring).

Some important reactions catalysed by NAD^+ and $NADP^+$ dehydrogenase:

Reaction **Location**

NAD-linked reactions:

$$A - KG + CoA + NAD^+ \rightleftharpoons \text{Succinyl CoA} + CO_2 + NADH + H + \text{Mitochondria}$$

$$\text{Malate} + NAD^+ \rightleftharpoons \text{Oxaloacetate} + NADH + H + \text{Mitochondria and cytosol}$$

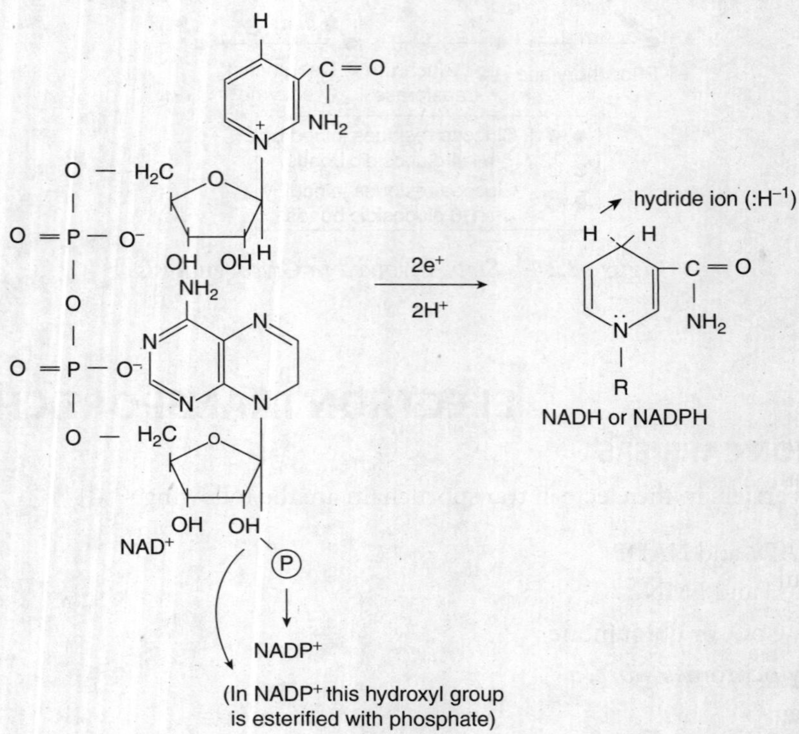

2. *Flavoproteins*
Flavoproteins contain a very tightly, sometimes covalently bound flavin nucleoticle either FMH or FAD. The oxidised flavin nucleotide can accept either one electron (H^+ and e^-) yielding semiquinone form (FADH˙, FADH˙) or two electrons ($2H^+$ and $2e^-$) yielding $FADH_2$ or $FMNH_2$.

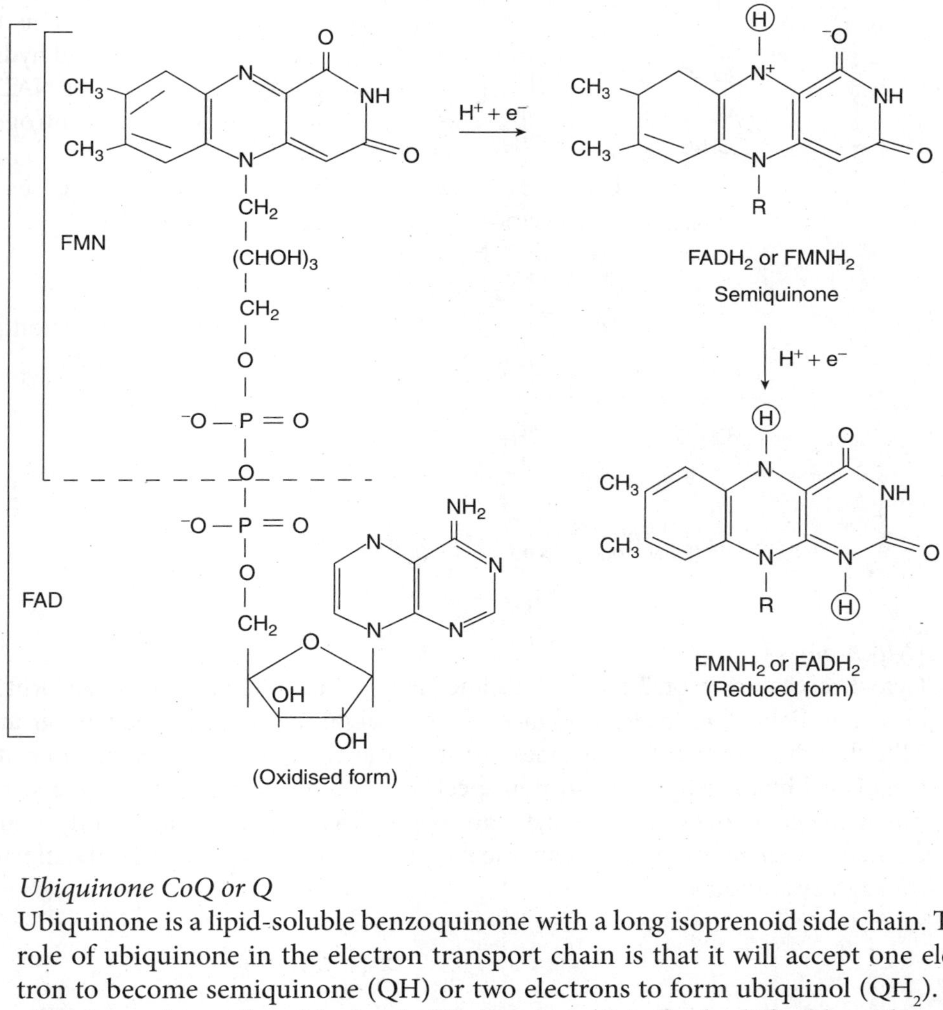

3. *Ubiquinone CoQ or Q*
Ubiquinone is a lipid-soluble benzoquinone with a long isoprenoid side chain. The role of ubiquinone in the electron transport chain is that it will accept one electron to become semiquinone (QH) or two electrons to form ubiquinol (QH_2). As ubiquinone is small and hydrophobic in nature, it is freely diffusible within the lipid bilayer of inner mitochondrial membrane and can shuttle, reducing equivalents between less-mobile electron carriers in the membrane. Since it carries both electrons and protons, it plays a central role in coupling electron flow to proton movement.

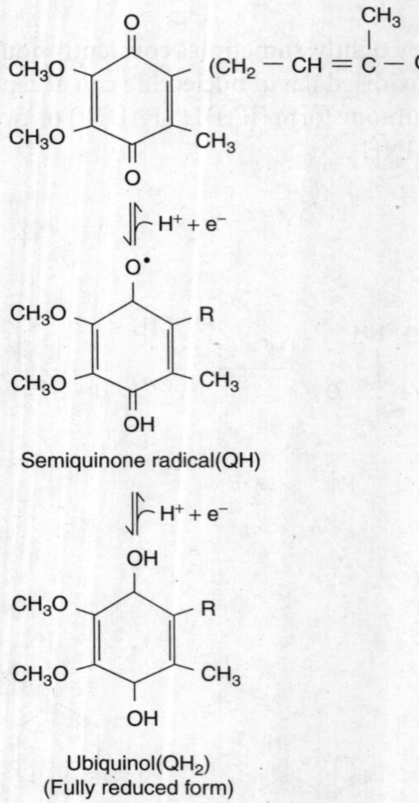

4. *Cytochromes*

Cytochromes are proteins in nature and have a characteristic strong absorption of visible light, due to the presence of iron-containing heme prosthetic groups. Mitochondria contain three classes of cytochromes designated as *a*, *b*, and *c*, distinguished by their light-absorption spectra: *a*-600 nm, *b*-560 nm, and *c*-550 nm. The heme cofactors of *a* and *b* cytochromes are tightly, but not covalently, bound to their associated proteins; the heme *c* type cytochromes are covalently attached through cys residues.

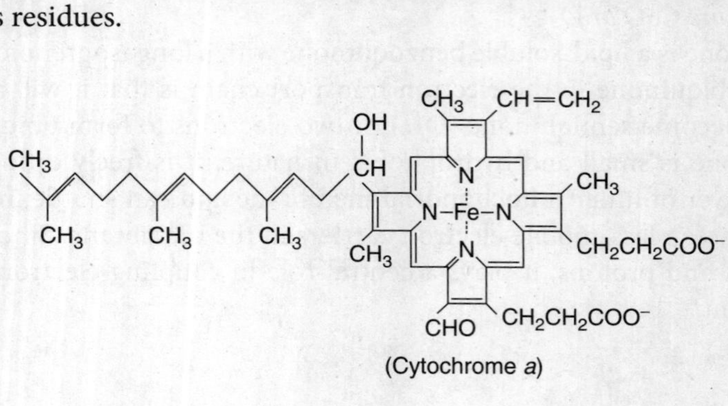

(Cytochrome *a*)

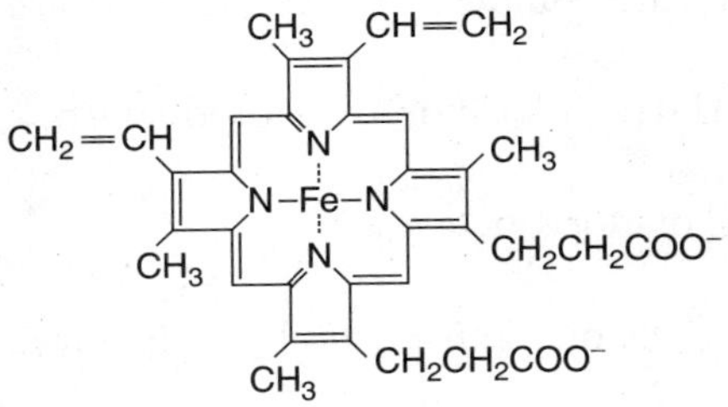

(Cytochrome *b*)

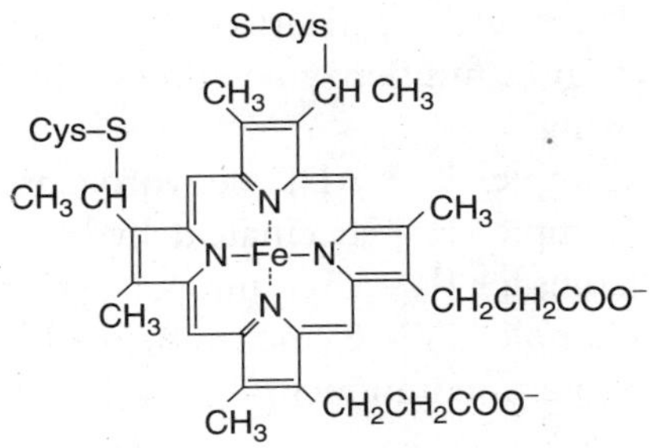

Heme-C (Cytochrome *c*)

IRON-SULPHUR PROTEINS

Helmut Beinert first discovered that the iron is present in heme and associated with inorganic sulphur atoms or with sulphur atoms of cysteine residues in the proteins or both. These iron sulphur (Fe-S) centres range from simple structures with single Fe atom coordinated with four cys-SH groups to more complex Fe-S centres with two or four atoms.

Rieske iron sulphur (named after the discoverer) is the iron atom that is coordinated with two of its residues rather than two cysteine groups. All iron-sulphur proteins participate in one electron transfer, among which one iron atom of the iron-sulphur cluster is oxidised or reduced as shown in Figure 8.15. At least eight Fe-S proteins function in mitochondrial electron transfer.

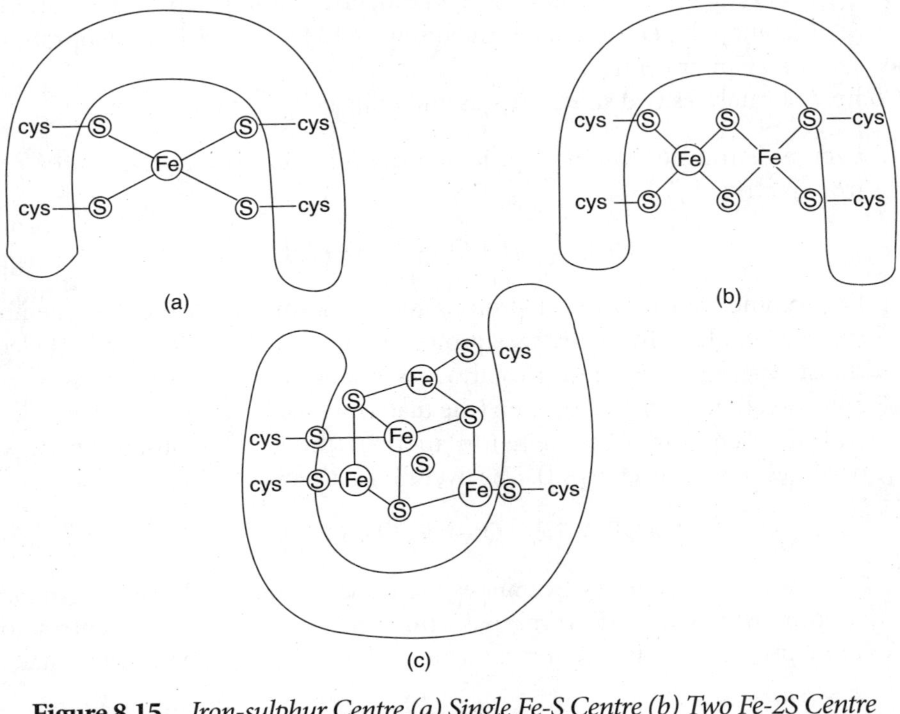

Figure 8.15 *Iron-sulphur Centre (a) Single Fe-S Centre (b) Two Fe-2S Centre (c) Four Fe-4S Centre*

The electron carriers of the respiratory chain or Electron Transport Chain (ETC) function in multienzyme complexes. Four enzyme complex involved in ETC are the following:

Complex I – NADH dehydrogenase or NADH: Ubiquinone oxidoreductase
Complex II – Succinate dehydrogenase
Complex III – Ubiquinone: Cytochrome C oxidoreductase
Complex IV – Cytochrome oxidase.

Protein components of mitochondrial electron transport chain is given in Table 8.2.

Table 8.2 *Protein Components of Mitochondrial Electron Transport Chain*

Enzyme complex	Number of subunits	Prosthetic group
I NADH DHase	42	FMN, Fe-S
II Succinate DHase	5	FAD, Fe-S
III Ubiquinone – cytochrome C oxidoreductase	11	Heme, Fe-S
IV Cytochrome oxidase	13	Heme, Cu A, Cu B

Complex I NADH to Ubiquinone

Complex I also called NADH. Ubiquinone reductase is a large enzyme composed of forty-two different polypeptide chains, including an FMA, containing flavoprotein and at least six iron sulphur centres.

Complex I catalyses two simultaneous and obligately coupled processes:

1. Exergonic transfer to ubiquinone of a hydride ion from NADH and a proton from matrix.

$$NADH + H + Q \rightarrow NAD + QH_2$$

2. Endergonic transfer of four protons forms the matrix to the intermembrane space. Complex I is, therefore, a proton pump driven by the energy of electron transfer, and the reaction it catalyses is vectorial. It moves protons in a specific direction from one location (the matrix which becomes negatively charged with the departure of proteins) to another (the intermembrane space which becomes positively charged). The overall reaction is

$$NADH + 5H + Q \rightarrow NAD + QH_2 + 4H^+$$

Figure 8.16 shows that complex I catalyses the transfer of a hydride ion from NADH to FMN, from which two electrons pass through a series of Fe-S centres to the iron-sulphur protein N-2 in the matrix arm of the complex. Electron transfer from

N-2 to ubiquinone on the membrane arm forms QH_2, which diffuses into the lipid bilayer.

It also drives the expulsion from the matrix of four protons per pair of electrons. This proton flux produces an electrochemical potential across the inner mitochondrial membrane (N-side portion), which conserves some of the energy released by the electron transfer reactions. This electrochemical potential drives ATP synthesis.

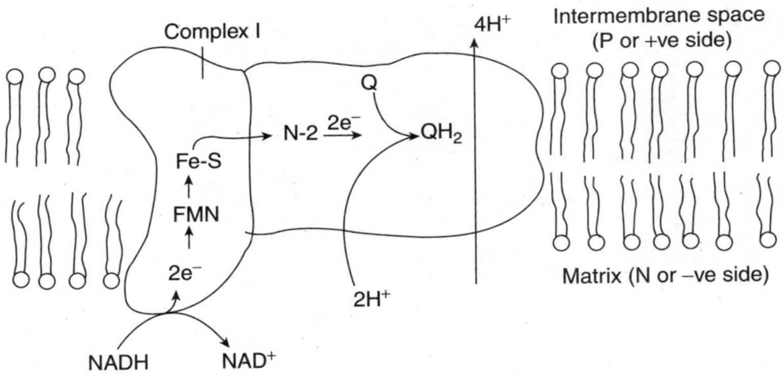

Figure 8.16 *NADH-Ubiquinone Oxidoreductase Complex I*

Complex II: Succinate to Ubiquinone

Complex II is called succinate dehydrogenase complex. It is the only membrane-bound enzyme in TCA. Complex II is smaller and simpler than complex I. However, it contains two types of prosthetic groups and at least four different proteins. To the first protein FAD is covalently bound and Fe-S centre with four Fe atoms. In addition, a second protein is present. Electrons pass form succinate to FAD and then through Fe-S centres to ubiquinone. Other substrates for mitochondrial dehydrogenase pass electrons into the respiratory chain at the level of ubiquinone but not through the complex II.

First step in β-oxidation of fatty acyl-CoA catalysed by flavoprotein acyl-CoA dehydrogenase involves transfer of electrons form the substrate to the FAD of the dehydrogenase and then to electron-transferring flavoprotein (ETF), which in turn passes its electrons to ETP, ubiquinone oxidoreductase. This enzyme passes electrons into the respiratory chain by reducing ubiquinone.

Glycerol-3-phosphate, formed either from glycerol released by triacylglycerol breakdown or by the reduction of dihydroxyacetone phosphate from glycolysis is oxidised by glycerol-3-phosphate dehydrogenase (Refer Figure 8.17). This enzyme is a flavoprotein, located on the outer face of the inner mitochondrial membrane, and it channels the electrons in respiratory chain by reducing ubiquinone.

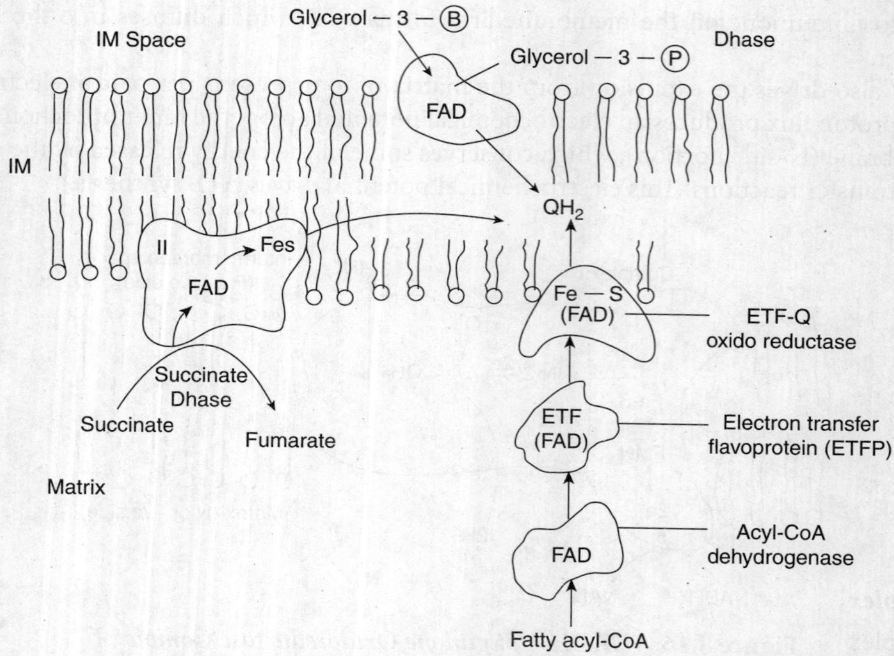

Figure 8.17 *Pathway of Electrons Flow from NADH, Succinate, Fatty acyl-CoA, Glycerol-3-phosphate to Ubiquinone*

Complex III: Ubiquinone to Cytochrome C

This is also called cytochrome bc1 complex or ubiquinone, cytochrome-oxidoreductase. This couples the transfer of electrons from ubiquinol (QH_2) to cytochrome c with the vectorial transport of protons from the matrix to the intermembrane space.

Net equation for the redox reactions of Q cycle:

$$QH_2 + 2Cyt\ c1\ (oxidised) + 2H + N \rightarrow Q + 2Cyt\ c1\ (reduced) + 4H + p$$

The Q cycle accommodates the switch between the two electron-carrier ubiquinones and the one electron-carrier's cytochrome b562, b566, c1, c.

Complex III contains cyt b562, cyt b566, cyt c1, and cyt c iron-sulphur protein and at least six other proteins subunits. These proteins are asymmetrically disposed in the inner mitochondrial membrane; cyt b spans the membrane.

Complex III functions as a proton pump, as a result of the asymmetric orientation of the complex. The protons produced when UQH_2 is oxidised to UQ are released to the intermembrane space, producing a transmembrane difference of proton concentration, a proton gradient is formed refer Figure 8.18. The importance of this proton gradient to mitochondrial ATP synthesis will soon become clear.

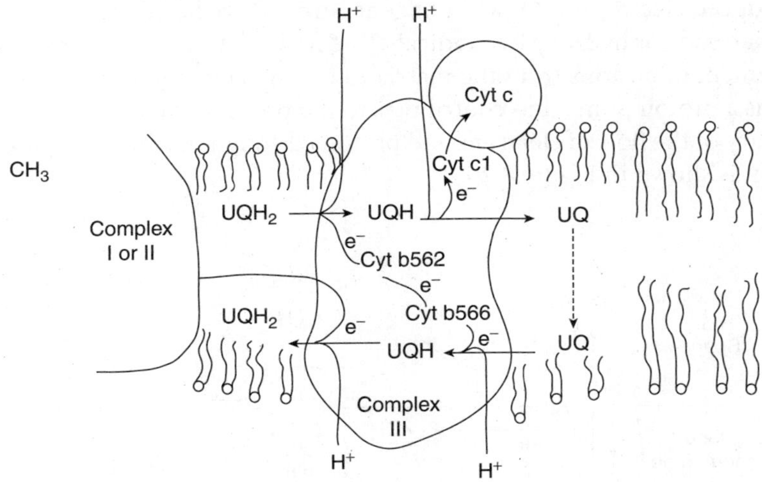

Figure 8.18 *Flow of Electrons through Complex III*

Complex IV: Reduction of O_2

Complex IV is also known as cytochrome oxidase and contains cytochromes a and a_3 as shown in Figure 8.19. The cytochromes are made up of two heme groups bound to different regions of the same large protein, which are therefore functionally and spectrally different. In addition, complex IV contains two copper ions, CuA and CuB, which are helpful in transferring the electrons to O_2. This complex is also involved in carrying

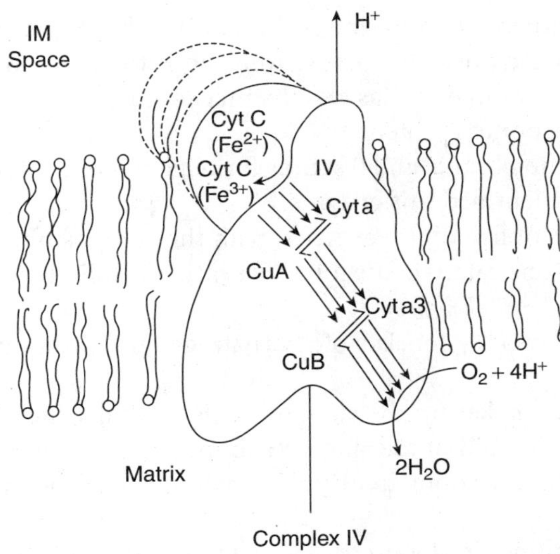

Figure 8.19 *Pathway of Electrons through Complex IV*

the four reduced electrons to O_2 without generating the reduced intermediates such as hydrogen peroxide or hydroxy free radicals. The flow of electrons in complex IV causes a net movement of protons from the matrix to the intermembrane space. Complex IV functions as a proton pump and contributes to the proton motive force.

Summary of the flow of electrons and protons through the four complexes of respiratory chain is shown in Figure 8.20.

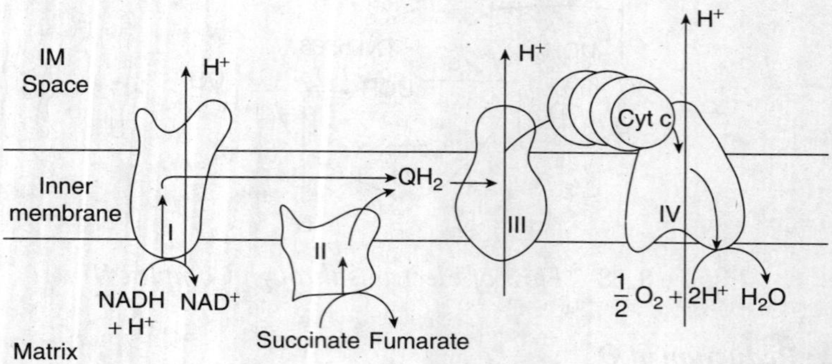

Figure 8.20 *Flow of Electrons and Protons through the Four Complexes of Respiratory Chain*

ATP SYNTHASE

Mitochondrial ATP synthase is an F-type ATPase. It catalyses the formation of ATP and Pi accompanied to the flow of protons from the P to N side of the membrane. ATP synthase also called complex V has two distinct components:

F_1 – a peripheral membrane protein

F_0 – integral to the membrane, and 0 stands for oligomycin sensitive

Efraim Racker first identified this enzyme.

F_1 has nine subunits of five different types, with the composition $\alpha_3\beta_3\gamma\delta\varepsilon$. Each of the three β subunits has one catalytic site for ATP synthesis. John E Walker determined the F_1 structure.

The knob-like portion of F_1 consists of alternating α and β subunits arranged like sections of an orange.

The polypeptide that makes up the stalk in the F_1 crystal structure is asymmetrically arranged with one domain of the single γ subunit, making up a central shaft that passes through F_1 and another domain of γ associated primarily with one of the three β subunits.

The F_0 complex, making up the proton pore, is composed of three subunits. a, b, and c in the proportion ab_2c_{12}. Subunit c is a small very hydrophobic polypeptide, consisting almost entirely of two transmembrane helices with a small loop extending from the matrix side of the membrane. The γ ε subunits of F_1 form a leg and foot that project from the bottom side of F_1 and stand as firm rings of (subunits as shown in Figure 8.21).

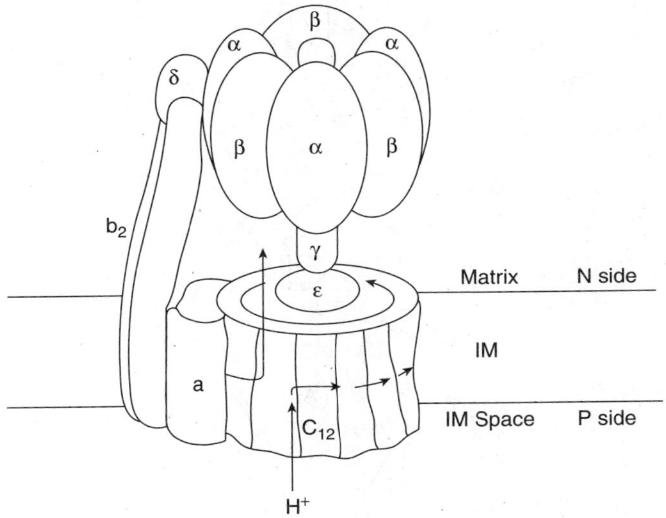

Figure 8.21 *Structure of ATP synthase*

MECHANISM FOR ATP SYNTHESIS

Rotational catalysis a key to the binding change mechanism for ATP synthesis. Paul Boyer proposed a mechanism in which the three active sites of F_1 take turns in the catalysis of ATP synthesis. A given β subunits start in the β-ADP conformation, which binds ADP and Pi from the surrounding medium. The subunit now changes conformation, assuming the β subunit changes to form that tightly binds and stabilises ATP. Finally, the subunit changes to β empty conformation, which has low affinity for ATP, and the newly synthesised ATP leaves the enzyme surface.

There is another round of analysis when this subunit again assumes the β-ADP forms and binds ADP and Pi. The conformational changes central to this mechanism are driven by the passage of protons through F_0 portion of ATP synthase. The streaming of protons through F_0 pore causes the cylinder of c subunits and the attached γ subunit to rotate about the long axis of γ, which is similar to the plane of membrane. The γ subunit passes through the centre of $\alpha_3\beta_3$ spheroid. With each rotation of 1200, γ comes into contact with a different β empty conformation.

The three β subunits interact in such a way that the first subunit assumes the empty configuration, the second subunit assumes the β-ADP form, and the third subunit the β-ATP form. Thus one complete rotation of the γ subunit causes each subunit to cycle through all the three possible conformations. Moreover, for each rotation, three ATPs are synthesised and released from the enzyme surface. The mode of operation of ATP synthase is shown in Figure 8.22.

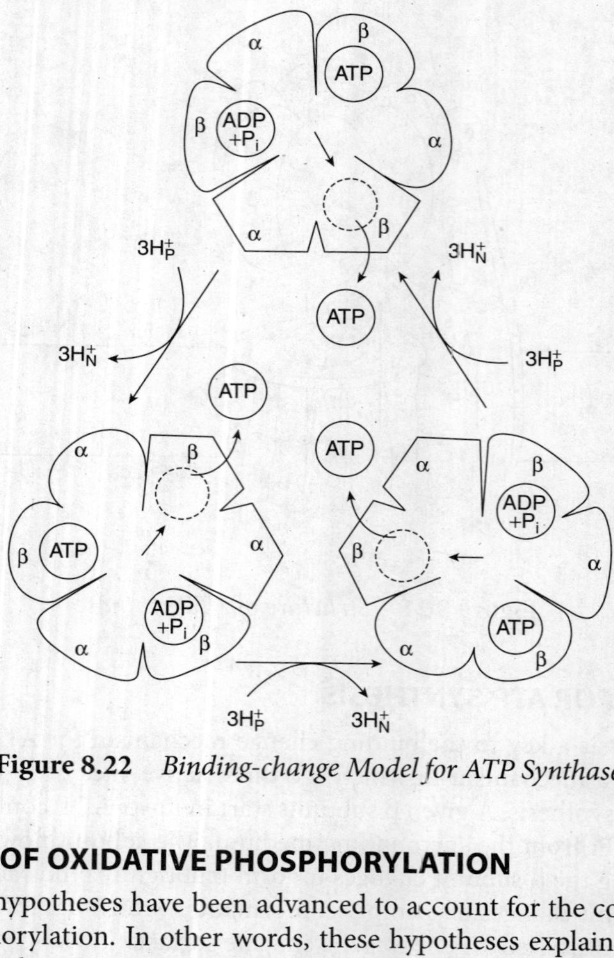

Figure 8.22 *Binding-change Model for ATP Synthase*

MECHANISM OF OXIDATIVE PHOSPHORYLATION

Three principal hypotheses have been advanced to account for the coupling of oxidation and phosphorylation. In other words, these hypotheses explain how the energy transfer between electron transport and ATP synthesis takes place.

1. Chemical coupling hypothesis
2. Conformational coupling hypothesis
3. Chemiosmotic coupling hypothesis

Chemical Coupling Hypothesis

This is the most accepted of the three hypotheses and proposes that electron transport is coupled to ATP synthesis by a sequence of consecutive reactions, in which a high energy covalent intermediate is formed by the electron transport and subsequently is cleaved and donates its energy to form ATP. The hypothesis explains the direct chemical coupling at all stages of the process. In a redox reaction between A reduced and B oxidised, and a factor I is incorporated into the formation of an energy-rich structure A oxidised ~ I, where ~ indicates a linkage having an energy rich nature.

$$\text{A red.} + \text{I} + \text{B oxi.} \rightarrow \text{A oxi.} - \text{I} + \text{B red.}$$

In subsequent reactions, an enzyme (E) replaces complex A oxidised~I to form an energy-rich E~I. Later, inorganic phosphate reacts with E~I to form phosphoenzyme complex E~P, containing the energy-rich enzyme–phosphate bond.

$$\text{A oxi}\sim\text{I} + \text{E} \rightleftharpoons \text{A oxi} + \text{E}\sim\text{I}$$
$$\text{E}\sim\text{I} + \text{Pi} \rightleftharpoons \text{E}\sim\text{P} + \text{I}$$

The enzyme phosphate component finally reacts with ADP to form ATP.

$$\text{E}\sim\text{P} + \text{ADP} \rightleftharpoons \text{E} + \text{ATP}$$

Conformational Coupling Synthesis

In mitochondria that are actively phosphorylating in the presence of an excess of ADP, the inner membrane pulls way from the outer membrane and assumes a condensed state. In the absence of ADP, the mitochondria have the normal structure or the swollen state, in which the cristae project into the large matrix.

According to this hypothesis, the energy released during the transport of electrons along the respiratory chain causes the conformational change. The inner mitochondrial membrane, that has this energy-condensed structure, in turn, is utilised for ATP synthesis as it changes to the energy-poor swollen conformation.

Chemiosmotic Coupling Hypothesis

Salient features

This is a simpler, radically different, and novel mechanism and was postulated by Peter Mitchell, a British biochemist in 1961. He proposed that "electron transport and ATP synthesis are coupled to high-energy intermediate or an activated protein" as shown in Figure 8.23.

According to this model, the transfer of electrons through the respiratory chain results in the pumping of protons (H^+) from the matrix side (M side) to the cytosol or cytoplasmic side (C-side) of inner mitochondrial membrane. The concentration of H^+ becomes positive on the cytoplasmic side. The hypothesis further proposes that H^+ ions ejected by electrons transport flow back into the matrix through a specific H^+ channel or 'pore' in the F_0F_1 ATPase molecule, driven by the concentration gradient of H^+. The free energy released as proton (H^+) flows back through the ATPase and causes the coupled synthesis of ATP from ADP and phosphate by ATP synthetase.

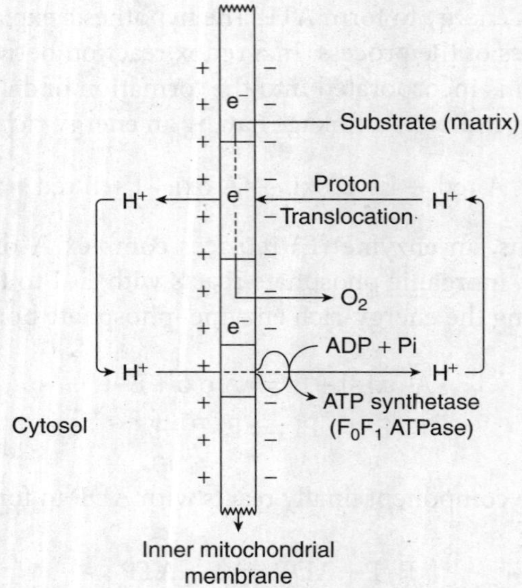

Figure 8.23 *Chemiosmotic Coupling Hypothesis*

RESPIRATORY INHIBITORS

The inhibitors bind to one of the components of ETC and block the transport of electrons. This causes the accumulation of reduced components before the inhibitor blockade step and oxidised components after the inhibitor step. The synthesis of ATP (phosphorylation) depends on electron transport. Hence, all the site-specific inhibitors of ETC also inhibit ATP formation. These possible action and inhibitors for the ETC are identified.

1. *Inhibitors of electron transport:*
   Rotenone, amytal, piericidin A, antimycin, dimercaprol cyanides, azide, H_2S, CO.
2. *Inhibitors of oxidative phosphorylation:*
   Oligomycin, Atractylate, Bongkrekic
3. *Uncouplers of oxidative phosphorylation:*
   2,4-Dinitrophenol (DNP), dicoumarol
4. *Ionophores of oxidative phosphorylation:*
   Valinomycin, gramicidin A, nigericin

Inhibitors of Electron Transport

The inhibitors involved in the electron transport arrest respiration by combing with membranes of the respiratory chain rather than with the enzymes that are involved in coupling respiration with ATP synthesis. They appear to act at three loci that may be identical to the energy transfer sites I, II, and III as stated in the below Figure.

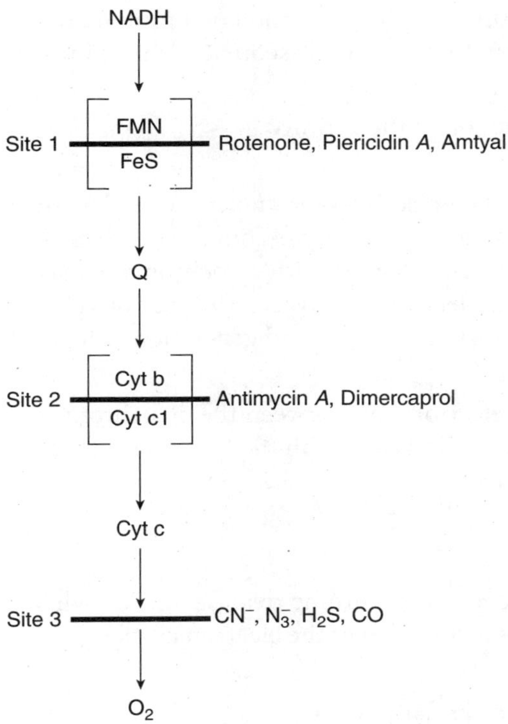

Site of Action of Inhibitors in ETC Chain

1. *Rotenone:*

   Rotenone is a compound extracted from the roots of tropical plant. It complexes avidly with NADH dehydrogenase and acts between Fe-S proteins and ubiquinone. Rotenone is relatively non-toxic to mammals because it is absorbed poorly although exposure of the lungs to dust is a little more dangerous.

2. *Piericidin A:*

   It is an antibiotic produced by streptomyces. It has an action similar to that of rotenone.

3. *Barbiturates:*

   Barbiturates block NADH dehydrogenase but it is required in much higher concentration for the purpose of sedative action on neural membranes and also for inhibition of respiration.

4. *Amytal:*

   Does not interfere with the oxidation of succinate because the electrons of these substrates enter the electron transport chain beyond the block of coenzyme Q.

5. *Antimycin:*

   It's an antibiotic produced by streptomyces. It inhibits the respiratory chain at or around site II and blocks electron flow between cytochromes b and c1, which prevent

ATP synthesis coupled to the generation of a proton gradient at site II. This block can be bypassed by the addition of ascorbate, which directly reduces cytochromes.

6. *Dimercaprol:*

   It is an identical action to the antimycin.

7. *Cyanides:*

   These are among the poisons better known by public. They enter the tissues very rapidly so that a sufficient quantity becomes lethal within a few minutes. The cyanide ion (CN^-) combines tightly with cytochrome oxidase aa_3^-, leading to the cessation of transfer of electrons to oxygen. The previous electron carriers in the chain accumulate in their reduced state, and generation of high energy phosphate stops.

8. *Azide:*

   It also blocks the electron flow between the cytochrome oxidase complex and O_2. Azide (N_3^-), also cyanide, reacts with the ferric form (Fe^{3+}) of this carrier.

9. *Hydrogen Sulfide (H_2S):*

   Its action is similar to HCN.

10. *Carbon Monoxide:*

    It also attacks between cytochrome oxidase and O_2 while cyanide and azide CO inhibit the ferrous form (Fe^{2+}) of the electron carrier.

OXIDATION-REDUCTION LOOP

Mitchell has explained the concept based on the fact that the reducing equivalents are transferred as H atoms by electron carriers (ubiquinone) and others (such as Fe-S centre and cytochrome). He explained that the hydrogen- and electron-carrying proteins are present in alternate in the respiratory chain to form three functional loops called the oxidation–reduction loops (= o/r loops). Each loop corresponds functionally to the coupling sites I, II, and III of chemical hypothesis.

PROTON TRANSPORT MECHANISM

According to Mitchell, the respiratory chain is folded into three oxidation reduction loops (o/r loops) as shown in Figure 8.24.

Each pair of electrons transferred from NADH to oxygen causes six protons to be translocated from inside to outside of the membrane. NADH first donates one H^+ and two electrons, which together with another H^+ from the internal medium reduces FMN to $FMNH_2$. FMN extends to full width of the coupling membrane so that it can release two to the exterior of the membrane and then return electrons to the inside via Fe-S proteins, which become reduced. Each reduced Fe-S complex donates one electron to ubiquinone (Q), which after accepting a proton (H^+) from inside, the membrane is reduced to QH_2 · QH_2, being small lipid soluble molecule, moves to the exterior of the membrane to discharge a pair of protons into the cytosol and donates two electron to moles of cytochrome b.

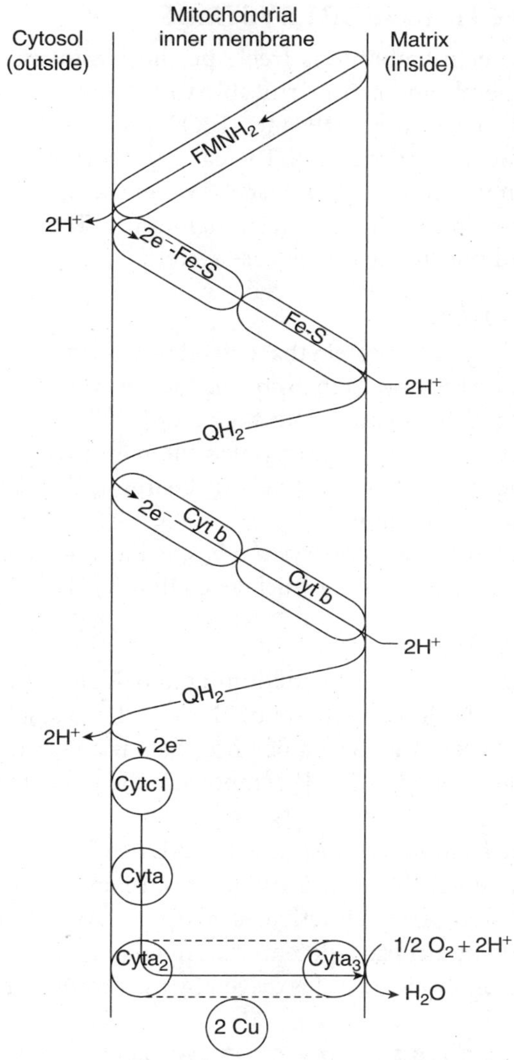

Figure 8.24 *Oxidation–Reduction Loop (Proton Transport Mechanism)*

Cytochrome b is thought to extend the mitochondrial membrane, enabling the electrons to join another molecule of ubiquinone along with two more protons from the internal medium. The QH_2, so produced moves to the outer surface to liberate two protons and two electrons are passed onto the two moles of cytochrome c. These electrons, passing through cytochrome a, then traverse the membrane to reach cytochrome a_3, which is located on the inner face of the membrane. Here two electrons combine with two protons from the internal medium and one O_2 atom to form a molecule of water.

INNER MEMBRANE TRANSPORT SYSTEMS

The mitochondrial outer membrane is freely permeable to most small solute molecules, but the inner membrane is impermeable to H^+, OA^-, K^+, and also many other ionic solutes. How, then, can the ADP^3 and HPO_4^{2-} produced in the cytosol enter the matrix and how can the newly formed ATP^{4-} leave again, since oxidative phosphorylation takes place within the inner matrix space? Two of the many specific transport systems are present in the inner mitochondrial membrane. They are adenine nucleotide translocase system and phosphate translocase system.

Adenine Translocase System

This system consists of a specific protein that extends across the inner membrane. It translocates one molecule of ADP inward in exchange for one molecule of ATP^{4-} coming out.

As the entrance of ADP is coupled to the exit of ATP, this system is better called ADP-ATP antiporter. Obviously, this transport system is moving more negative charges out than it is bringing in, so it is effectively discharging the outside positive electric potential across the inner membrane.

The adenine nucleotide transport system is specific since it carries ADP and ATP and not AMP or any other nucleotides such as GDP or GTP.

Phosphate Translocase System

This is the second membrane system functioning in oxidative phosphorylation. It promotes transport of $H_2PO_4^-$ along with that of H^+ from the cytosol into the matrix compartments as the entrance of the inorganic phosphate is coupled to the entrance of H^+. This system is aptly designated as Pi-H^+ symporter. The phosphate translocase system is specific for phosphate.

Thus, the combined action of phosphate and ADP-ATP translocase allows external phosphate and ADP to enter the matrix and the resulting ATP to return to the cytosol, where most of the ATP-requiring cell activities take place. An antiporter exchanges ADP^3 and ATP^{4-} and is driven by electrical potential across the membrane, where a symporter carries a proton and monobasic Pi and is driven by the proton concentration gradient.

SHUTTLE SYSTEMS: MITOCHONDRIAL OXIDATION OF CYTOSOLIC NADH

The NADH dehydrogenase of the inner mitochondrial membrane of animal cells can accept electrons only from NADH in the matrix. The inner membrane is not permeable to NADH. This has to be reoxidised to NAD^+ by O_2 via the respiratory chain. This can be overcome by the special shuttle systems carry reducing equivalents from cytosolic NADH into mitochondria by an indirect route.

1. Malate-aspartate shuttle system
2. Glycerol 3-phosphate shuttle system

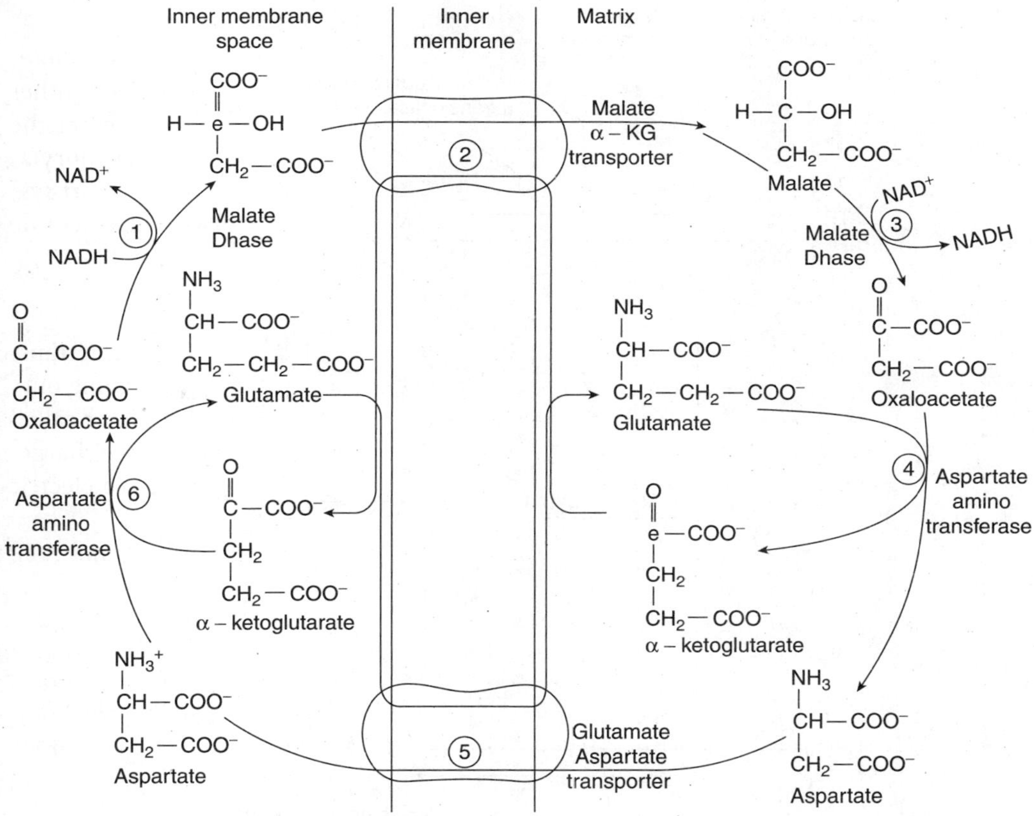

MALATE-ASPARTATE SHUTTLE SYSTEM

The most active NADH shuttle, which functions in liver, kidney, and heart mitochondria, is the *malate-aspartate shuttle.* The cytosolic reducing equivalents of NADH are first transferred to the cytosolic oxaloacetate to yield malate, and this is catalysed by cytosolic malate dehydrogenase. The malate thus formed passes through the inner membrane via the malate–ketoglutarate transporter. Within the matrix, the reducing equivalents are passed to NAD by the action of matrix malate dehydrogenase, forming NADH; this NADH can pass electrons directly to the respiratory chain. Cytosolic oxaloacetate must be regenerated by transamination reactions and the activity of membrane transporters to start another cycle of the shuttle.

Glycerol 3-phosphate Shuttle System

Skeletal muscle and brain use a different NADH shuttle, the *glycerol 3-phosphate shuttle.* It differs from the malate-aspartate shuttle in that it delivers the reducing equivalents from NADH to ubiquinone and thus into complex III, not complex I, providing only enough energy to synthesise 1.5 ATP molecules per pair of electrons.

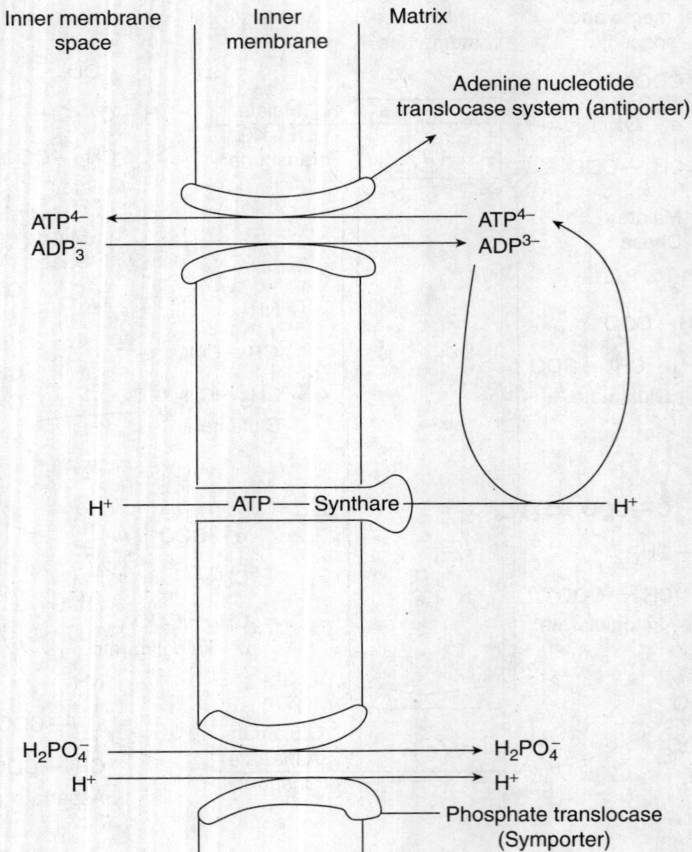

Summing up the reaction sequence:

$$\text{Glucose} + 38\ \text{Pi} + 38\ \text{ADP} + 6O_2 \rightarrow 6CO_2 + 38\ \text{ATP} + 44H_2O$$

Thus, one mole of glucose on complete oxidation to CO_2 and H_2O in the heart, liver, and kidney, where the malate-aspartate shuttle operates, leads to the production of 38 moles of ATP as shown in Table 8.3.

In skeletal muscles, where glycerol-3-phosphate shuttle operates in which a total of 36 ATP moles is generated when per mole of glucose gets oxidised.

Table 8.3 *ATP Yield from Complete Ordination of Glucose*

S.No	Reaction Sequence	ATP/GLUCOSE
	Glycolysis	+2
1	Dephosphorylation of 2 moles of 1,3 DPG	+2
2	Dephosphorylation of 2 moles of PEP	
3	Conversion of pyruvate to acetyl-CoA (2 NADH are formed)	
4	*TCA*	
5	2 moles of GTP are formed from 2 moles of succinyl CoA	+2
6	6 NADH are formed in oxidation of 2 moles each isocitrate, α-KG and mate.	
7	2 FADH2 are formed in oxidation of 2 moles of succinate	
8	Oxidative phosphorylation	
9	2 NADH (glycolysis) 2*3	+6
10	2 NADH (pyruvate to acetyl-CoA) 2*3	+6
11	2 FADH (TCA) 2*2	+4
12	6 NADH (TCA) 6*3	+18
13		40 – 2 = 38

P:O ratio

P:O ratio is defined as the number of moles of inorganic phosphate incorporated into organic form per atom of oxygen consumed.

$$\text{Malate-aspartate shuttle P:O ratio} = 38\ \text{ATP}/12\ O_2 = 3.16$$

$$\text{Glycerol 3-phosphate shuttle P:O ratio} = 36\ \text{ATP}/12\ O_2 = 3$$

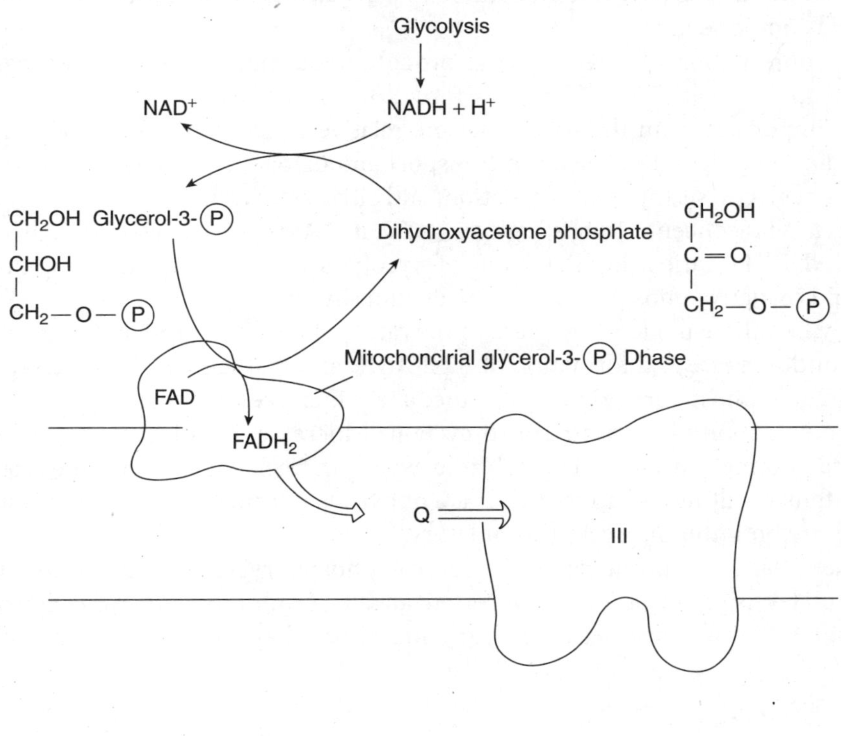

REGULATION OF OXIDATIVE PHOSPHORYLATION

The rate of respiration (O_2 consumption) is tightly regulated in the mitochondria. It is generally regulated by the availability of ADP as a basic substrate for phosphorylation. Depending on the rate of consumption of O_2 and also on the availability of Pi acceptor, this kind of control is called acceptor control of respiration. Availability of ADP is the one of the measures of the energy status of cells, [ATP]/[ADP][Pi]. Usually this ratio is very high, so the ATP-ADP system is fully saturated.

In some reactions, the energy-requiring rate is more; for example, protein synthesis, and the rate of breakdown of ATP to ADP and Pi increases, lowering the mass action ratio. With more ADP available for oxidative phosphorylation, the rate of respiration increases, causing the regeneration of ATP. This process continues until the ratio slows again. The rate of oxidation of cellular fuel is regulated with the availability of [ATP]/[ADP][Pi] ratio. In most of the tissues, even during extreme variations of energy demand this ratio gets fluctuated only slightly. In short, ATP is formed as fast as it is used in energy-requiring cellular activities.

Respiratory control among glycolysis, TCA, and oxidative phosphorylation.

The three energy-yielding stages in carbohydrate metabolism are glycolysis, TCA, and oxidative phosphorylation. Each stage is so regulated as to satisfy the time-to-time need of the cell for its products.

These three stages are coordinated with each other in such a way that they function most economically in a self-regulated way. They produce ATP and certain specific intermediates such as pyruvate and citrate that act as precursors for the biosynthesis of other cell components.

The coordination of three stages is brought about by the interlocking regulatory mechanism.

It is important from the figure that the relative concentrations of ATP and ADP control not only the rate of electron transport and oxidative phosphorylation but also the rates of glycolysis, pyruvate oxidation, and citric acid cycle.

When ATP concentration is high and ATP and AMP are correspondingly low (i.e. [ATP]/[ADP][Pi] ratio is high), the rate of glycolysis, pyruvate oxidation, the citric acid cycle, and oxidative phosphorylation are at a minimum.

But when there is a large increase in the rate of ATP utilisation by the cell, with the corresponding increased formation of ATP, AMP, and Pi, the rate of electron transport and oxidative phosphorylation will immediately increase. Simultaneously, the rate of pyruvate oxidation via the citric acid cycle will increase, thus increasing the flow of electrons into the respiratory chain. These events, in turn, will enhance the rate of glycolysis, thus resulting in an increased rate of pyruvate formation. Thus, the regulatory controls are both inhibitory and stimulatory.

Whenever ATP produced by oxidative phosphorylation and citrate increase to higher levels, they produce concerted allosteric inhibition of phosphofructokinase (PFK), the two together being more inhibitory than the sum of their individual effects.

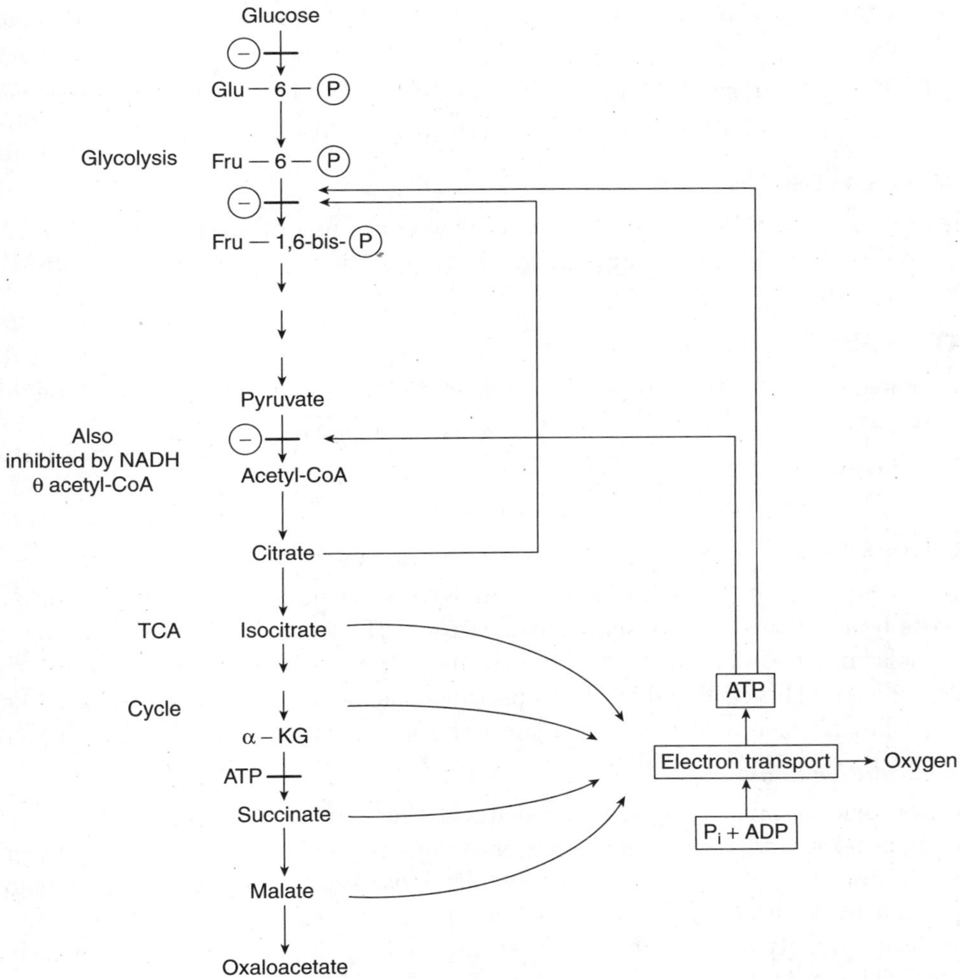

Figure 8.25 *Regulation of Oxidation Phosphorylation*

In addition, increased levels of NADH and acetyl-CoA also inhibit the oxidation of pyruvate to acetyl-CoA.

In nutshell, interlocking and regulatory mechanism control glycolysis so that pyruvate is produced at a rate at which it is required by the citric acid cycle. A process that donates electrons for oxidative phosphorylation is shown in Figure 8.25.

HEXOSE MONOPHOSPHATE PATHWAY

INTRODUCTION

Hexose monophosphate pathway or HMP shunt is also called pentose phosphate or phosphogluconate pathway. This is an alternative pathway of glycolysis and TCA cycle for oxidation of glucose. It does not generate ATP but has the following two major functions:

1. The generation of NADPH for reductive synthesis such as fatty acids and steroids
2. The provision of ribose residues for the nucleotide biosynthesis

LOCATION OF PATHWAY

The enzymes of HMP shunt are located in the cytosol tissues mostly heavily involved in lipid biosynthesis (liver, mammary gland, adipose tissue, adrenal cortex, testis, and erythrocytes).

PATHWAY

The sequence of reactions of HMP shunt is divided into two phases as shown in Figure 8.26.

1. Oxidative
2. Non-oxidative

Oxidative Phase

Glucose 6-phosphate dehydrogenase is an NADP-dependent enzyme that converts glucose 6-phosphate to 6-phosphogluconolactone. The later is then hydrolysed by the glucanolactone hydrolase to 6-phosphogluconate. The next reaction involving the synthesis of NADPH is catalysed by 6-phosphogluconate dehydrogenase to produce 3-keto 6-phosphogluconate, which then undergoes decarboxylation to give ribulose 5-phosphate.

Non-oxidative Phase

The non-oxidative reactions are concerned with the inter-conversion of 3–, 4–, 5–, and 7-carbon monosaccharides. Ribulose 5-phosphate is acted upon by an enzyme epimerase to produce xylulose 5-phosphate while ribose ketoisomerase converts ribulose 5-phosphate to ribose 5-phosphate. The enzyme transketolase catalyse the transfer of two carbon moiety from xylulose 5-phosphate to ribose 5-phosphate to give 3-carbon glyceraldehydes 3-phosphate and a 7-carbon sedoheptulose 7-phosphate. Transketolase depends on the coenzyme thiamine pyrophosphate (TTP) and Mg^{2+}.

Transaldolase brings about the transfer of a 3-carbon fragment from sedoheptulose 7-phosphate to glyceraldehydes 3-phosphate to give fructose 6-phosphate and 4-carbon erythrose 4-phosphate. Transketolase acts on xylulose 5-phosphate and transfers 2-carbon fragment (glyceraldehydes) from it to erythrose 4-phosphate to generate fructose 6-phosphate and glyceraldehyde 3-phosphate. Fructose 6-phosphate and glyceraldehyde 3-phosphate can be further catabolised through glycolysis and TCA. Glucose may also be synthesised from these two compounds.

SIGNIFICANCE OF HMP SHUNT

HMP shunt is unique in generating two important products: pentoses and NADPH needed for the biosynthetic reactions and other functions. The sequence of reactions of HMP shunt is divided into two phases as shown in Figure 8.26.

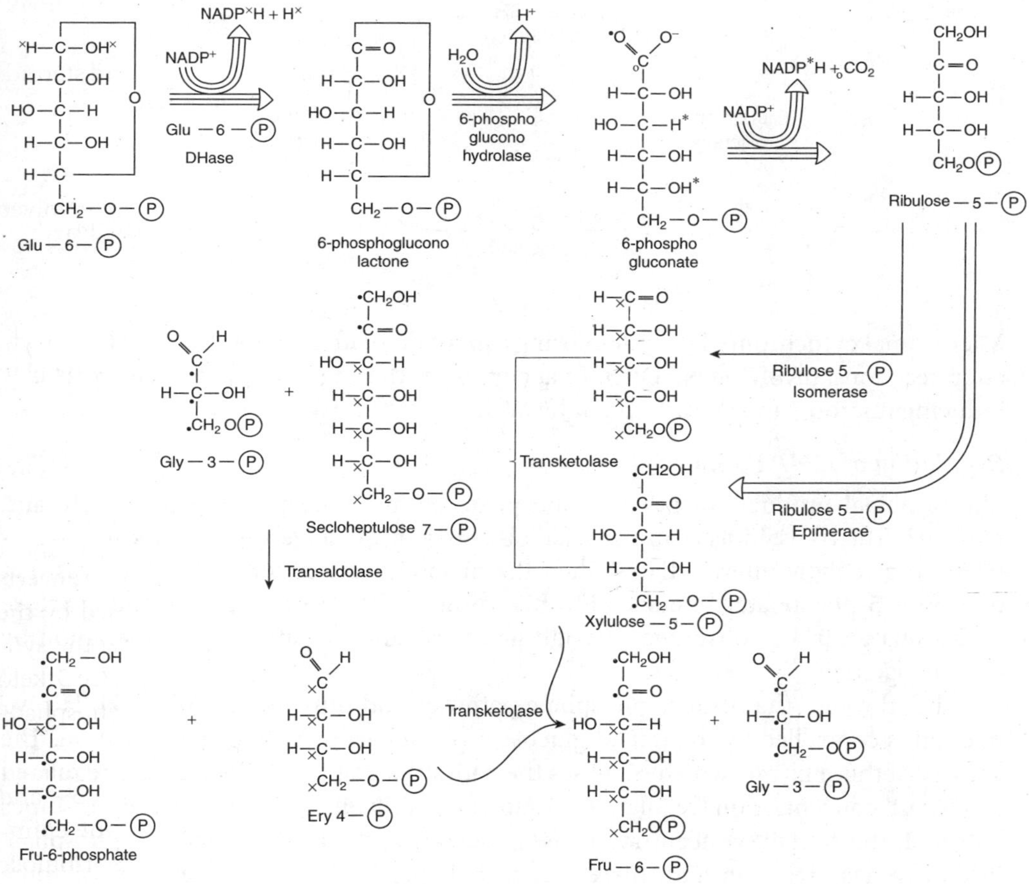

Figure 8.26 *Hexose Monophosphate Pathway*

Importance of Pentoses

In the HMP shunt, hexoses are converted into pentoses, the most important being ribose 5-phosphate. It is useful for the synthesis of nucleic acids (RNA and DNA) and many nucleotides such as ATP, NAD^+, FAD, and CoA.

Importance of NADPH

NADPH is needed for the reductive biosynthesis of fatty acids and steroids, and therefore, HMP occurs more actively in the tissues concerned with lipogenesis, for example, adipose tissue and liver. NADPH is used in the synthesis of amino acids involving the enzyme glutamate dehydrogenase.

There is a continuous production of H_2O_2 in living cells, which is chemically damaged unsaturated lipids, proteins, and DNA. This is prevented to a large extent through antioxidant reactions involving NADPH, glutathione-mediated reactions of H_2O_2.

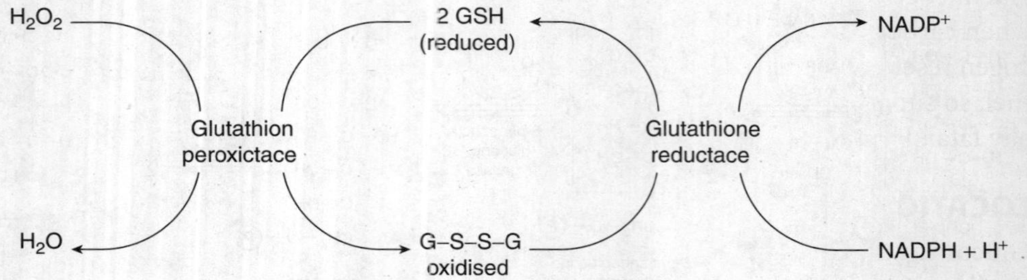

Microsomal cytochrome P_{450} system brings about detoxification of drugs by hydroxylation reactions, involving NADPH. Phagocytosis is the engulfment of foreign particles inducing microorganism, carried out by WBC.

Regulation of HMP Pathway

The principal products of pentose phosphate pathway are ribose 5-phosphate and NADPH. The transaldolase and transketolase reactions convert excess ribose 5-phosphate to glycolytic intermediates when the metabolic needs for NADPH excess that of ribose 5-phosphate in nucleotide biosynthesis. The resulting gap and fructose 6-phosphate can be consumed through glycolysis and oxidative phosphorylation by gluconeogenesis.

Flux through the pentose phosphate pathway, and thus the rate of NADPH production is controlled by the rate of glucose 6-phosphate dehydrogenase reaction. The activity of this enzyme, which catalyses the pathway's first committed step, is regulated by $NADP^+$ concentration (regulation by substrate availability). When the cell consumes NADPH, the $NADP^+$ concentration rises, increasing the rate of glucose 6-phosphate dehydrogenase reaction and, therefore, stimulating NADPH regeneration. In some tissues, the amount of enzyme synthesised also appears to be under hormonal control. Deficient of glucose 6-phosphate dehydrogenase results in syndrome of hemolytic anemia, and Wernicke-Korsakoff syndrome is associated with transketolase activity (genetic disorder).

GLUCONEOGENESIS

INTRODUCTION

The synthesis of glucose of glycogen from a non-carbohydrate compounds is known as gluconeogenesis. The major substrates precursors for gluconeogenesis are lactate and pyruvate, glucogenic amino acids, propionate, and glycerol.

BIOMEDICAL IMPORTANCE

Gluconeogenesis is the term used to include all pathways responsible for converting all non-carbohydrate precursors to glucose or glycogen. Liver and kidney are the major gluconeogenic tissues. Gluconeogenesis meets the glucose requirement of the body when carbohydrate is not available in the required amount from the diet or from glycogen reserves. Erythrocytes and nervous systems especially require glucose as energy fuel, so supply of glucose is a must for these tissues. Failure of gluconeogenesis is usually fatal. Failure of gluconeogenesis is usually fatal.

LOCATION

Gluconeogenesis mainly occurs in the cytosol. Gluconeogenesis mostly takes place in the liver and, to some extent, in the kidney matrix.

REACTIONS OF GLUCONEOGENESIS

Gluconeogenesis closely resembles the reversed pathway of glycolysis, the citric acid cycle, and some special reactions as shown in Figure 8.27. The three stages bypassed by alternate enzymes specific to gluconeogenesis are discussed below:

1. The conversion of fructose 1, 6 bis phosphate to fructose-6-p
2. The conversion of glucose 6-phosphate to glucose
3. The conversion of pyruvate to phosphoenol pyruvate

The following are the enzymes involved in gluconeogenesis:

- The conversion of fructose 1,6-bisphosphate
- Pyruvate carboxylase, phosphoenol pyruvate carboxylase
- Fructose α1-6 linkage bisphosphate
- Glucose 6-phosphate

Conversion of Pyruvate to Phosphoenol Pyruvate

This takes place in two steps. Pyruvate carboxylase is a biotin-dependent mitochondrial enzyme that converts pyruvate to oxaloacetate in the presence of ATP and CO_2. This enzyme regulates gluconeogenesis. Oxaloacetate is synthesised in the mitochondrial matrix. It has to be transported to cytosol to be used in gluconeogenesis, where the rest of the pathway occurs. The reversible conversion of oxaloacetate to malate is catalysed by malate dehydrogenase, an enzyme present in both mitochondria and cytosol. In the cytosol, phosphoenol pyruvate carboxylase converts oxaloacetate to phosphoenol pyruvate.

Conversion of Fructose 1,6-bisphosphate to Fructose 6-Phosphate

Phosphoenol pyruvate undergoes the reversal of glycolysis until fructose 1,6-bisphosphate is produced. The enzyme fructose 1,6-bisphosphatase converts fructose 1,6 bisphosphate to fructose 6-phosphate.

Conversion of Glucose 6-phosphate to Glucose

Glucose 6-phosphatase catalyses the conversion of glucose 6-phosphate to glucose.

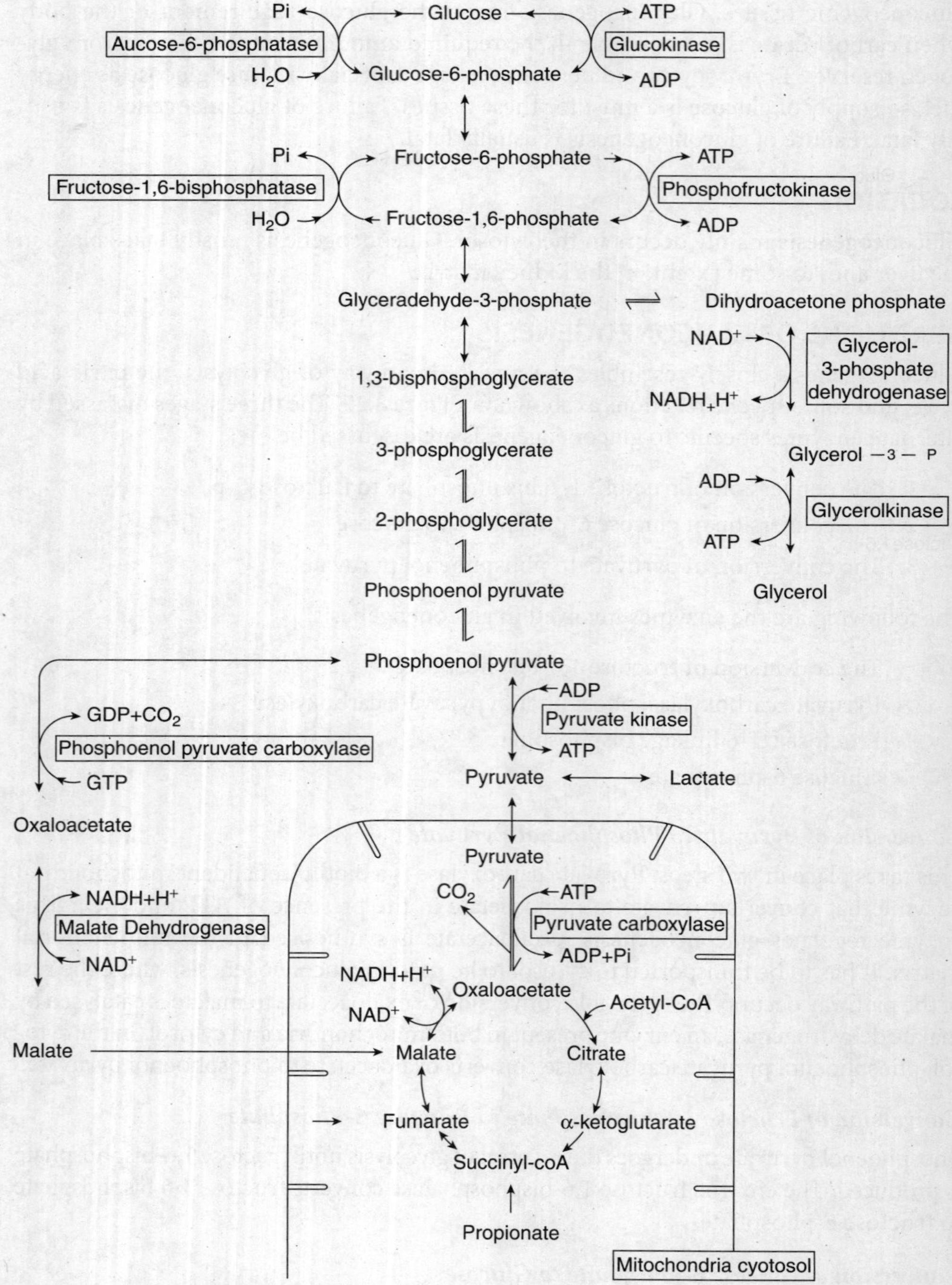

Figure 8.27 *Pathway of Gluconeogenesis and Glycolysis in Liver*

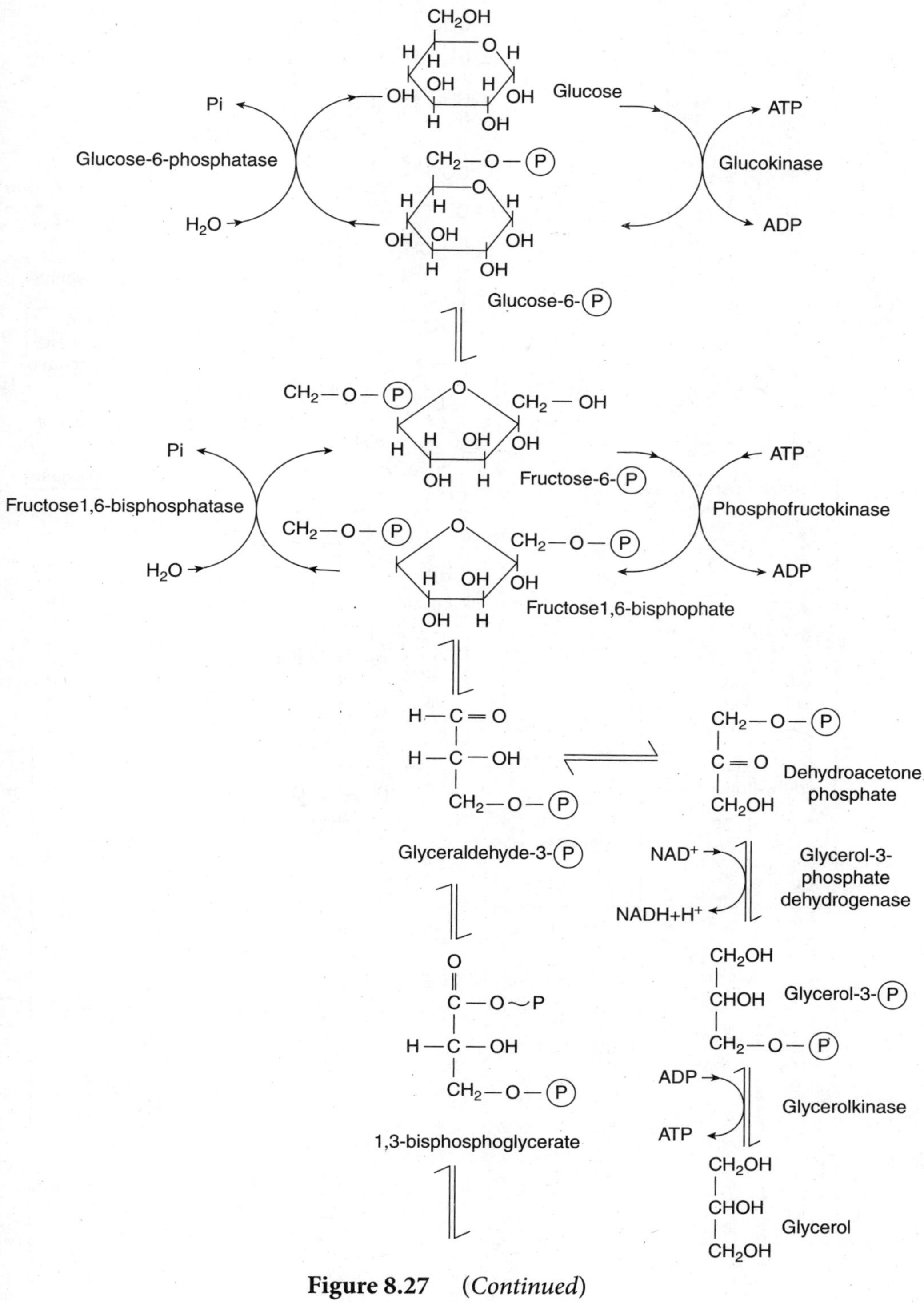

Figure 8.27 *(Continued)*

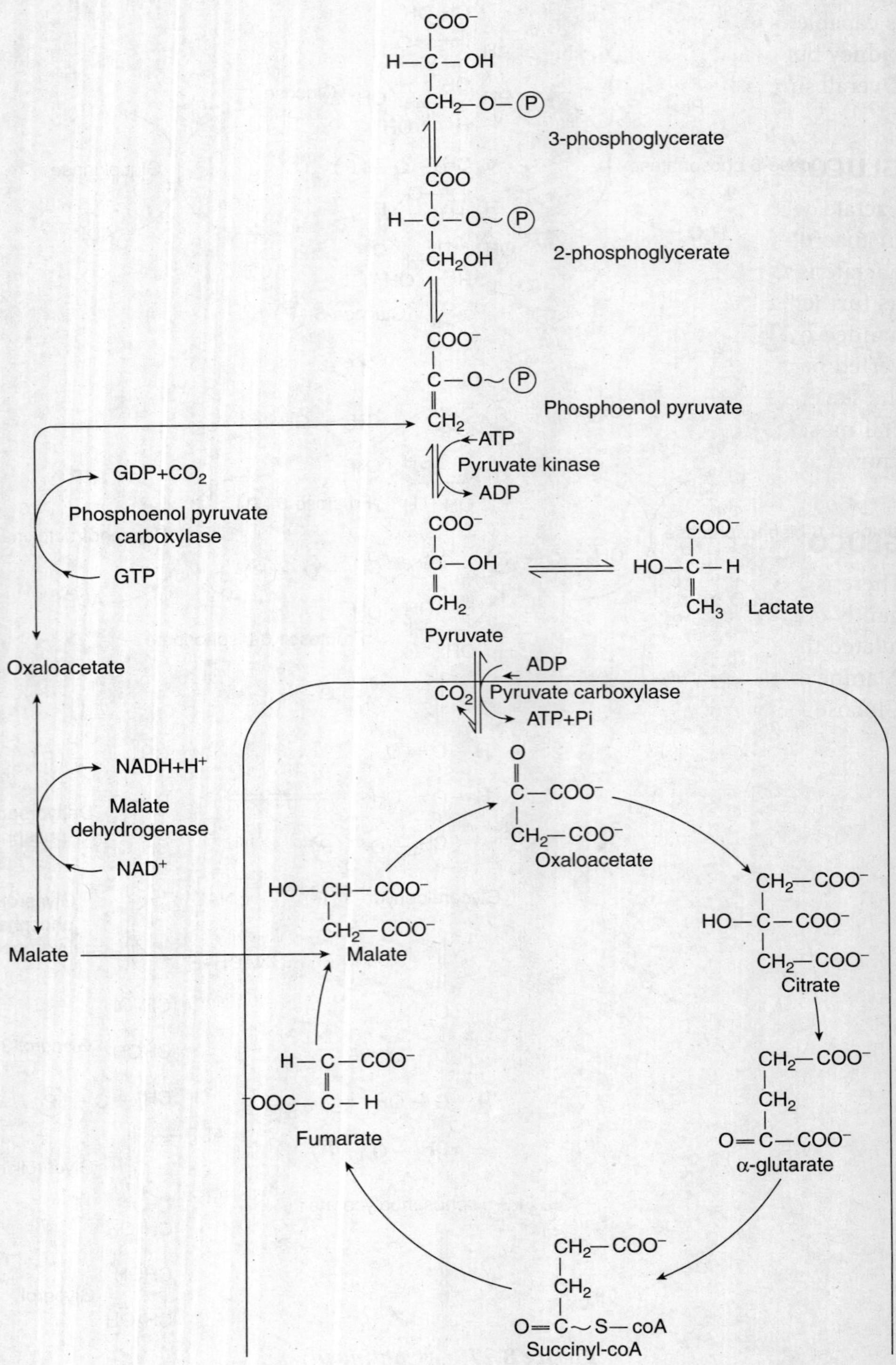

3-phosphoglycerate
2-phosphoglycerate
Phosphoenol pyruvate
ATP
Pyruvate kinase
ADP
GDP+CO2
Phosphoenol pyruvate carboxylase
GTP
Pyruvate
Lactate
Oxaloacetate
ADP
CO2
Pyruvate carboxylase
ATP+Pi
NADH+H+
Malate dehydrogenase
NAD+
Oxaloacetate
Malate
Malate
Citrate
Fumarate
α-glutarate
Succinyl-coA

The presence or absence of this enzyme in a tissue determines whether the tissue is capable of contributing glucose to the blood or not. It is mostly present in liver and kidney but not in muscle, brain, and adipose tissue.
Overall summary of gluconeogenesis:-

GLUCONEOGENESIS FROM CORI CYCLE

Lactate, which is produced in the skeletal muscle, is the precursor for gluconeogenesis. In anaerobic conditions, pyruvate is reduced to lactate by using lactate dehydrogenase. Lactate is the end product of glycolysis, since it must be reconverted to pyruvate for its further metabolism. Lactate is carried from the skeletal muscle through blood and handed over to the liver, where it is oxidised to pyruvate. Pyruvate so formed is converted back to glucose through gluconeogenesis pathway and further transported to the skeletal muscle. The cycle involving the synthesis of glucose in the liver from skeletal muscle lactate and the reuse of glucose by the muscle for the energy purpose is known as 'Cori cycle'.

GLUCONEOGENESIS FROM GLUCOSE-ALANINE CYCLE

There is a continuous transport of amino acids from muscle to liver, which predominantly occurs during starvation. Alanine dominates among the amino acids. It is postulated that pyruvate in skeletal muscle undergoes transamination to produce alanine. Alanine is transported to liver and used for gluconeogenesis. This cycle is referred to as Glucose Alanine Cycle (Figure 8.28).

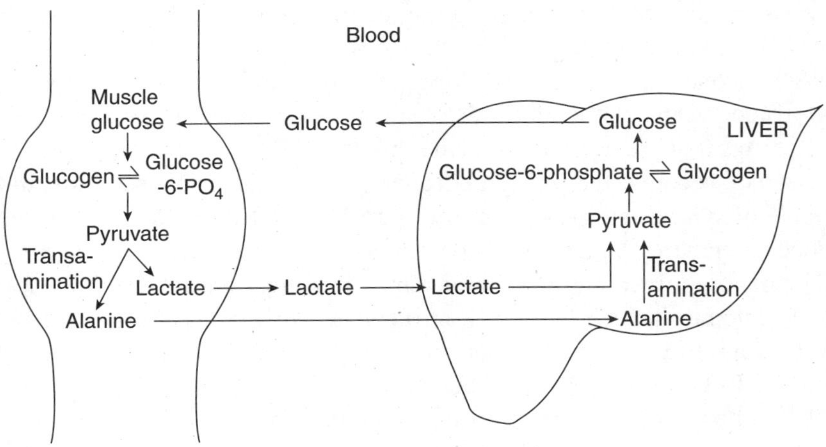

Figure 8.28 *The Cori Cycle or (Glucose Alanine Cycle)*

REGULATION OF THE PATHWAY

The hormone glucagons and the availability of substrates mainly regulate gluconeogenesis as shown in Figure 8.29. Glucagon is a hormone secreted by α cells of the pancreatic islets. It stimulates gluconeogenesis. Availability of substrates among the various substrates, glucogenic amino acids, has stimulating influence on gluconeogenesis. This is particularly important in a condition like diabetes mellitus where amino acids are mobilised from muscle protein for the purpose of gluconeogenesis.

SIGNIFICANCE OF THE PATHWAY

Fructose 2,6-bisphosphate plays a unique role in regulation of gluconeogenesis in liver. The concentration of the blood glucose is regulated within narrow limits. Since glycolysis and gluconeogenesis share the same pathway but operate in opposite directions, their activities are regulated reciprocally. The pathway of gluconeogenesis in liver and kidney utilises those reactions in glycolysis, which are reversible, plus four additional reactions that circumvent the irreversible non-equilibrium reactions. Glucagon is secreted as a response to hypoglycemia and activates both gluconeogenesis and glucogenolysis in the liver, causing release of glucose into the blood.

URONIC ACID PATHWAY

In the liver, glucose is converted to glucuronic acid and ascorbic acid (except in human beings and other species in which ascorbate and pentose are produced) through uronic acid pathway. It is also considered as an alternative pathway that does not lead to the formation of ATP. Glucose 6-phosphate reacts with uridine triose phosphate to form uridine diphosphate glucose (UDPGlc) in a reaction catalysed by UDPGlc pyrophosphorylase, as in glycogen synthesis. UDP-glucuronate occurs in a two-step reaction catalysed by UDPGlc dehydrogenase.

UDP-glucuronate is the precursor for glucuronate and is involved in its incorporation into proteoglycans or for synthesis of substrates such as steroid hormones, and a number of drugs that are excreted in urine or bile as UDP-glucuronide conjugates.

Glucuronate is converted to L-glucuronate; the precursor of ascorbate, in the animals capable of synthesising this vitamin in an NADPH-dependent reaction. Due to the absence of enzyme L-gulonolactone oxidase, ascorbic acid is not synthesised in humans, primates, guinea pigs, bats, and some birds and fishes. L-gulonate is oxidised to 3-keto-L-gulonate, which is then decarboxylated to L-xylulose. L-xylulose is converted to D-isomer by an NADPH-dependent reduction of xylitol. Furthermore, xylitol is oxidised to D-xylulose in an NAD-dependent reaction. After the conversion to D-xylulose 5-phosphate, it is metabolised via pentose phosphate pathway as shown in Figure 8.30.

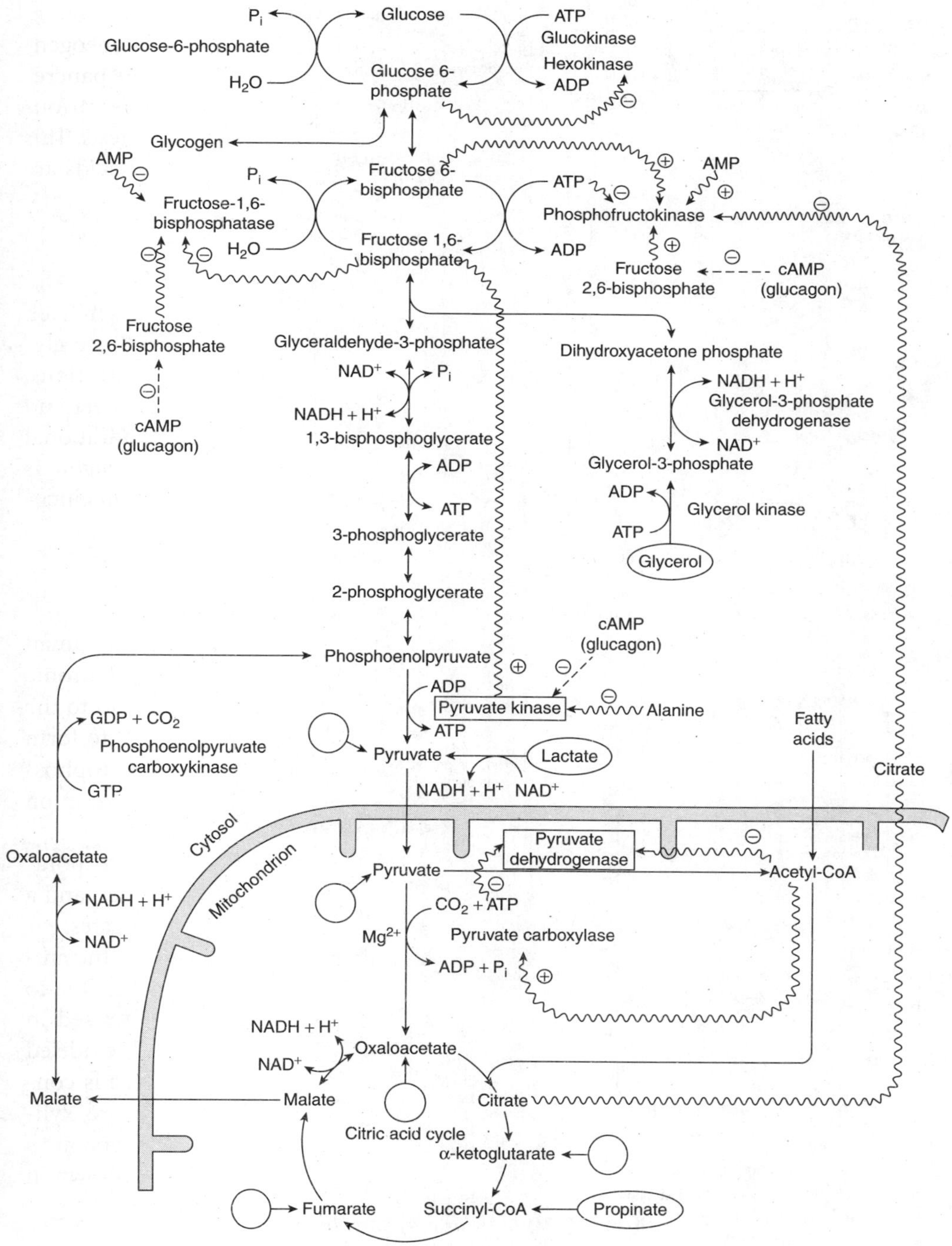

Figure 8.29 *Biosynthetic Pathway and Regulation of Gluconeogenesis*

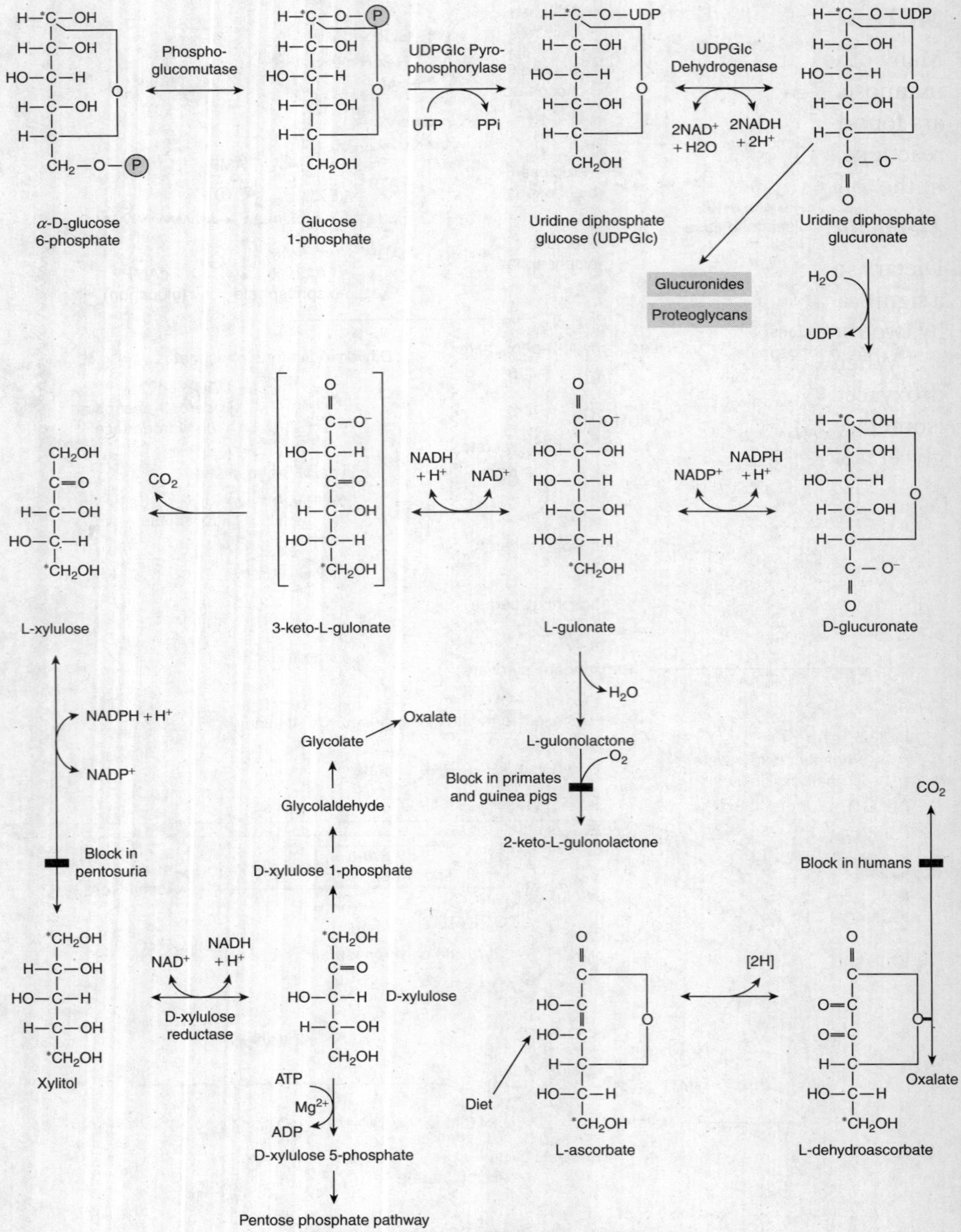

Figure 8.30 *Uronic Acid Pathway*

METABOLISM OF OTHER SUGARS

Many other sugars are consumed by human beings. They are fructose, galactose, and mannose. Beside glucose, fructose, galactose, and mannose are the most common sugars found in oligosaccharides and polysaccharides. They are also energy sources. The reactions by which these sugars are converted into glycolytic intermediates are shown in the following pathways.

Metabolism of Fructose

Dietary sources of fructose include fruit, honey, and the disaccharide sucrose. Fructose is a significant source of carbohydrate in the human diet. It can enter the glycolytic pathway by two routes. In liver, fructose is converted to fructose-1-phosphate by fructose kinase.

When fructose-1-phosphate enters the flycolytic pathway, it is first split into dihydroxyacetone phosphate (DHAP) and glyceraldehyde-3-phosphate by triose phosphate isomerase. Glyceraldehyde-3-phosphate is generated from glyceraldehyde and ATP by glyceraldehyde kinase as shown in Figure 8.31.

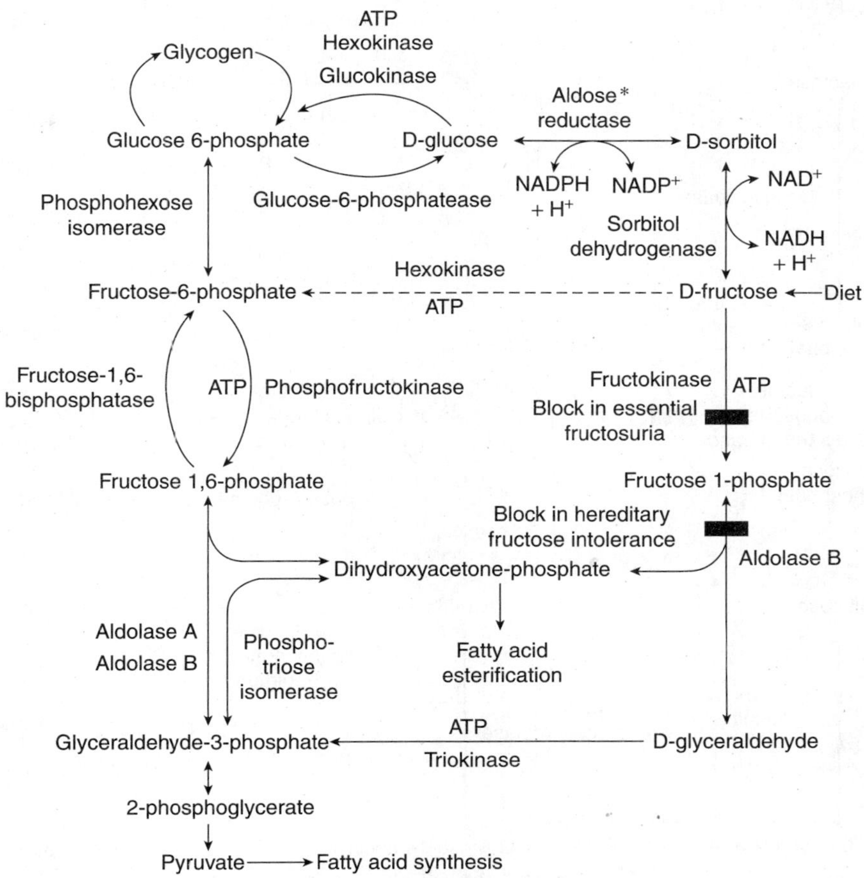

Figure 8.31 *Metabolism of Fructose*

The conversion of fructose-1-phosphate into glycolytic intermediates bypasses two regulatory steps by hexokinase and PFK-1. Thus, fructose is metabolised more quickly than glucose.

In muscle and adipose tissue, fructose is converted into glycolytic intermediate fructose-6-phosphate by hexokinase. As the hexokinase has a low affinity for fructose, this reaction is of minor importance unless fructose consumption is exceptionally high.

Galactose Metabolism

The galactose and glucose have similar structures; that is, they are epimerase. Several reactions are required for this sugar to enter the glycolytic pathway. Galactose is initially converted to galactose-1-phosphate by galactokinase.

The galactose-1-phosphate is transformed into the nucleotide derivative UDP galactose. During fetal development and childhood, the first step in this conversion is catalysed by galactose-1-phosphate by uridyltransferase. Beginning in adolescence, UDP-galactose is produced in a reaction caralysed by UDP-galactose pyrophosphorylase as shown in Figure 8.32.

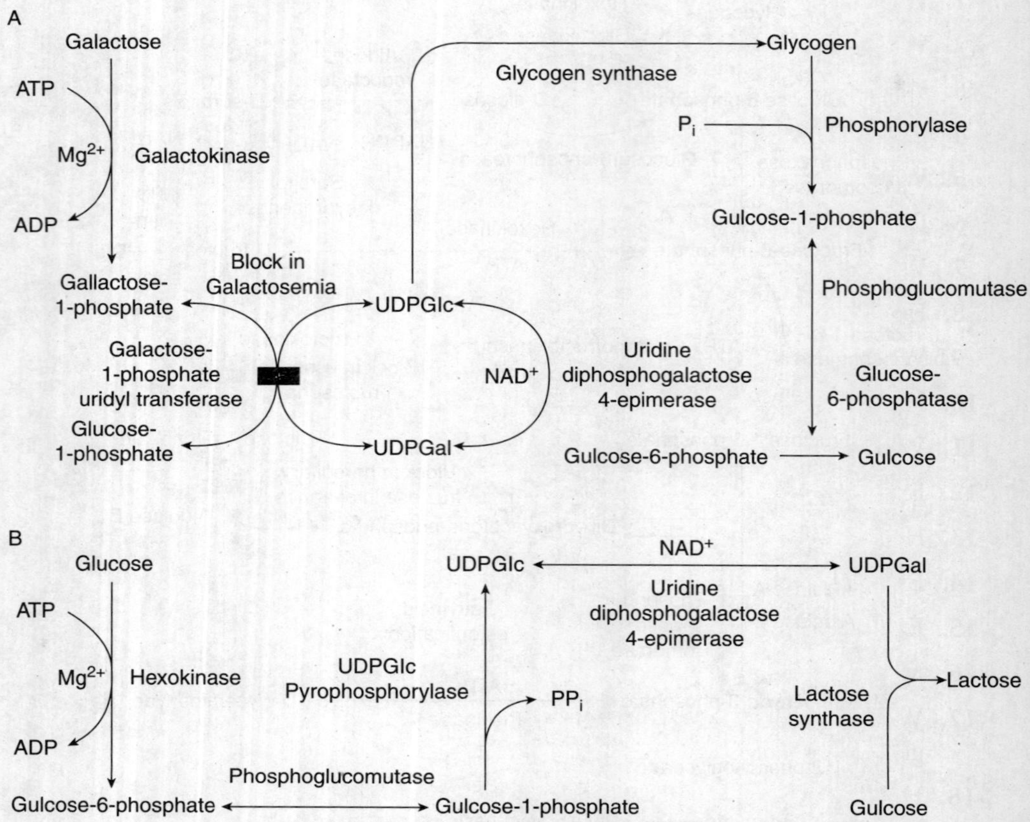

Figure 8.32 *Metabolism of Galactose*

Then UDP -glucose is formed by the isomerisation of galactose catalysed by UDP-glucose-4-epimerase. Depending on the cell's metabolic needs, UDP-glucose is used directly in glycogen synthesis or is converted to glucose-1-phosphate by UDP-glucose pyrophosphorylase. Glucose-1-phosphate enters the glycolytic pathway after its conversion to glucose-6-phosphate by phosphoglucomutase.

Metabolism of Mannose

Mannose is an important component of the oligosaccharides that are found in glycoproteins. As it is a minor component in the diet, mannose enters the glycolytic pathway as fructose-6-phosphate.

QUESTION BANK

SECTION A

1. What is PO ratio?
2. What is substrate level phosphorylation?
3. What are anaplerotic reactions?
4. Explain Cori cycle.
5. Write about debrancing enzyme.
6. Define cellular respiration.
7. What are glucogenic amino acids?
8. Explain the action of ATPase?
9. What is the importance of malate shuttle?
10. How is starch degraded?
11. What is the significance of LDH reaction?
12. Explain the mechanism of action of α-KG dehydrogenase.
13. How is lactose synthesised?
14. Explain the action of isocitrate lyase.
15. Explain the action of phosphoenol pyruvate carboxykinase.
16. Muscle glycogen cannot directly contribute to blood glucose. Why?
17. What are the important intermediates of TCA cycle that can be converted into glucose?
18. How is fructose converted to glucose?
19. When does acetyl-CoA enter into ketogenesis?

SECTION B

1. Discus the malate/aspartate shuttle.
2. Explain how glycolysis and gluconeogenesis are reciprocally regulated.
3. How many ATPs are obtained when one molecule of glucose is metabolised in aerobic glycolytic pathway?
4. What is an amphibolic pathway? Explain the amphibolic nature of the TCA cycle.
5. Discuss the glyoxylate pathways. Explain the significance of glyoxylate pathways.
6. Explain the role of acetyl-CoA as an intermediary metabolite.
7. What is the amount of energy released during the hydrolysis of ATP?
8. Explain aerobic glycolysis.
9. Explain the various steps involved in the process of glycogenolysis.
10. How does acetyl-CoA act as the intermediary compound of lipid and carbohydrate metabolism?
11. Explain the role of cytochrome oxidase in the electron transport chain.
12. What is the role of liver in maintaining blood glucose level?
13. How is glucose converted to ribose in HMP shunt?
14. How is galactose metabolised?
15. How is propoinonic acid converted to TCA cycle intermediate compound?

SECTION C

1. Explain the sequence of electron transport through the electron transport chain.
2. Discus the reactions and regulation of glycolytic pathways.
3. What is gluconeogenesis? Explain the synthesis of glucose from non-carbohydrate processors.
4. Discus hexose monophosphate shunt. Detail its importance.
5. What are the different members of the electron transport chain? Explain their action.
6. What are the important steps of the TCA cycle? How is it regulated?
7. How is glycogen synthesised? Explain the role of cAMP in the regulation of glycolysis metabolism.
8. Explain the chemiosmotic theory of oxidative phosphorylation.

CHAPTER

9

AMINO ACID METABOLISM

Concept Outline

- ❑ Introduction
- ❑ Catabolism of Amino Acids
- ❑ Decarboxylation
- ❑ Urea Cycle
- ❑ Phenylalanine and Tyrosine
- ❑ Synthesis of Melanin
- ❑ Synthesis of Epinephrine
- ❑ Synthesis of Thyroid Hormones
- ❑ Synthesis of Creatine

INTRODUCTION

In vertebrates, deamination takes place particularly in liver cells, where the excess of amino acid supplied by eaten food is deaminated. Ammonia produced in this way is highly toxic and must be disposed of. Elimination often occurs in the form of uncharged ammonia or a more complex nitrogenous substance, which is either uric acid or urea.

Based on the nature of the nitrogenous excretory end product, animals are classified into three main categories:

1. Ammonotelic
2. Uricotelic
3. Ureotelic

AMMONOTELIC

Animals which excrete their nitrogenous wastes as ammonia are said to be ammonotelic. Excretion of ammonia, which is presumed to be the primitive material, is especially suited to an aquatic habit. Ammonia is highly soluble in water, with which it forms ammonium hydroxide (NH_4OH). The latter passes readily across the cell membranes and is dispersed into surrounding water. Ammonia is found as chief excretory product in the majority of primitively aquatic animals and certain protozoa, annelids, crustaceans, some mollusks, and teleost fishes. Anuran (amphibian) tadpoles excrete mainly ammonia, whereas the adults produce urea.

URICOTELIC

Animals which excrete their nitrogen predominantly in the form of uric acid are called uricotelic. All genuinely terrestrial animals, including insects, terrestrial gastropods, terrestrial reptiles (snakes and lizards), and birds excrete mainly uric acid. Insects and land snails excrete solid pellets of uric acid, whereas reptiles and birds void a thick paste of uric acid crystals.

UREOTELIC

Animals which excrete urea as the major excretory product are said to be ureotelic. Urea is a useful product for a semi-terrestrial animal that is not exposed to scarcity of water but is not necessarily surrounded by large quantities of it. It is, therefore, the main product of some earthworms, shore gastropods, adult amphibians, and semi-terrestrial tortoises, turtles, and mammals. In humans, blood contains about 0.04 g/100 ml of urea, whereas the level of urea in the bladder is 2 g/100 ml of urine.

CATABOLISM OF AMINO ACIDS

INTRODUCTION

The amino acid obtained from dietary source or body protein turnover is utilised for protein biosynthesis and the production of a wide range nitrogen-containing compounds (creatine, amines, phorphyrin, and so on).

The amino parts of amino acid are utilised for the biosynthesis of urea; the remaining carbon skeletons are converted to various intermediates of TCA cycle. An overview of amino acid metabolism is shown in Figure 9.1.

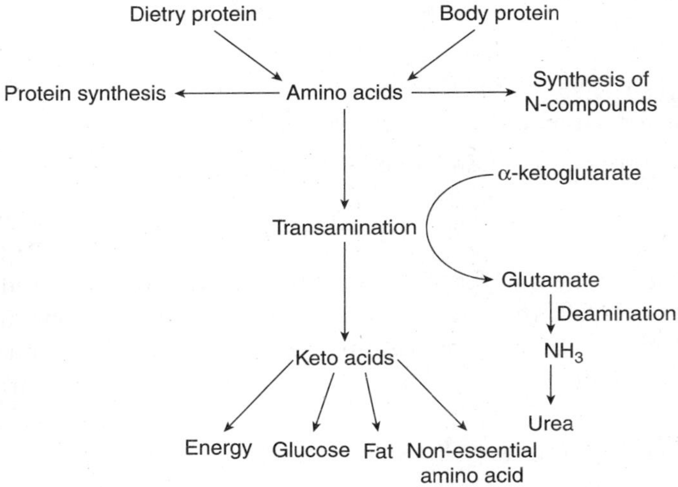

Figure 9.1 *An Overview of Amino Acid Metabolism*

The structure given below (Figure 9.2) shows the products formed from carbon skeleton of amino acids.

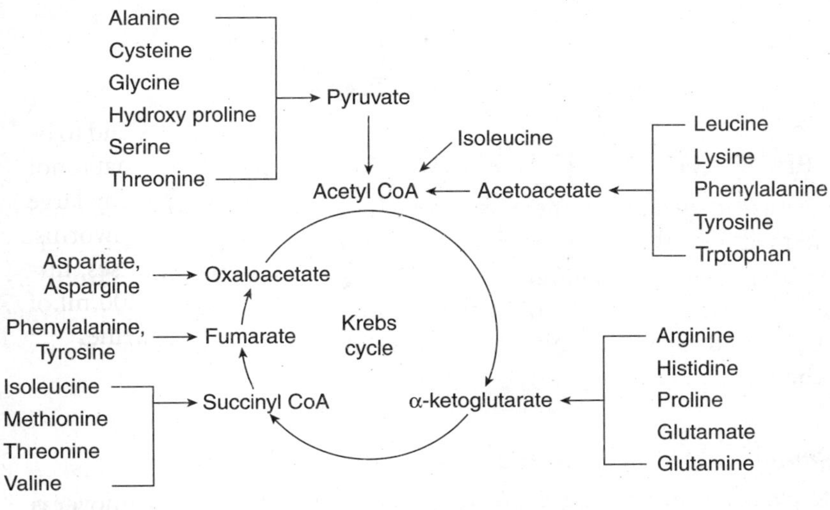

Figure 9.2 *Summary of the Products Formed from Carbon skeleton of Amino Acids*

Amino acids undergo common reactions like transamination followed by deamination for the liberation of ammonia. The amino group of the amino acids is utilised for the formation of urea, which is an excretory end product of protein metabolism. The carbon skeleton of the amino acid is first converted into keto acids (by transamination), which meet one or more of the following fates:

1. Utilised to generate energy
2. Used for the synthesis of energy
3. Diverted for the formation of fat or ketone bodies
4. Involved in the production of non-essential amino acid

TRANSAMINATION

Transamination is defined as the transfer of amino group ($-NH_2$) from an amino acid to keto acid. This process entails the inter-conversion of a pair of amino acids and a pair of keto acids catalysed by a group of enzymes known as transaminases (amino transferases) as shown in Figure 9.3.

TRANSAMINATION REACTION

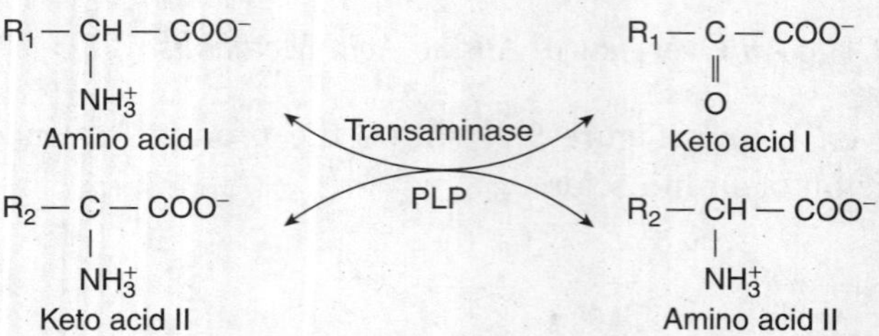

Figure 9.3 *Transamination Reaction*

SUBSTRATE SPECIFICITY OF AMINO TRANSFERASES

Each amino transferase is specific for one or at most a few amino group donors. Amino transferases are named after the specific amino group donors because the acceptor of the amino acid is almost always α-ketoglutarate. The most important amino transferase reactions are catalysed by alanine amino transferase or serum glutamate pyruvate transaminase (SGPT) and aspartate amino transferase (AST) or serum glutamate oxaloacetate transaminase (SGOT).

SGPT (Serum Glutamate Pyruvate Transferase)

Serum glutamate pyruvate transferase is present mostly in liver. This enzyme catalyses the transfer of the amino group from alanine to α-ketoglutarate, resulting in the formation of pyruvate and glutamate. This reaction is reversible, so during amino acid metabolism the enzyme proceeds towards the direction of glutamate synthesis. Thus, glutamate is considered to be the collector of nitrogen from alanine.

SGOT (Serum Glutamate Oxaloacetate Transferase)

Serum glutamate oxaloacetate transferase, also known as aspartate transaminase (AST), is the exception to the rule that amino transferase funnels amino groups to form glutamate. AST transfers amino groups from glutamate to oxaloacetate, thus forming aspartate, which acts as a source of nitrogen in the urea cycle.

MECHANISM OF ACTION OF AMINO TRANSFERASE

All amino transferase requires the coenzyme pyridoxal phosphate (a derivative of vitamin B6), which is covalently linked to an e-amino group of a specific lysine residue at the active site of the enzyme (Schiff base) as shown in Figure 9.4.

Amino transferase transfers the amino group from an amino acid to the pyridoxal part of the coenzyme to produce pyridoxamine phosphate. The pyridoxamine form of coenzyme then reacts with an α-keto acid to form an amino acid and returns to its original form of coenzyme.

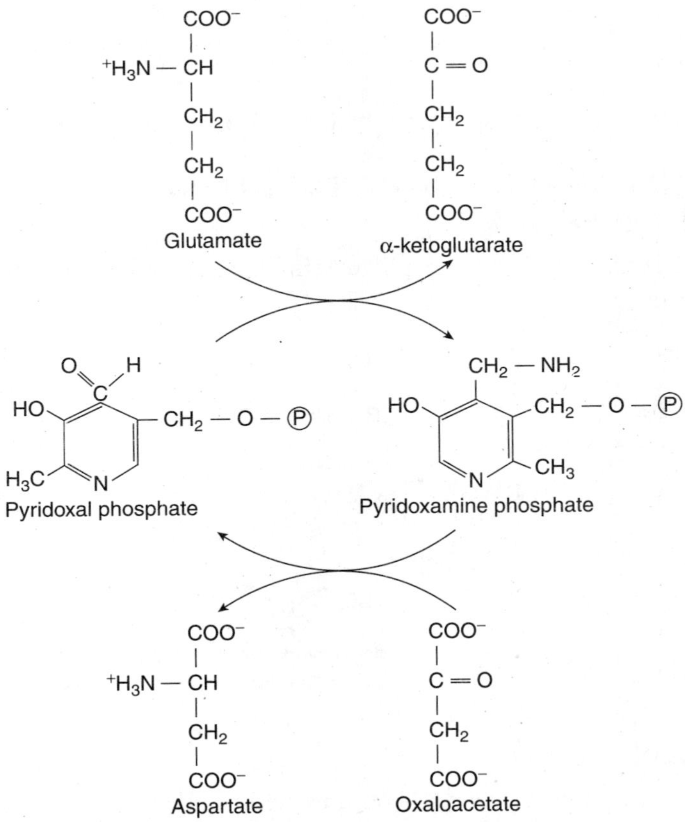

Figure 9.4 *Mechanism of Transamination*

MECHANISM OF TRANSAMINATION

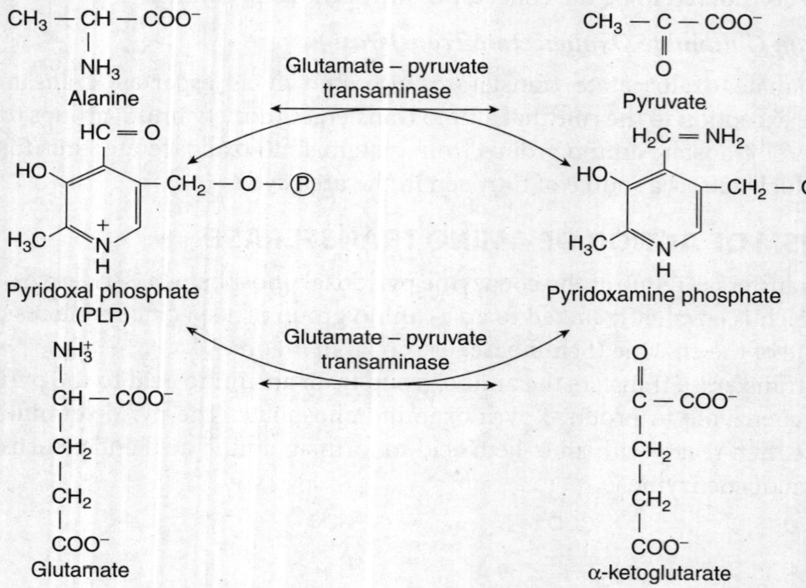

Involvement of pyridoxal phosphate in the transfer of amino group.

TRANSAMINATION REACTION DEPENDENT ON PYRIDOXAL PHOSPHATE

Formation of enzyme pyridoxal phosphate-Schiff base and amino add-pyridoxal phosphate Schiff base.

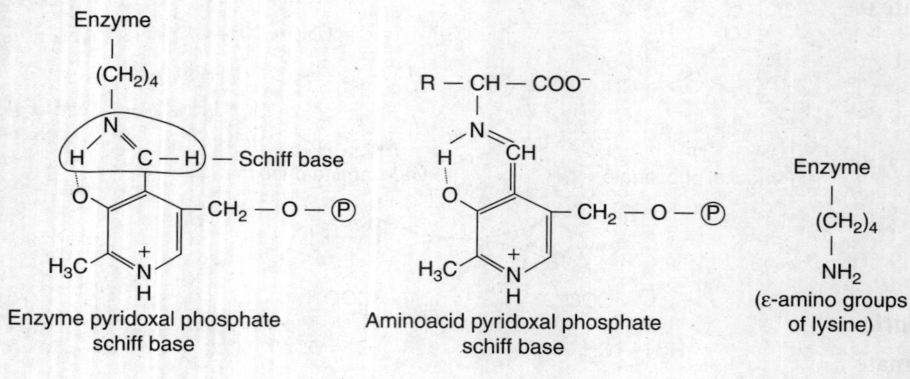

DEAMINATION

The removal of amino group from the amino acid as NH_3 is termed as *deamination*. Transamination involves only the shuffling of amino groups among the amino acids. On the other hand, deamination results in the liberation of ammonia for urea synthesis.

These reactions occur primarily in liver and kidney and provide α-keto acid (which can enter the central pathway of energy metabolism). Deamination may be oxidative or non-oxidative.

Oxidative Deamination

Oxidative deamination is the liberation of free ammonia from the amino group of the amino acid coupled with oxidation. This takes place mostly in liver and kidney. The purpose of oxidative deamination is to provide NH_3 for urea synthesis and α-keto acid for a variety of reactions, including energy generation.

Glutamate Dehydrogenase

In the process of transamination, the amino groups of most amino acid are transferred to α-ketoglutarate to produce glutamate. The glutamate serves as a collection *centre* for amino groups in the biological system. Glutamate rapidly undergoes oxidative deamination catalysed by glutamate dehydrogenase (GDH) to liberate ammonia. This enzyme is unique, in that it can utilise either NAD^+ or NAD^+ as coenzyme. Glutamate dehydrogenase catalysed reaction is important as it reversibly links up glutamate metabolism with TCA cycle through α-ketoglutarate. GDH is involved in both catabolic and anabolic reactions.

Reaction

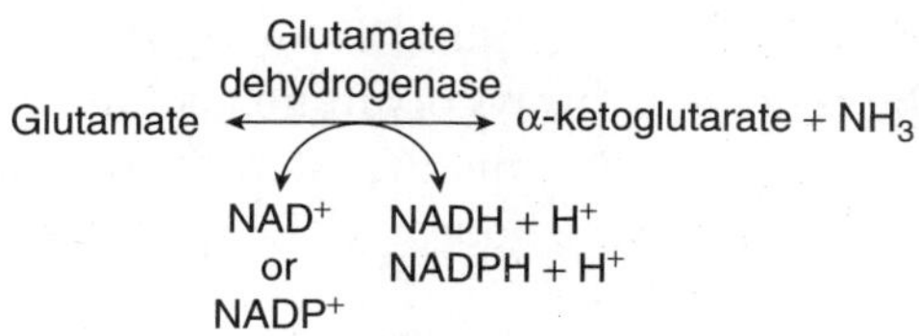

Conversion of glutamate to α-ketoglutarate occurs through the formation of an intermediate α-imminoglutarate.

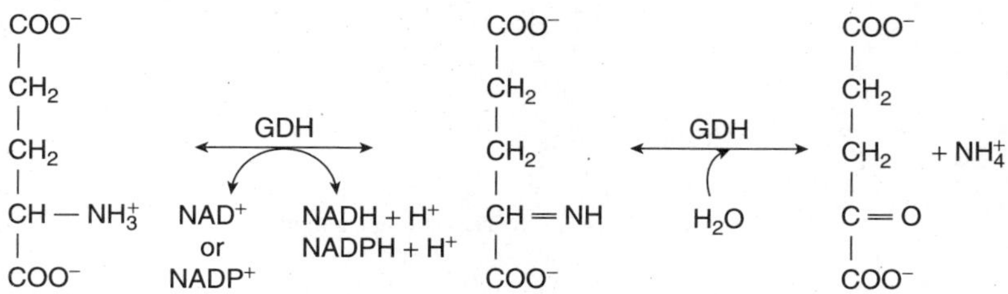

Regulation of Glutamate Dehydrogenase

Glutamate dehydrogenase is zinc, containing mitochondrial enzyme. GDH is controlled by allosteric regulation. ATP and GTP inhibit GDH, whereas ATP and GTP activate GDH.

Protein-rich diet: Glutamate level is elevated, which is converted to α-ketoglutarate with liberation of NH_3. When the energy levels are low, the dehydration of glutamate is increased to provide α-ketoglutarate, which enters TCA cycle to liberate energy.

Oxidative deamination by amino acid oxidase

α-amino acid oxidase and D-amino acid oxidase are flavour proteins, processing FMN and FAD, respectively. They act on amino acids (L and D) to produce α-keto acids and NH_3. In this reaction, O_2 is reduced to H_2O_2, which is later decomposed by catalase.

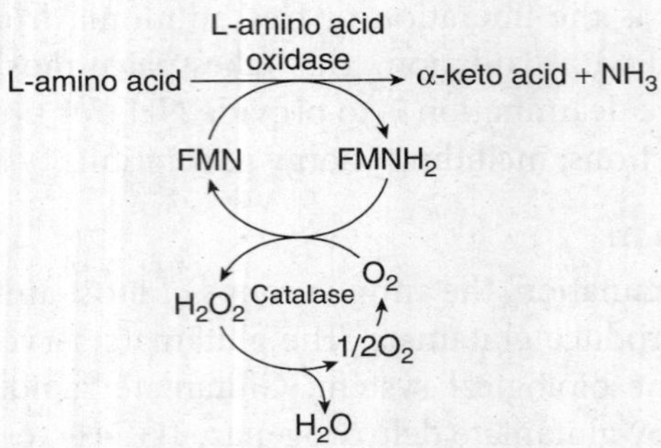

Fate of D-amino acids

D-amino acids are present in the plants and cell walls of microorganisms and are not used in the biosynthesis of mammalian proteins. They are present in the diet and are metabolised by liver. D-amino acid is an FAD-dependent enzyme that catalyses the oxidative deamination of these unnatural isomers of amino acids. The resulting α-keto acid can enter the general pathways of amino acid metabolism. Thus, D-amino acid oxidase is important as it initiates the first step for the conversion of unnatural D-amino acids to α-amino acids in the body.

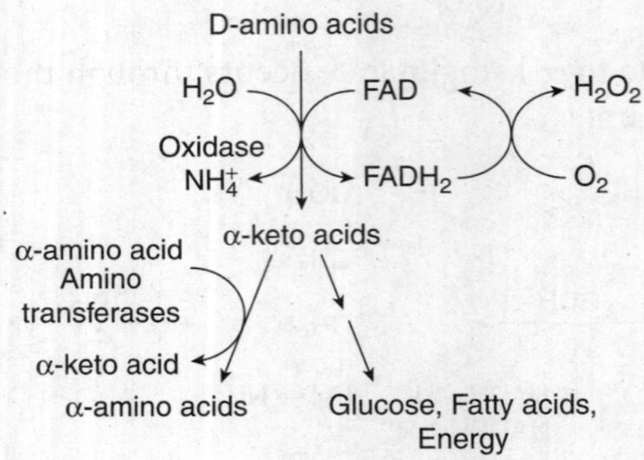

Non-Oxidative Deamination

Amino acid which is deaminated to liberate without undergoing oxidation is termed as non-oxidative deamination.

Amino acid dehydrses

Serine, threonine, and homoserine are the hydroxyl amino acids; they undergo non-oxidative deamination catalysed by PLP-dependent hydrases.

$$\left.\begin{array}{c}\text{Serine}\\\text{Threonine}\\\text{Hommoserine}\end{array}\right.\xrightarrow{\text{dehydratase}}\alpha\text{-keto acids}+NH_3$$

Amino acid desulphydrases

The sulphur amino acids, namely cysteine and homocysteine, undergo deamination coupled with desulphydration to give keto acids.

$$\text{Cysteine}\xrightarrow{\text{Desulphydrases}}\text{Pyruvate}+NH_3$$

Histidase

The enzyme histidase will act on histidine to liberate NH_3 by the non-oxidative deamination process.

$$\text{Histidine}\xrightarrow{\text{Histidase}}\text{Uroconate}+NH_3$$

DECARBOXYLATION

Amino acids are converted to the respective amines by decarboxylation process. This is carried out by a group of enzymes called decarboxylase. The coenzyme involved in the reaction is PLP (pyridoxal phosphate). Biogenic amines, which have important functions, are synthesised by PLP decarboxylation.

Serotonin – Neurotransmitter regulates sleep, behaviour, and blood pressure.

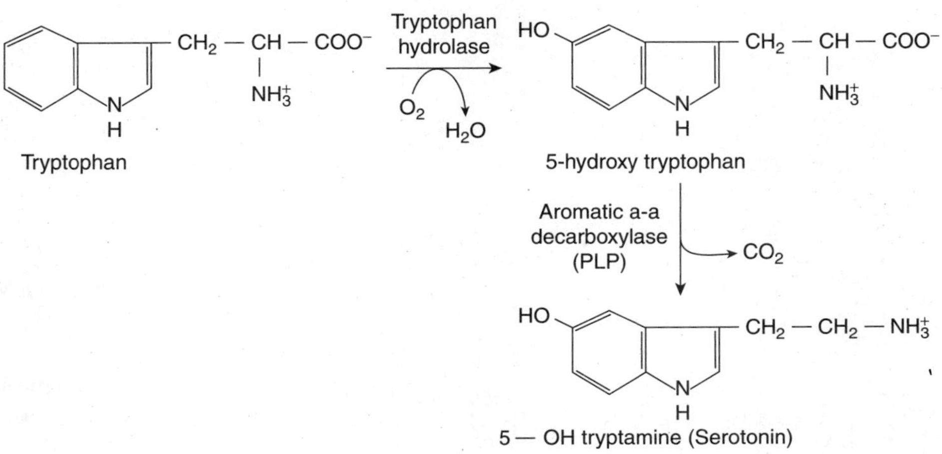

Histamine – Vasodilator lowers BP, stimulates gastric HCl secretion, and involves in inflammation and allergic reactions.

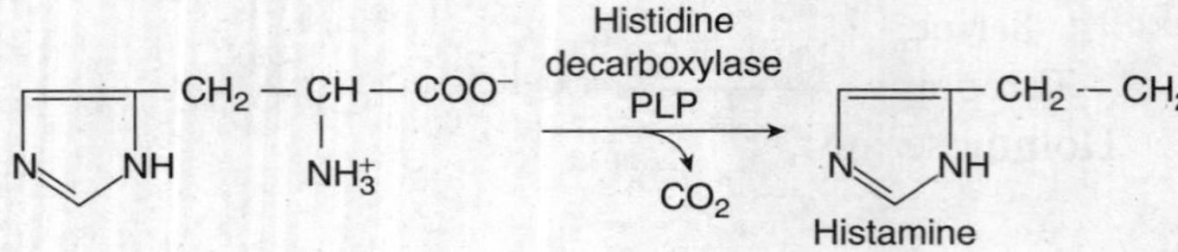

GABA (γ-amino butyric acid) – Inhibitory neurotransmitter.

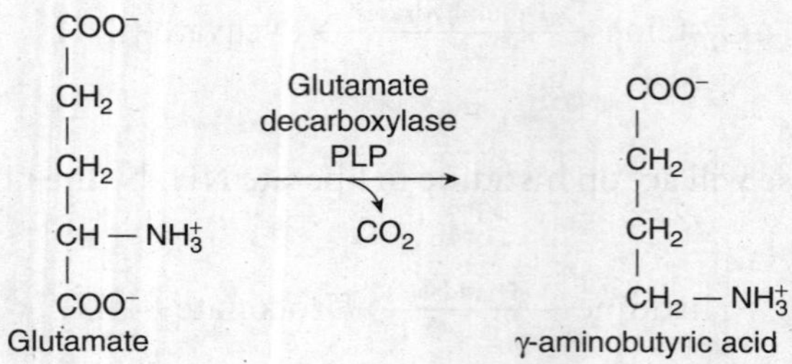

Catecholamines (dopamine, norepinephrine, and epinephrine) – Involved in metabolic and nervous regulations.

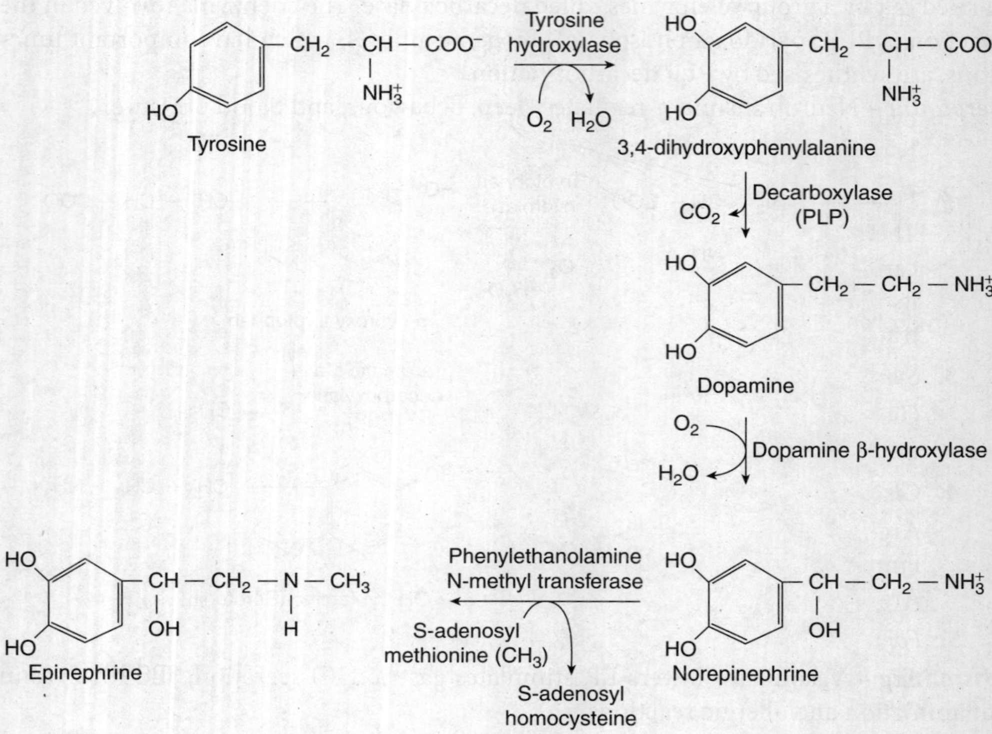

UREA CYCLE

Urea is the end product of protein metabolism. The nitrogen of amino acids converted to ammonia is toxic to the body. It is converted to urea and detoxified. As such, urea accounts for 80%–90% of the nitrogen-containing substances excreted in urine.

SITE

Urea is processed in two locations; the synthesis occurs in liver and it is transported to kidney for excretion. Urea cycle was first described by Hans Krebs and Kurt Henseleit in 1932. Hence, this cycle is also known as Krebs–Henseleit cycle.

MECHANISM

The skeleton of urea consists of two amino ($-NH_2$) groups and one carbon atom. The first ($-NH_2$) group is derived from glutamate reservoir ($-NH_2$) and the second ($-NH_2$) group is derived from aspartate. The carbon atom is derived from CO_2. The urea cycle occurs in five steps with five enzymes. The first two enzymes are located in mitochondria while the rest are located in cytosol.

1. *Synthesis of carbamoyl phosphate:*

   Carbamoyl phosphate synthase I (CPS I) in the mitochondria condenses the ammonium ions (NH_4^+) with CO_2 to form carbamoyl phosphate. During this step, two moles of ATP are consumed and this is an irreversible and rate-limiting step. N-acetyl glutamate (NAG) is involved in this process. Carbamoyl phosphate synthase II is involved in pyrimidine synthesis and is present in cytosol. CPS II accepts amino group from glutamine and does not require N-acetylglutamate for its activity.

2. *Formation of citrulline:*

   The second step in urea cycle is the formation of citrulline. It is produced from carbamoyl phosphate and ornithine. Ornithine is reused in the urea cycle. So its role is comparable to oxaloacetate in the citric acid cycle. Citrulline is then transferred to cytosol by the transporter system.

3. *Synthesis of argininosuccinate:*

   Citrulline condenses with aspartate and produces argininosuccinate. The second amino group is incorporated in this reaction and this step requires ATP.

4. *Cleavage of argininosuccinate:*

   Argininosuccinase cleaves argininosuccinate to fumarate and arginine. Fumarate acts as the connecting link with TCA cycle and gluconeogenesis. Arginine is the next precursor for urea production.

5. *Formation of urea:*

   Arginase cleaves arginine and produces the final products urea and ornithine. Ornithine so produced enters mitochondria for its reuse in the urea cycle.

Arginine is induced by CO_2 and Mn^{2+}. Ornithine and lysine compete with Arginase (competitive inhibition). Arginase is present in liver, whereas the rest of the enzymes are present in other tissues. Hence, arginine synthesis may occur in various tissues but only the liver can finally produce urea. The synthesis of urea is presented in the structure given below:

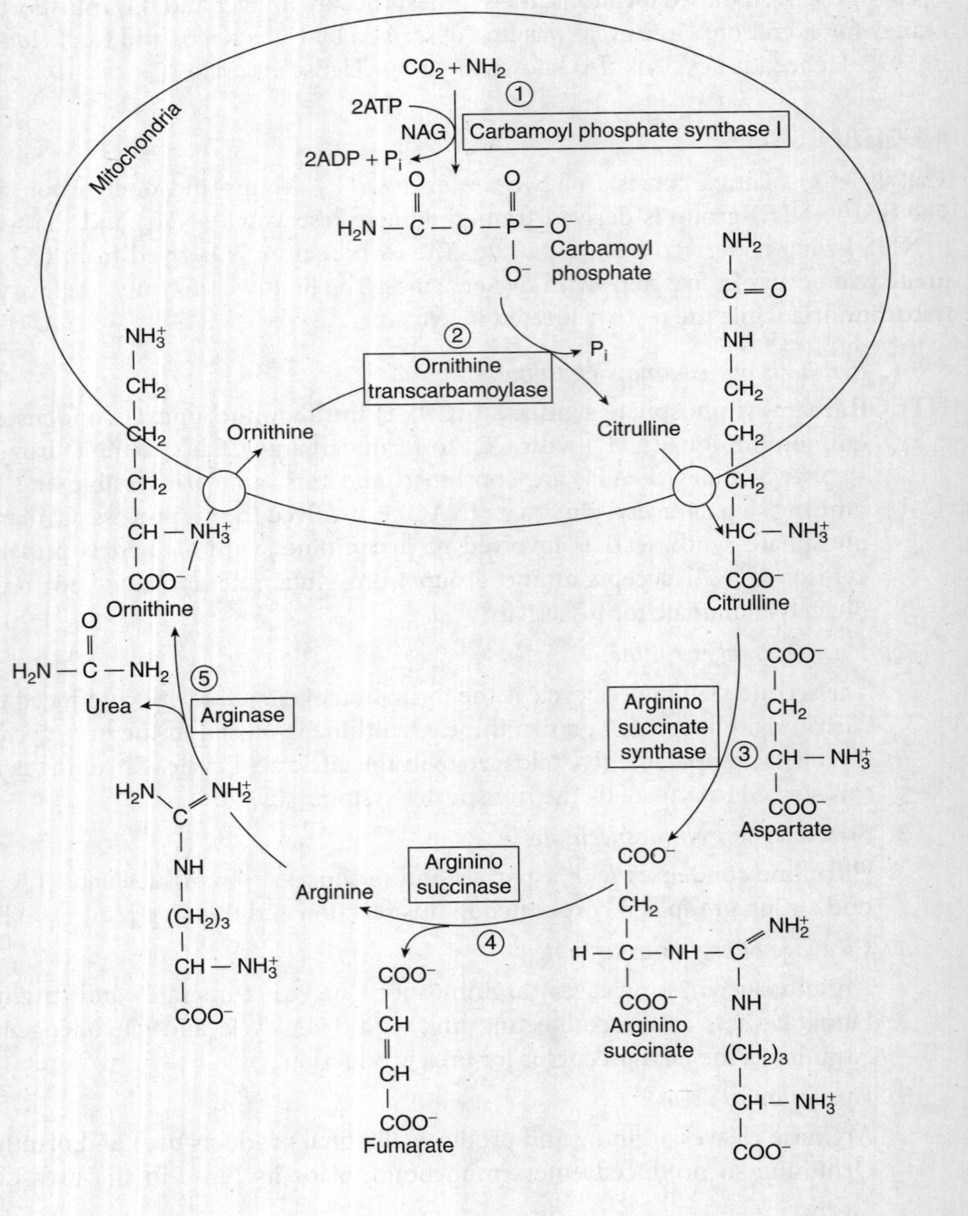

REGULATION OF UREA CYCLE

The first reaction catalysed by CPS I is the rate-limiting enzyme reaction or committed step in urea synthesis. CPS I is allosterically activated by NAG. It is synthesised from glutamate and acetyl CoA by synthase and degraded by a hydrolase.

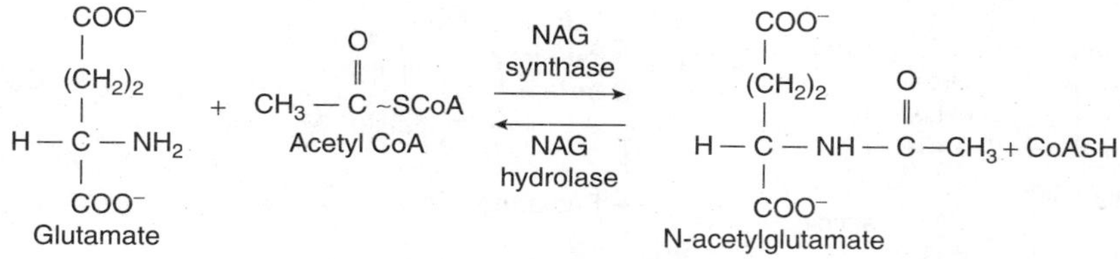

The rate of urea synthesis in liver is correlated with the concentration of NAG. The consumption of protein-rich meal increases the level of NAG in liver, leading to enhanced urea synthesis.

CPS I and glutamate dehydrogenase are localised in the mitochondria. They coordinate with each other in the formation of NH_3 and its utilisation for the synthesis of carbamoyl phosphate. The remaining four enzymes of urea cycle are mostly controlled by the concentration of their respective substrates.

INTEGRATION BETWEEN UREA CYCLE AND TCA

Urea cycle is linked with TCA cycle in three different ways:

1. Production of fumarate in urea cycle is the most important integrative point with TCA cycle. Fumarate is converted to malate and then to oxaloacetate in TCA cycle. Oxaloacetate undergoes transamination to produce aspartate, which enters urea cycle. Here, it combines with citrulline to produce argininosuccinate. Oxaloacetate is an important metabolite which can combine with acetyl CoA to form citrate and get finally oxidised. Oxaloacetate can also serve as a precursor for the synthesis of glucose (gluconeogenesis).
2. ATP (12) is generated in TCA cycle, while ATP (4) is utilised for urea synthesis.
3. Citric acid cycle is an important metabolic pathway for complete oxidation of various metabolites to CO_2 and H_2O. The CO_2 liberated in TCA cycle (in the mitochondria) can be utilised in urea cycle.

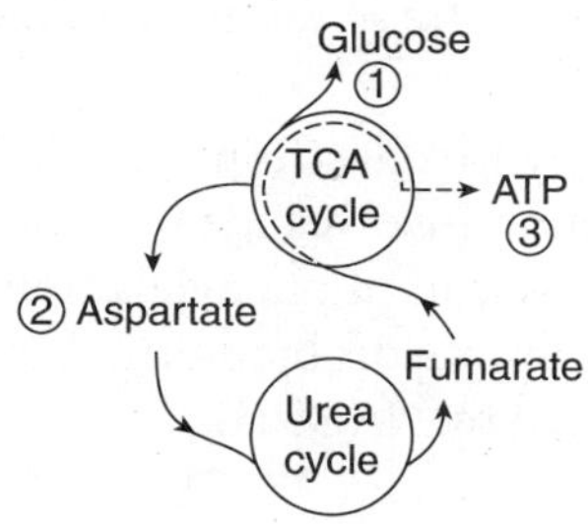

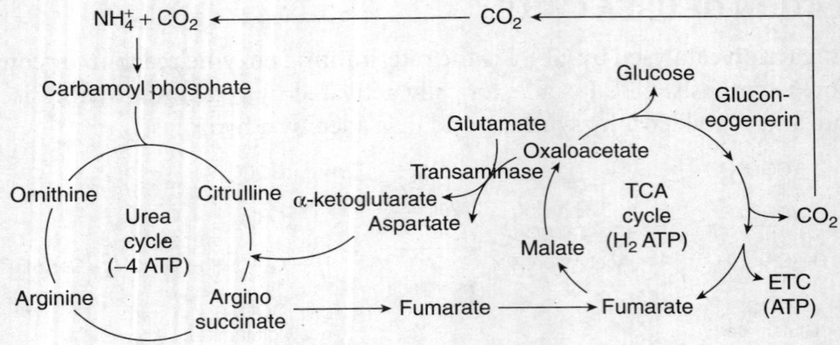

PHENYLALANINE AND TYROSINE

Phenylalanine and tyrosine are structurally related aromatic amino acids. Phenylalanine is an essential amino acid, while tyrosine is non-essential. Besides its incorporation into proteins, the only function of phenylalanine is its conversion to tyrosine. For this reason, ingestion of tyrosine can reduce the dietary requirement of phenylalanine. This phenomenon is referred to as 'sparing action' of tyrosine on phenylalanine.

The predominant metabolism of phenylalanine occurs through tyrosine. Tyrosine is incorporated into proteins and is involved in the synthesis of a variety of biologically important compounds epinephrine, norepinephrine, dopamine (catecholamine), thyroid hormones, and the pigment melanin. During the course of degradation of phenylalanine fumarate and acetoacetate are formed which serve as precursors for the synthesis of glucose and fat. Hence, these amino acids (phenylalanine and tyrosine) are both ketogenic and glucogenic in nature.

CONVERSION OF PHENYLALANINE TO TYROSINE

Phenylalanine is hydroxylated at para position by phenylalanine hydroxylase to produce tyrosine. This is an irreversible reaction and requires the participation of a specific coenzyme biopterin. The active forms of biopterin is tetrahydrobiopterin, in the reaction it is converted to dihydrobiopterin. In the phenylalanine hydroxylase reaction, tetrahydrobiopterin. Tetrahydrobiopterin is then regenerated by an NADPH-dependent dihydrobiopterin reductase.

The enzyme phenylalanine hydroxylase is present in liver. In the conversion of phenylalanine to tyrosine, the reaction involves the incorporation of one atom of molecular oxygen (O_2) into the para position of phenylalanine while the other atom of O_2 is reduced to form water. It is the tetrahydrobiopterin that supplies the reducing equivalents, which, in turn, are provided by NADPH.

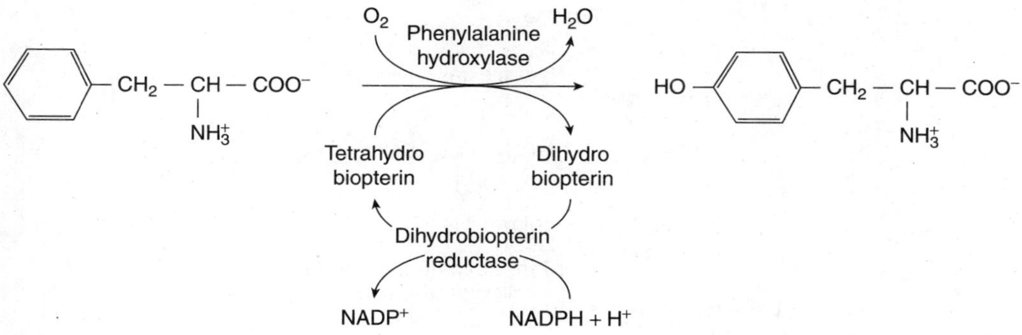

DEGRADATION OF TYROSINE

As phenylalanine is converted to tyrosine, a single pathway is responsible for the degradation of both these amino acids, which occurs mostly in liver. In a sequence of reactions, tyrosine is converted to fumarate and acetoacetate. Tyrosine first undergoes transamination to 5-p-hydroxyphenyl pyruvate. P-hydroxyphenyl pyruvate dioxygenase is a copper-containing enzyme. It catalyses oxidative decarboxylation as well as hydroxylation of the phenyl ring of p-hydroxyl-phenyl pyruvate to produce homogentisate. This reaction involves a shift in hydroxyl group at para position to meta position and incorporates a new hydroxyl group at para position. This step in tyrosine metabolism requires ascorbic acid.

Homogentisate oxidase cleaves the benzene ring of homogentisate to form 4-maleylacetoacetate. Molecular oxygen is required for this reaction to break the aromatic ring. Maleylacetoacetate undergoes isomerisation to form 4-fumarylacetoacetate, and this reaction is catalysed by maleylacetoacetate isomerase. Fumarylacetoacetase brings about the hydrolysis of fumarylacetoacetate to liberate fumarate and acetoacetate.

Fumarate is an intermediate of citric acid cycle and also serves as precursor for gluconeogenesis. Acetoacetate is a ketone body, from which fat can be synthesised.

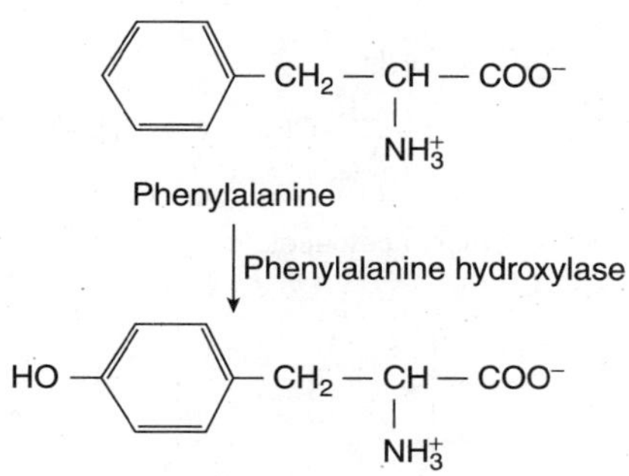

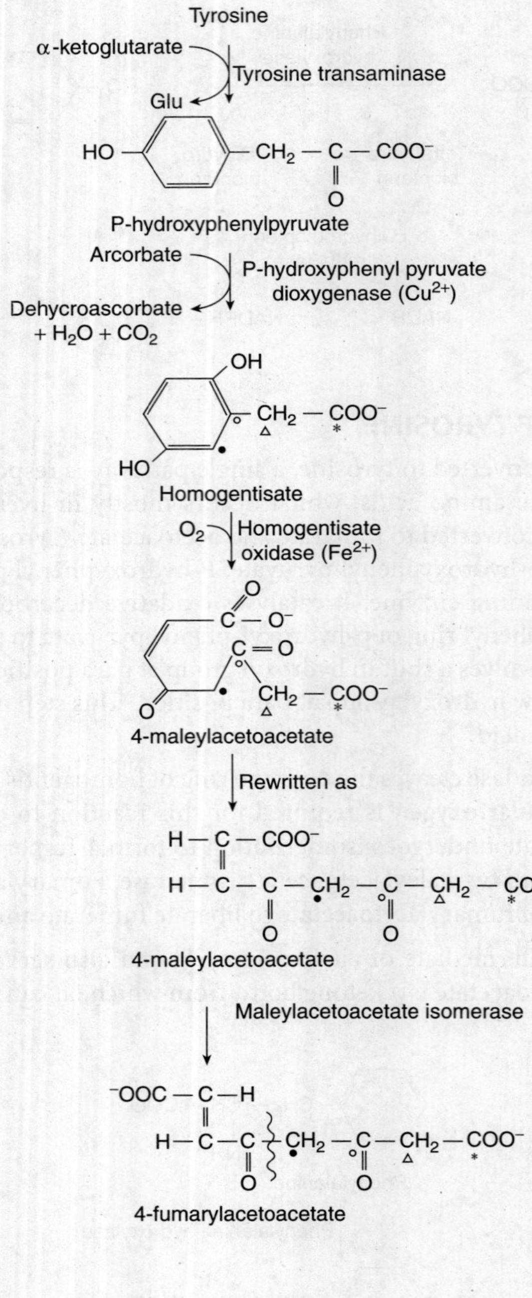
Tyrosine
α-ketoglutarate
Tyrosine transaminase
Glu
HO
CH2 — C — COO⁻
O
P-hydroxyphenylpyruvate
Arcorbate
P-hydroxyphenyl pyruvate
dioxygenase (Cu2+)
Dehycroascorbate
+ H2O + CO2
OH
CH2 — COO⁻
HO
Homogentisate
O2
Homogentisate
oxidase (Fe2+)
C — O⁻
C = O
CH2 — COO⁻
4-maleylacetoacetate
Rewritten as
H — C — COO⁻
H — C — C — CH2 — C — CH2 — COO⁻
4-maleylacetoacetate
Maleylacetoacetate isomerase
⁻OOC — C — H
H — C — C — CH2 — C — CH2 — COO⁻
4-fumarylacetoacetate

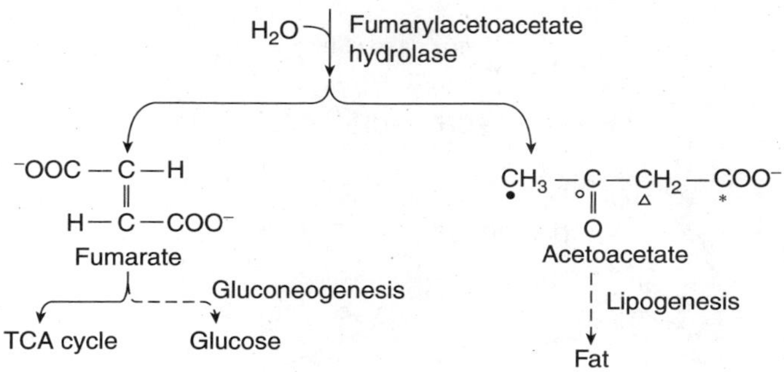

SYNTHESIS OF MELANIN

Melanin is the pigment of skin, hair, and eye. The synthesis of melanin occurs in melanosomes present in melanocytes, the pigment-producing cells. Tyrosine is the precursor for melanin and only one enzyme, namely tyrosinase, is involved in its formation. Tyrosinase hydroxylates tyrosine to form 3,4-dihydroxyphenylalanine (DOPA). DOPA can act as a cofactor for tyrosinase. The next reaction is also catalysed by tyrosinase, in which DOPA is converted to dopaquinone. It is believed that the subsequent couple of reactions occur spontaneously, forming leuco dopachrome followed by 5,6-dihydroxy indole. The oxidation of 5,6-dihydroxy indole to indole 5,6-quinone is catalysed by tyrosinase, and DOPA serves as a cofactor. This reaction, inhibited by tyrosine, regulates the synthesis of melanin. Melanochromes are formed from indole quinone, which, on polymerisation, is converted to black melanin.

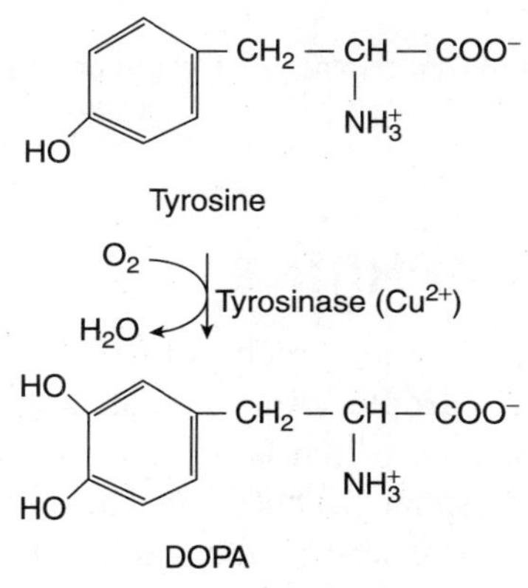

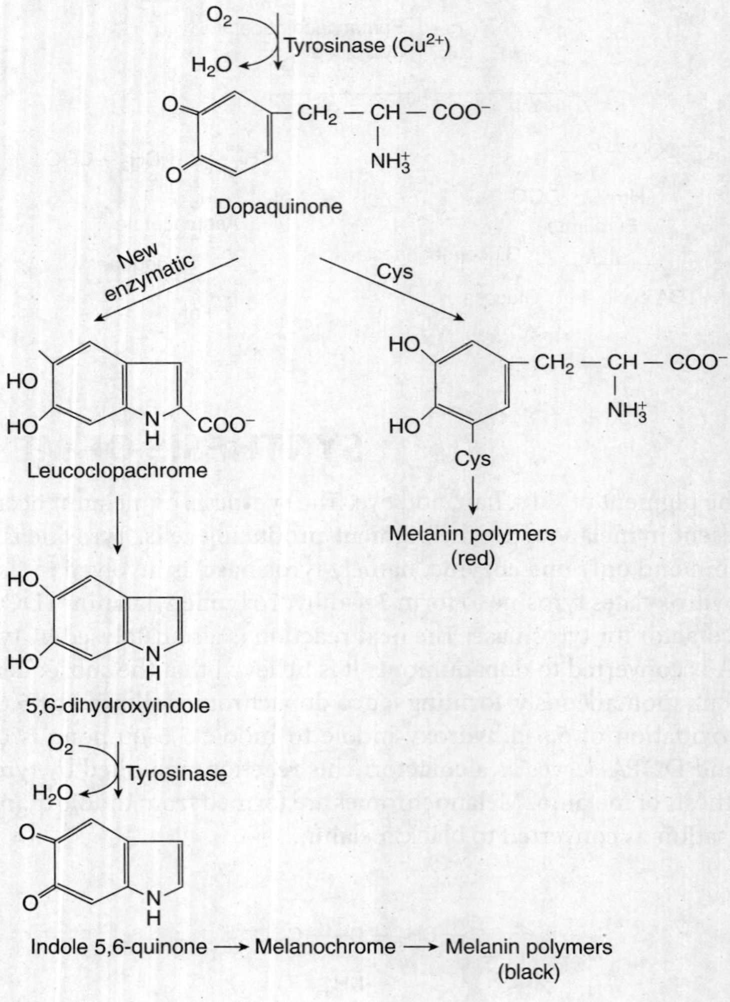

SYNTHESIS OF EPINEPHRINE

Phenylalanine is converted into tyrosine, which is further converted into DOPA by tyrosine hydroxylase with tetrahydropteridine as cofactor. DOPA is decarboxylated to dopamine by a decarboxylase, which is present in many tissues, including a drenal medulla with pyridoxal phosphate as cofactor. The hydroxylation of dopamine is carried out by dopamine β-hydroxylase to form norepinephrine in the presence of ascorbic acid and molecular oxygen. The methylation from methionine converts norepinephrine to epinephrine.

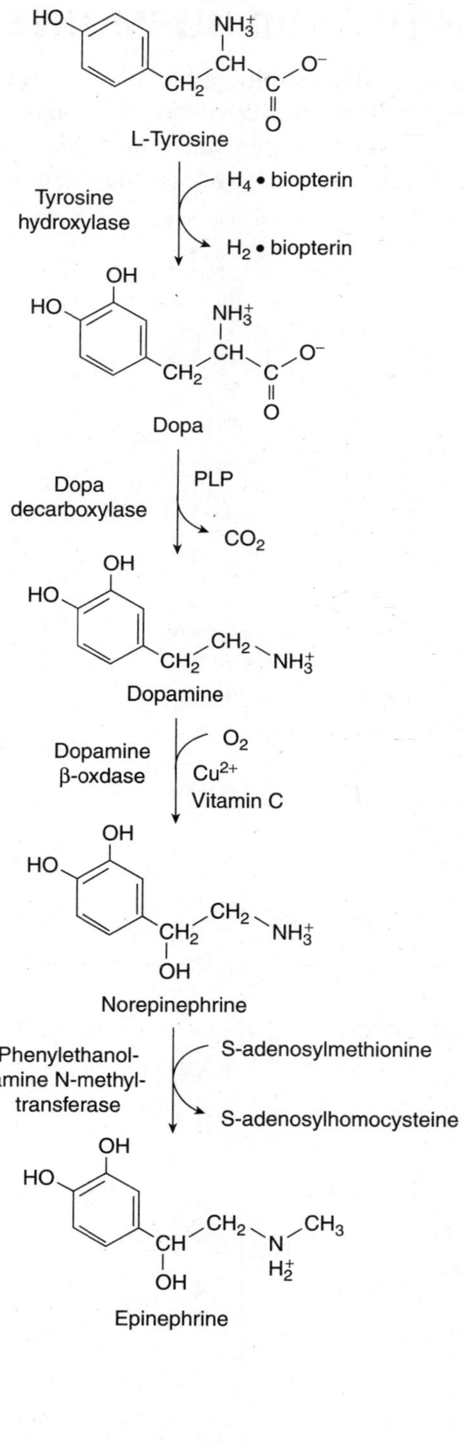
L-Tyrosine
Tyrosine hydroxylase
H4 • biopterin
H2 • biopterin
Dopa
Dopa decarboxylase
PLP
CO2
Dopamine
Dopamine β-oxdase
O2
Cu2+
Vitamin C
Norepinephrine
Phenylethanolamine N-methyltransferase
S-adenosylmethionine
S-adenosylhomocysteine
Epinephrine

SYNTHESIS OF THYROID HORMONES

Thyroid hormones, thyroxine and triiodothyronine, are synthesised from the tyrosine residues of the protein thyroglobulin and activated iodine. Iodination of tyrosine ring occurs to produce mono-iodination and diiodotyrosine, from which triiodothyronine (T_3) and thyroxine undergo proteolytic breakdown to release the free hormones, T_3 and T_4.

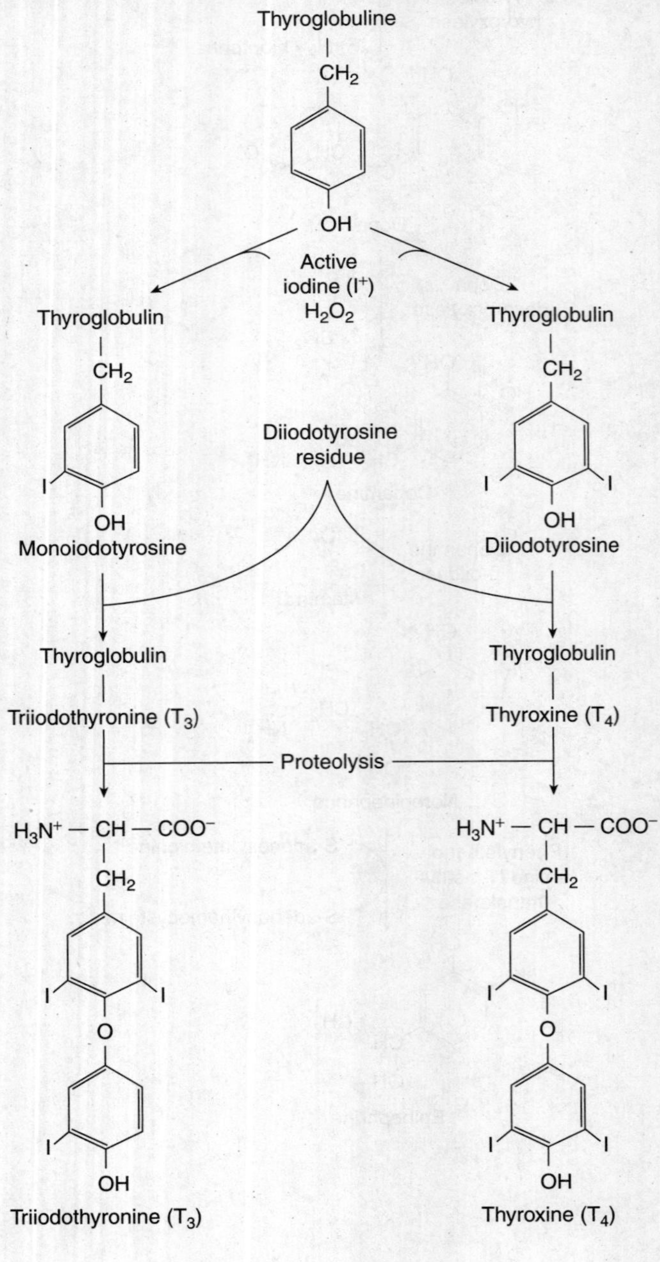

SYNTHESIS OF CREATINE

Creatine is present in the tissues (muscle, brain, blood, and so on) as the high energy compound, phosphocreatine, and as free creatine. Three amino acids – glycine, arginine, and methionine – are required for the formation of creatine. The first reaction occurs in kidney. It involves the transfer of guanidine group of arginine to glycine, catalysed by arginine-glycine transmidinase to produce guanidoacetate. S-adenosylmethionine donates methyl group to glycocyamine to produce creatine. This reaction occurs in liver. Creatine is reversibly phosphorylated to phosphocreatine by creatine kinase. It is stored in muscle as high energy phosphate. Creatinine is an anhydride of creatine. It is formed by spontaneous cyclisation of creatine or creatinine in human serum and urine.

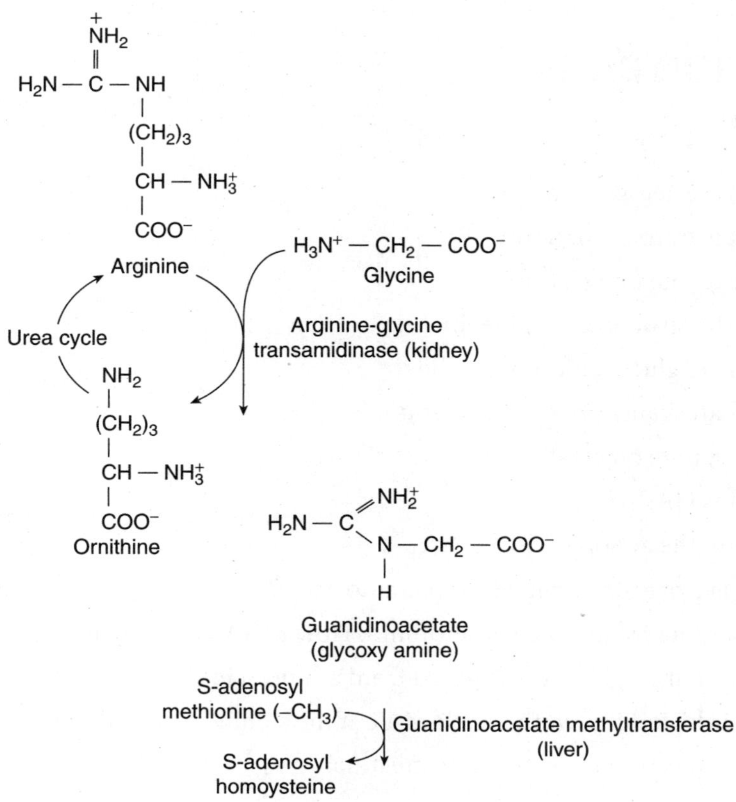

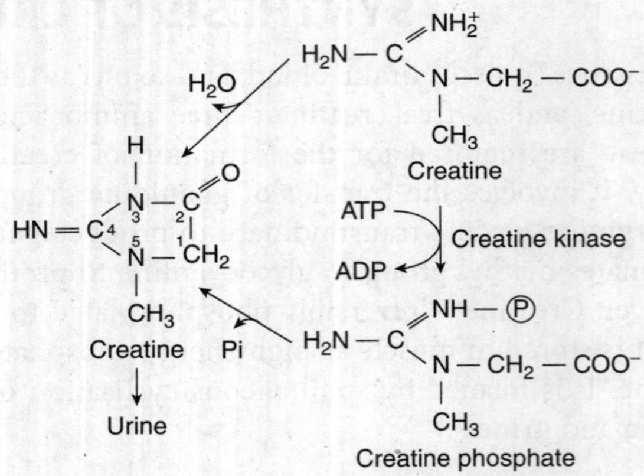

QUESTION BANK

SECTION A

1. What are ketogenic amino acids?
2. What is transamination?
3. Write a note on creatine.
4. Give the structure of epinephrine.
5. What are glucogenic amino acids?
6. What are transminases? Give example.
7. Write a note on creatinine.
8. What is GABA?
9. Explain the action of deaminase.
10. How is pyruvate obtained from amino acids?
11. Explain the significance of serotonin. How is GABA formed?
12. How is α-ketoglutarate obtained from amino acids?
13. How is histamine synthesised? List out the significance of the same.
14. How is creatinine converted to creatine?

SECTION B

1. Explain decarboxylation of amino acids.
2. Discuss the synthesis of epinephrine.

3. How are dopamines synthesised?
4. Explain the action of glutamate dehydrogenase.
5. How is epinephrine formed?

SECTION C

1. What is transamination? Explain how amino acids are degraded by transamination.
2. Explain how ammonia is detoxified to urea.
3. Write short notes on (i) serotonin (ii) GABA (iii) dopamine (iv) epinephrine.
4. Explain the mechanism of transamination reaction.
5. Explain the urea cycle. Explain the regulation of the pathway.

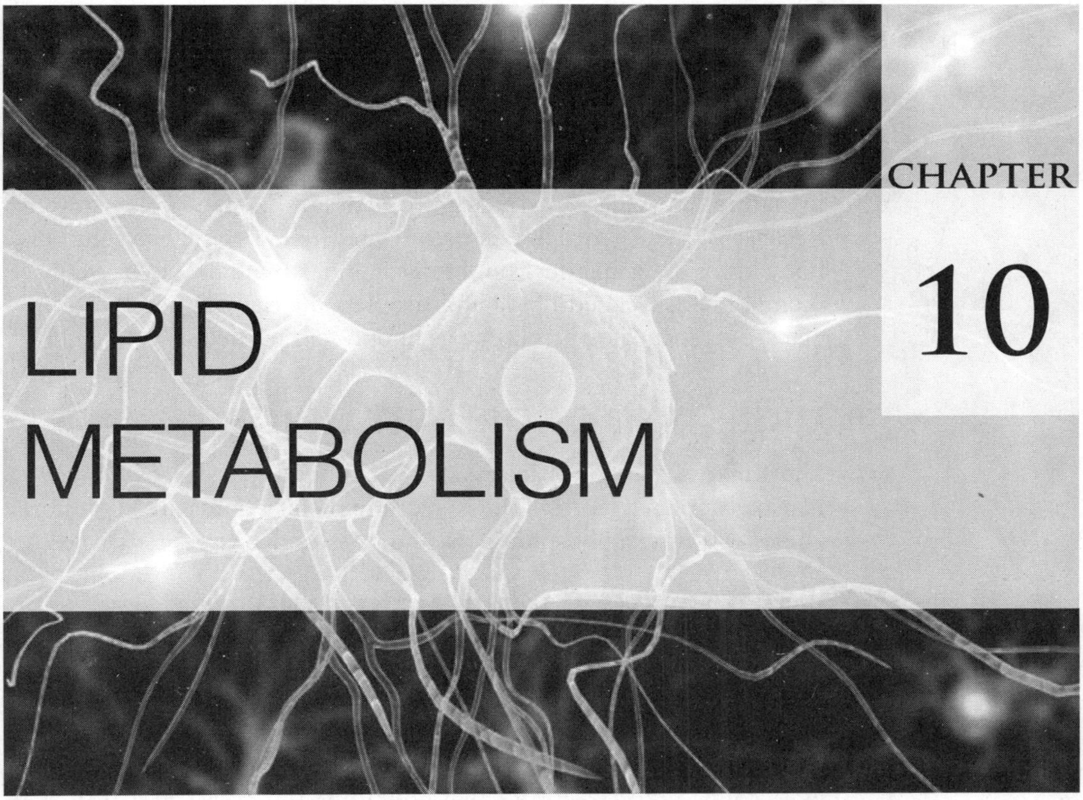

CHAPTER 10

LIPID METABOLISM

Concept Outline

- Introduction
- Digestion of Dietary Lipids
- Degradation of Triacylglycerol
- Cholesteryl Ester Degradation
- Phospholipid Degradation
- Absorption of Lipids
- Resynthesis of Triacylglycerol and Cholesteryl Esters by the Intestinal Cells
- Secretion of Lipids from Intestinal Cells
- Secretion of Chylomicrons
- Use of Dietary Lipids by the Tissues
- Fate of Glycerol
- Fate of Chylomicron Components
- Fatty Acid Oxidation
- Biosynthesis of Fatty Acid
- Acetyl-CoA Carboxylase
- Cholesterol Biosynthesis
- Degradation of Cholesterol
- Biosynthesis of Acylglycerol
- Complex Lipids
- Cardiolipin
- Plasmalogens
- Degradation of Sphingomyelin
- Niemann-Pick Disease
- Farber's Disease
- Ketone Bodies

INTRODUCTION

Lipids are water-insoluble organic molecules, which are heterogeneous in nature and can be extracted from tissues by non-polar solvents. Because of their hydrophobic nature (insoluble in aqueous solutions), body lipids are generally found to be compartmentalised, for example, membrane-associated lipids and droplets of triacylglycerol in adipocytes, or transported by plasma as lipoprotein particles.

DIGESTION OF DIETARY LIPIDS

The intake of an adult is about 60–150 g of lipids per day, of which 90% is usually triacylglycerol. The remaining dietary lipids are cholesterol, cholesteryl esters, phospholipids, and unesterified (free) fatty acids.

FATE OF DIETARY LIPID IN MOUTH AND STOMACH (LIMITED PROCESSING)

The digestion of lipids starts in the stomach, catalysed by an acid-stable lipase, which originates in the glands at the back of the tongue, known as lingual lipase. However, the rate of hydrolysis is very slow because the lipid is not emulsified for the digestion process. Lipase can hydrolyse or degrade triacylglycerol only in the lipid–water interface . However, the rate of hydrolysis is slow because the lipid does not get emulsified, and the enzyme can only degrade the triacylglycerol in the lipid water interface.

The triacylglycerol molecule consists of fatty acids of short or medium chain length (milk), which can be catalysed by a different enzyme called gastric lipase. The optimum activity of gastric lipase is found to be in neutral pH and therefore this enzyme is not of much use. In adults, the pH in gastric juice is low. However, in infants, whose gastric juice pH is nearer neutral and whose diet is only milk lipids, gastric lipase may digest the milk lipids in such neutral pH. Overall, in adults, dietary lipids are not digested in mouth and stomach and the digestion starts only in the small intestine. The overview of the lipid digestion is shown in Figure 10.1.

EMULSIFICATION OF DIETARY LIPIDS IN SMALL INTESTINE

In duodenum, the critical process of emulsification of the dietary lipids occurs. Because these compounds are practically insoluble in water, the enzymic hydrolysis of lipids of triacylglycerol can occur only on the surface of a lipid droplet, which is at the interface between the lipid droplet and the surrounding aqueous solution.

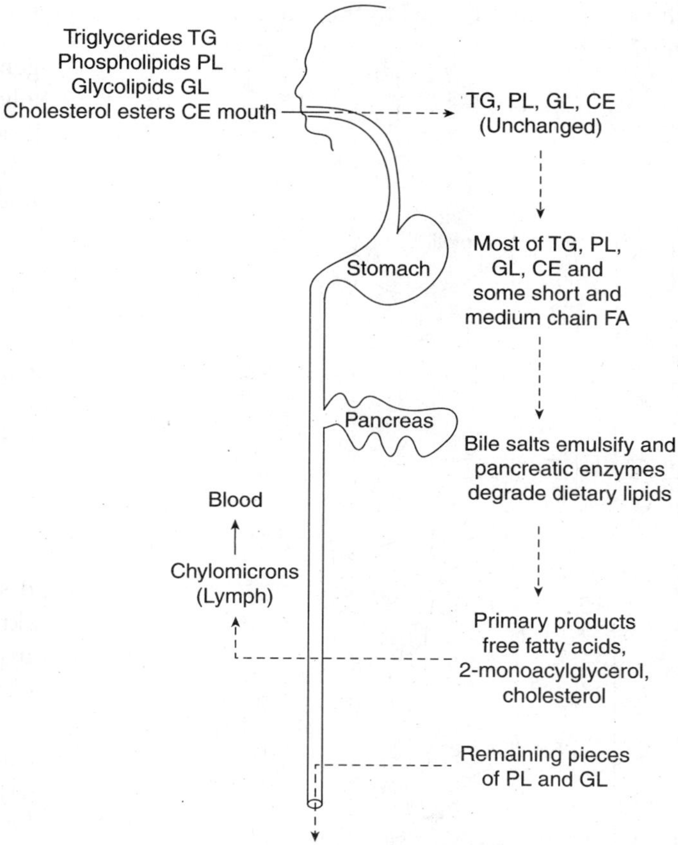

Figure 10.1 *Overview of Lipid Digestion*

ENZYMIC DEGRADATION OF DIETARY LIPIDS

The hydrolytic enzymes that degrade the dietary lipids in the intestine are secreted by the pancreatic juice in the pancreas. The secretion of this enzyme is under the control of hormones. A small peptide hormone cholecystokinin, also known as pancreozymin, is produced by the cells in the mucosa of jejunum and lower duodenum. This hormone is produced in response to the presence of lipids and partially digested proteins present in the upper small intestine. The role of this hormone is to act on the exocrine cells of the pancreas, which makes them to release digestive enzyme. It also decreases gastric motility, thereby resulting in a slower release of gastric content into the small intestine as shown in Figure 10.2 .

Another small peptide hormone, secretin, is produced by the intestinal cells in response to the low pH of the chime entering the intestine. Secretin makes the pancreas to release a watery solution, rich in bicarbonate, which helps to neutralise the pH of the intestine, thus enabling the pH to aid in the enzymic digestive activity.

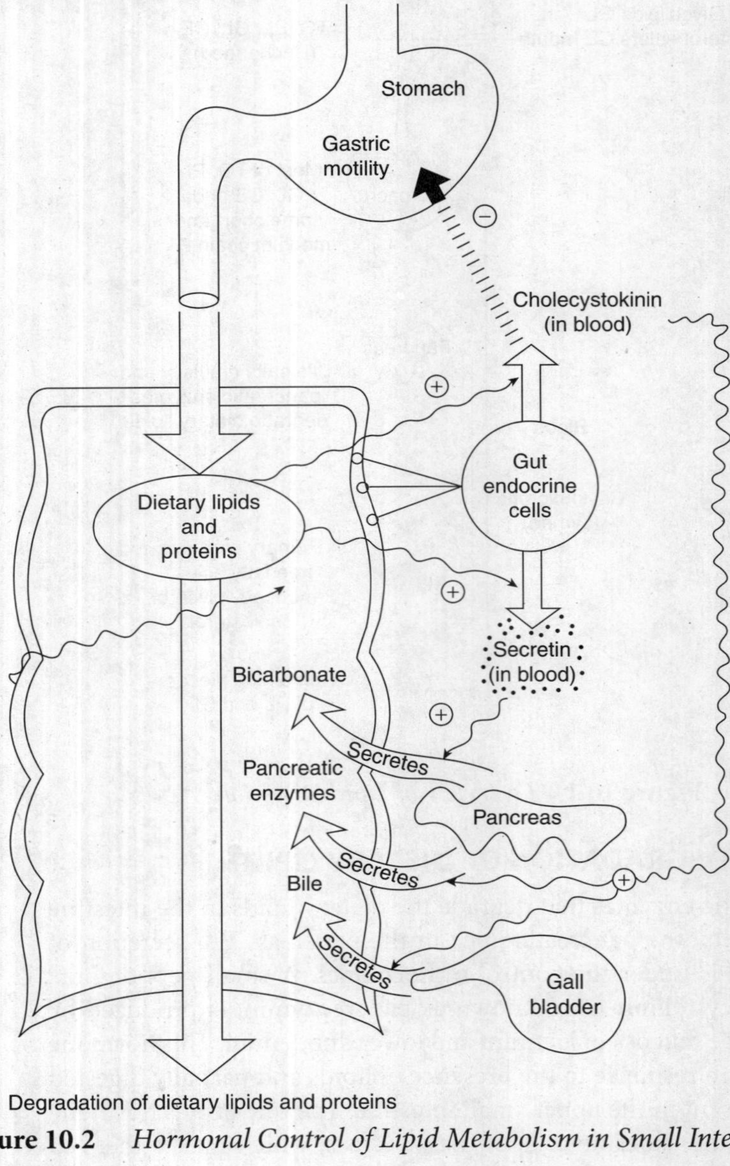

Figure 10.2 *Hormonal Control of Lipid Metabolism in Small Intestine*

DEGRADATION OF TRIACYLGLYCEROL

Triacylglycerols are too large to be taken by the mucosal cells of the intestinal villi. Hence, they are acted upon by an esterase, pancreatic lipase, which removes the fatty acids at carbon 1 and 3. The resulting products of hydrolysis are a mixture of 2-monoacylglycerol and free fatty acids. A second protein, colipase is secreted by the pancreas, which also helps to hold and stabilise the lipase at the lipid–aqueous interface.

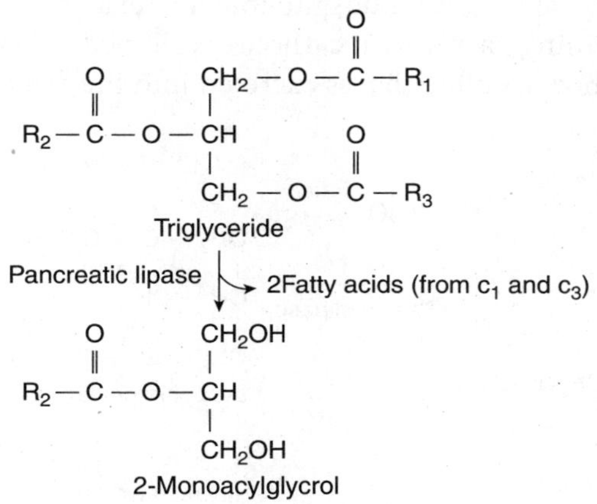

CHOLESTERYL ESTER DEGRADATION

Cholesteryl esters are hydrolysed by cholesteryl hydrolase, resulting in cholesterol and free fatty acids.

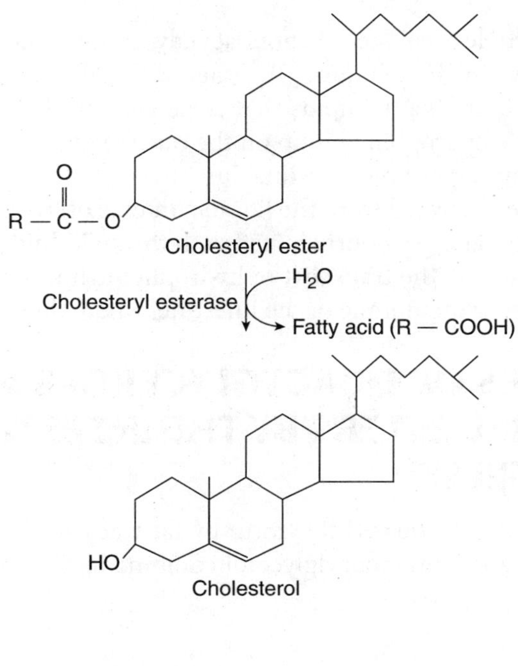

PHOSPHOLIPID DEGRADATION

Pancreatic juice is rich in the proenzymes of phospholipase A_2, which is converted to active form with the help of trypsin. Phospholipase A_2 converts the fatty acid at carbon 2 of the phospholipids, leaving a lysophospholipid. For example, lecithin is converted to lysolecithin. The remaining fatty acid at carbon 1 is cleaved by lysophospholipase, producing glycerylphosphorylcholine that is excreted into the faeces or further degraded and absorbed.

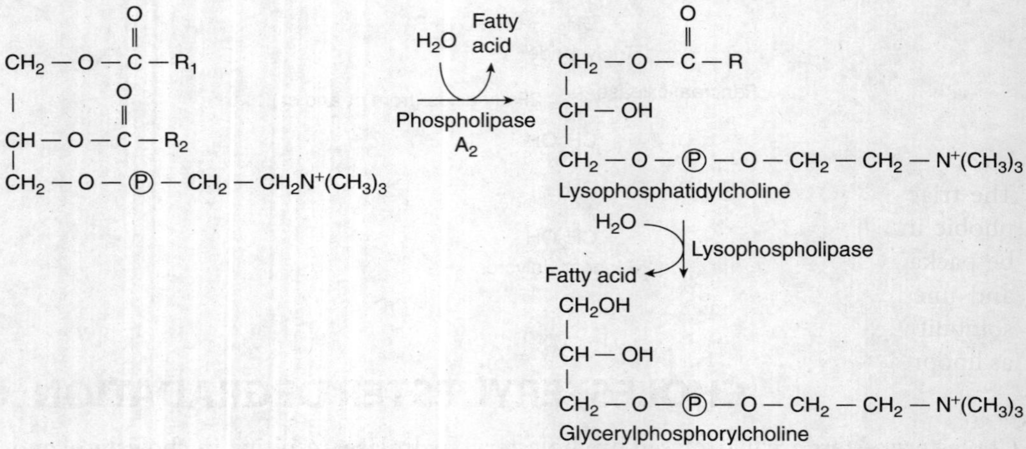

ABSORPTION OF LIPIDS

Free fatty acid, free cholesterol, and 2-monoacylglycerol are the primary products of dietary lipid digestion in the jejunum. Together with bile salts, they form micelles, which are clusters of amphipathic lipids that coalesce with their hydrophobic groups on the inside and their hydrophilic group on the outside of the cluster and are soluble in the aqueous environment of the intestinal lumen.

These micelles are separated from the liquid contents of the intestinal lumen by an unstirred layer of water and get poorly mixed with the bulk fluid. The hydrophilic surface of the micelles favours the transport of hydrophobic lipids through the unstirred layer to the brush border membrane of the intestine, where they get absorbed.

RESYNTHESIS OF TRIACYGLYCEROLS AND CHOLESTERYL ESTERS BY THE INTESTINAL MUCOSAL CELLS

Fatty acids are converted to their active form by fatty acyl-CoA synthase and thiokinase. Fatty acyl-CoA and 2-monoacylglycerol combine in the intestinal mucosal cells and are converted to triacylglycerol with the help of acyl transferases that recognise fatty acyl-CoA of specific chain length.

$$\text{2monoacylglycerol} + \text{2fatty acyl-CoA} \xrightarrow{\text{Acyl transferase}} \text{Triacylglycerol} + \text{2CoA}$$

SECRETION OF LIPIDS FROM INTESTINAL CELLS

The triacylglycerol and the cholesteryl esters that are resynthesised are very hydrophobic in nature and aggregate in an aqueous environment. Therefore, they need to be packaged as lipid droplets surrounded by a thin layer of proteins, phospholipids, and unesterified cholesterol. These packages stabilise the lipids and increase their solubility. The lipids that are packaged with the proteins and phospholipids are known as lipoproteins.

SECRETION OF CHYLOMICRONS

Chylomicrons are processed out by exocytosis from the intestinal mucosal cells into the intestinal lacteals and then to the lymphatic system to the thoracic duct; they then enter into the left subclavian vein, where they enter the blood.

USE OF DIETARY LIPIDS BY THE TISSUES

Triacylglycerols packed in chylomicrons are converted by the skeletal muscle and adipose tissues and also in heart, kidney, and liver to free fatty acids and glycerol by lipoprotein lipase. This enzyme is synthesised in adipose tissues and muscle cells. It is associated with the luminal surface of the endothelial cells of the capillary beds of the peripheral tissues.

FATE OF GLYCEROL

Glycerol formed from triacylglycerol is utilised by the liver to produce glycerol-3-phosphate, which can enter glycolysis or gluconeogenesis by oxidation into dihydroxyacetone phosphate.

FATE OF CHYLOMICRON COMPONENTS

The chylomicron remnants, which contain cholesteryl esters, phospholipoproteins, and some triacylglycerols, are consumed by the liver, where they are further hydrolysed to their component parts. Cholesterol and the nitrogen bases of phospholipids are reused in the body.

FATTY ACID OXIDATION

INTRODUCTION

The fatty acids are oxidised in the system mainly through β-oxidation. The oxidation of the fatty acids on the beta carbon atom is known as β-oxidation. This results in the sequential of a two carbon fragment, acetyl-CoA. Knoop (1904) proposed that these fatty acids are degraded by oxidation of the β-carbon.

$$\underset{\omega}{R} - (CH_2)_n - \underset{\beta}{CH_2} - \underset{\alpha}{CH_2} - COOH$$

Fatty acid

LOCATION

The fatty acid oxidation takes place in mitochondria. Fatty acids are oxidised by most of the tissues in the body. However, brain, erythrocytes, and adrenal medulla cannot utilise fatty acids for energy requirement.

The β-oxidation of fatty acids involves the following three stages:

1. Activation of fatty acid occurs in the cytosol
2. Transport of fatty acids into mitochondria
3. Beta oxidation in the mitochondrial matrix

TRANSPORTATION

Fatty acids refer to fatty acids that are in unesterified state – unesterified fatty acids (UFA) or non-esterified fatty acids (NEFA). In plasma, Free Fatty Acids (FFA) of longer-chain fatty acids are combined with albumin, and in cell, they are attached to a fatty-acids-binding protein so that in fact, they are never really free.

ACTIVATION OF FATTY ACIDS

Fatty acids are activated to acyl-CoA by thiokinase or acyl-CoA synthetase. The reaction occurs in two steps and requires ATP, CoA, and Mg^{2+}.

Fatty acid reacts with ATP to form acyl-adenylate, which then combines with CoA to produce acyl-CoA.

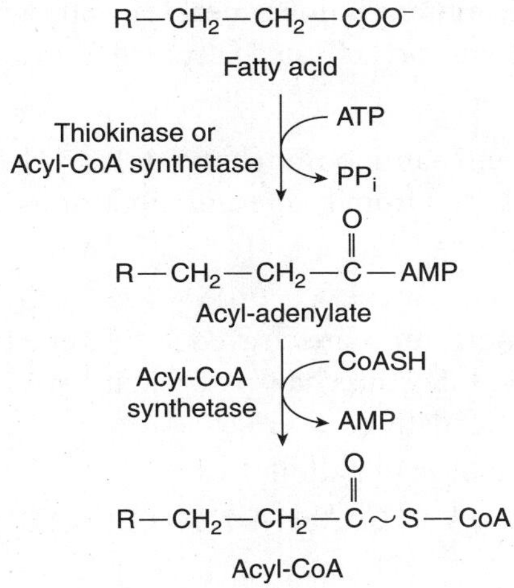

TRANSPORT OF ACYL-CoA INTO MITOCHONDRIA

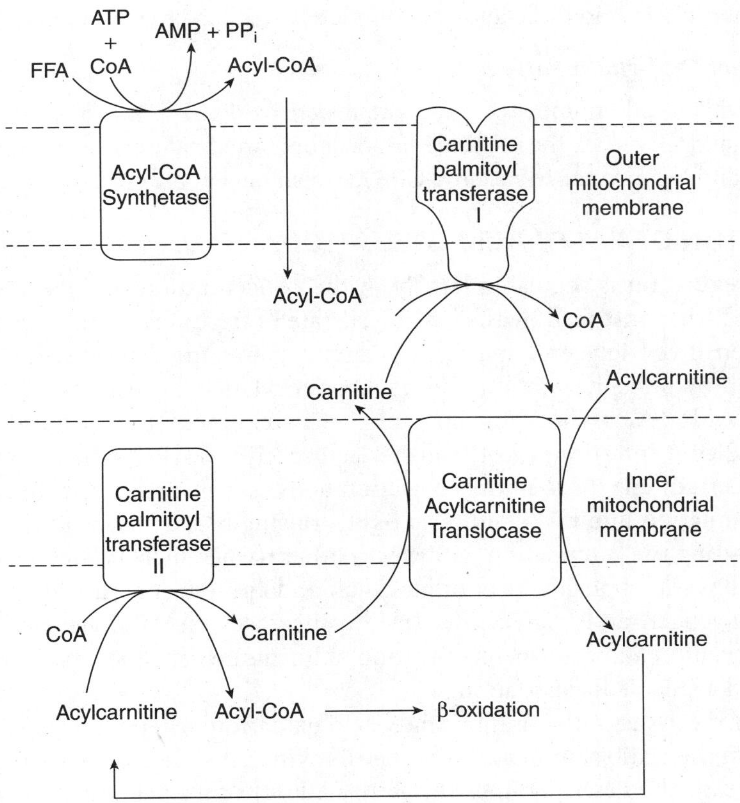

The inner mitochondrial membrane is impermeable to fatty acids. A specialised carnitine carrier system operates to transport activated fatty acid from cytosol to the mitochondria.

Carnitine

Carnitine (β-hydroxy trimethylammonium butyrate) is widely distributed and is abundant in muscle. It is synthesised from lysine and methionine in liver and kidney.

β-OXIDATION

In beta oxidation, two carbon atoms are cleaved at the time from acyl-CoA molecules, starting at the carboxyl end. So the chain is broken between α(2) and β(3) carbon atoms. Hence, the name β-oxidation.

Overall equation: Oxidation of palmityl-CoA

$$\text{Palmitoyl-CoA} + 7\text{CoASH} + 7\text{FAD} + 7\text{NAD}^+ + 7\text{H}_2\text{O} \\ \downarrow \\ 8\ \text{Acetyl-CoA} + 7\ \text{FADH}_2 + 7\ \text{NADH} + \text{H}^+$$

Palmityl-CoA undergoes seven cycles of β-oxidation to yield 8 acetyl-CoA. Acetyl-CoA can enter citric acid cycle and get completely oxidised to CO_2 and H_2O.

Oxidations of Odd-chain Fatty Acids

Fatty acids with an odd number of carbon atoms are oxidised by β-oxidation and produce acetyl-CoA until they reach the last three carbon atoms and finally end with propionyl-CoA. This compound is converted to succinyl-CoA, a constituent of the citric acid cycle.

REGULATION OF FATTY ACID OXIDATION

Fatty acid regulation is regulated largely in the concentration of fatty acid in blood, which is, in turn, controlled by the hydrolysis rate of triglycerides in adipose tissue by hormone-sensitive triglycerol lipase. This enzyme is so named because it is susceptible to regulation by phosphorylation and dephosphorylation in response to hormonally controlled cAMP levels. Glucagon, epinephrine, and norepinephrine increases adipose tissue cAMP concentration; cAMP allosterically activates APK, which, in turn, phosphorylates certain enzymes; phosphorylation activates hormone-sensitive lipase and, thereby, stimulation lipolysis in adipose tissue, raising blood fatty acid levels and ultimately activating the β-oxidation pathway in other tissues such as liver and muscle.

Insulin has the opposite effect of glucagon and epinephrine. It stimulates the formation of glycogen and triglyceride. Insulin increases cAMP levels, leading to the dephosphorylation and thus the inactivation of hormone-sensitive lipase. This reduces the fatty acid available for oxidation.

Another mechanism that inhibits fatty acid oxidation, when fatty acid syntheses, is stimulated in the inhibition of carnitine palmitoyltransferase 1 by malonyl-CoA. This inhibition keeps the newly synthesised fatty acid out of mitochondria and thus away from β-oxidation system as shown in Figure 10.3.

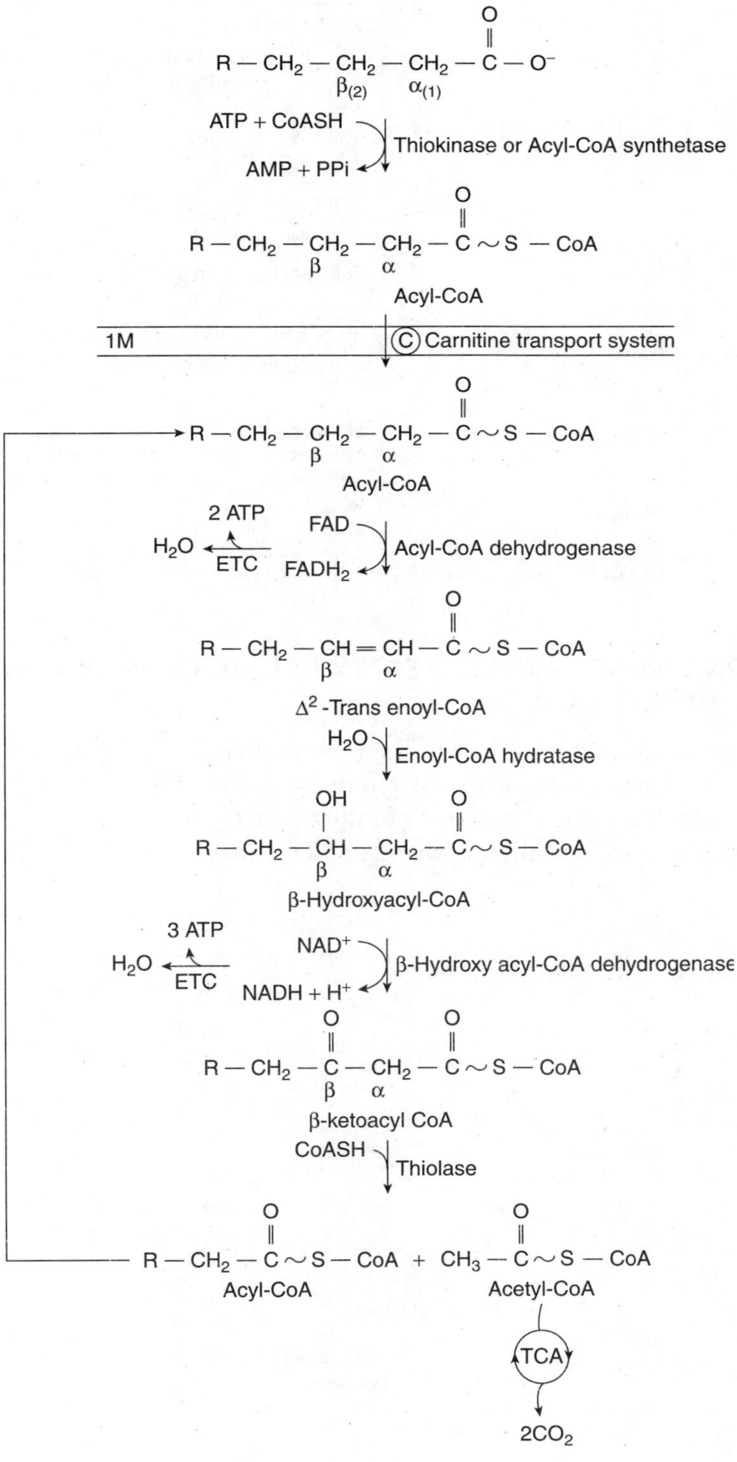
R — CH2 — CH2 — CH2 — C — O⁻
β(2) α(1)
ATP + CoASH
AMP + PPi
Thiokinase or Acyl-CoA synthetase
R — CH2 — CH2 — CH2 — C ~ S — CoA
β α
Acyl-CoA
1M
C Carnitine transport system
R — CH2 — CH2 — CH2 — C ~ S — CoA
β α
Acyl-CoA
2 ATP
H2O
ETC
FAD
FADH2
Acyl-CoA dehydrogenase
R — CH2 — CH = CH — C ~ S — CoA
β α
Δ2 -Trans enoyl-CoA
H2O
Enoyl-CoA hydratase
OH
R — CH2 — CH — CH2 — C ~ S — CoA
β α
β-Hydroxyacyl-CoA
3 ATP
H2O
ETC
NAD+
NADH + H+
β-Hydroxy acyl-CoA dehydrogenase
R — CH2 — C — CH2 — C ~ S — CoA
β α
β-ketoacyl CoA
CoASH
Thiolase
R — CH2 — C ~ S — CoA + CH3 — C ~ S — CoA
Acyl-CoA
Acetyl-CoA
TCA
2CO2

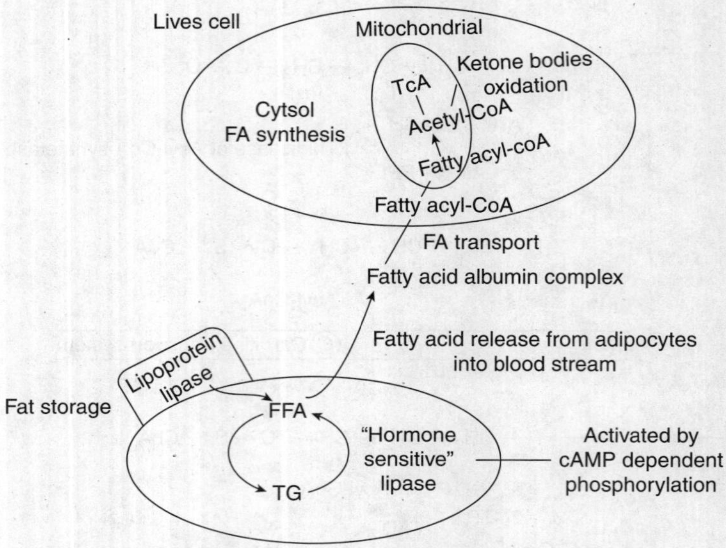

Figure 10.3 *Regulation of Fatty Acid Oxidation*

OXIDATION OF UNSATURATED FATTY ACID OCCURS BY A MODIFIED β-OXIDATION PATHWAY

Due to the presence of double bonds, the unsaturated fatty acids are not reduced to some extent as saturated fatty acids. Therefore, oxidation of unsaturated fatty acids, in general, provides less energy than that of saturated fatty acids. This oxidation is overcome by two additional enzymes, isomerase and epimerase.

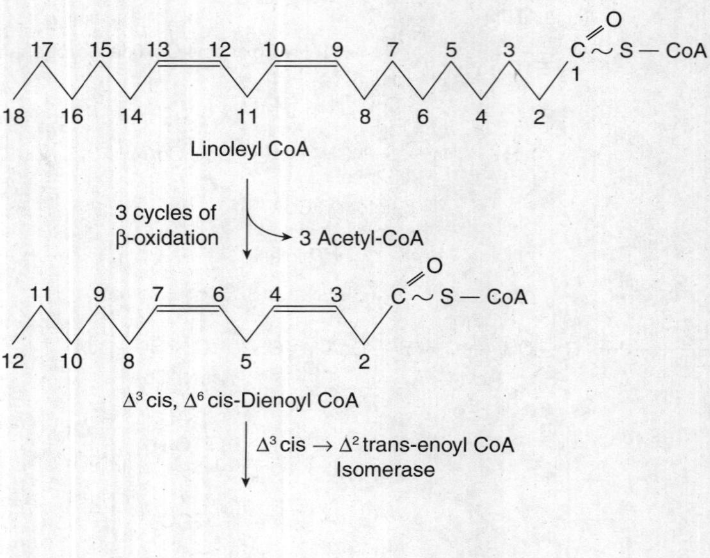

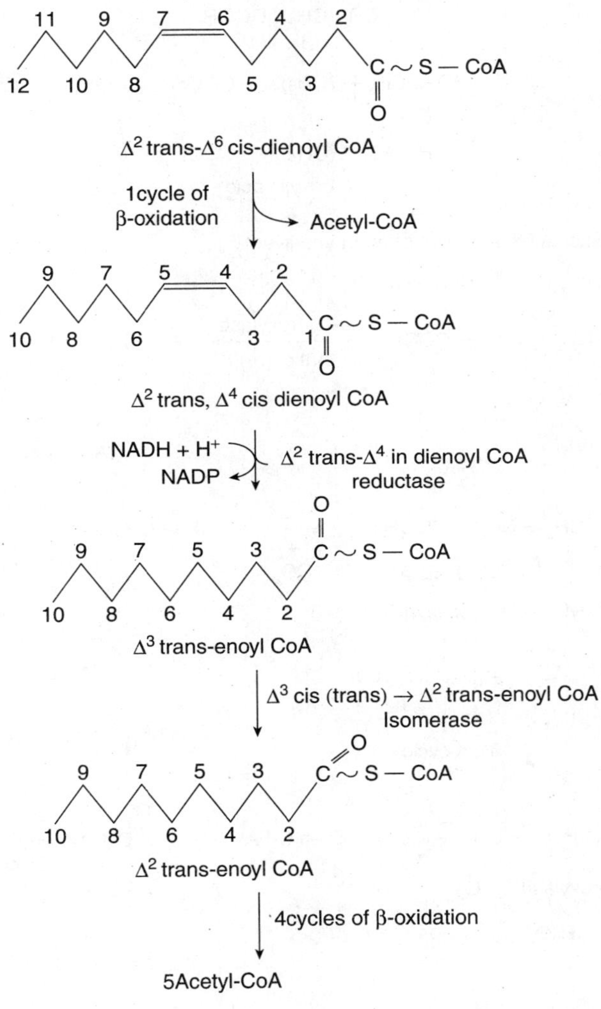

α-OXIDATION OF FATTY ACIDS

β-oxidation is the most predominant pathway for fatty acid degradation. However, the removal of one carbon unit at a time by oxidation of α-carbon atom of fatty acid is known as α-oxidation.

This type of oxidation has been detected in brain tissues. It does not require CoA intermediates and does not require high energy phosphates.

ω-oxidation is a minor pathway and is brought about by hydrolase enzymes involvng cytochrome p450 in the endoplasmic reticulum. The CH_3 group is converted to CH_2OH group that is subsequently oxidised to COOH, thus forming dicarboxylic acid. This is β-oxidised usually to adipic and suberic acids which are excreted in urine.

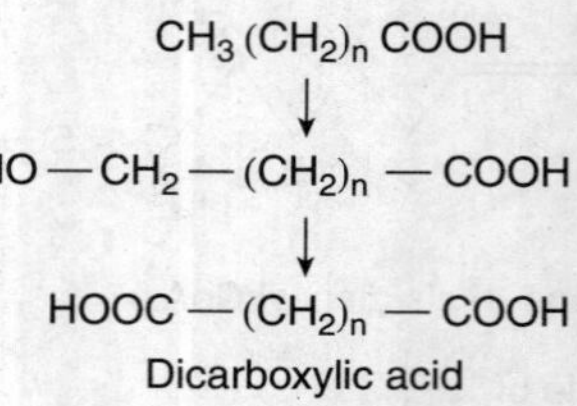

α-Oxidation:

Sites – Microsomes of brain and liver

Enzyme–α-hydroxylase

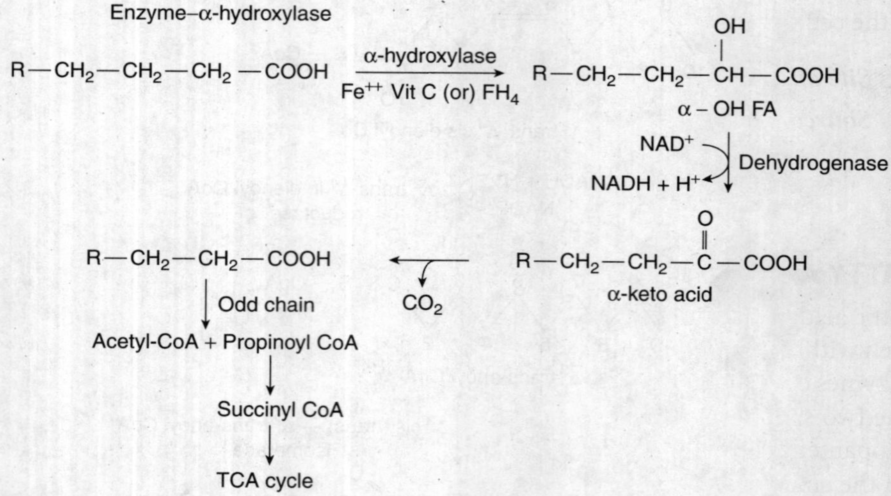

ω-Oxidation:

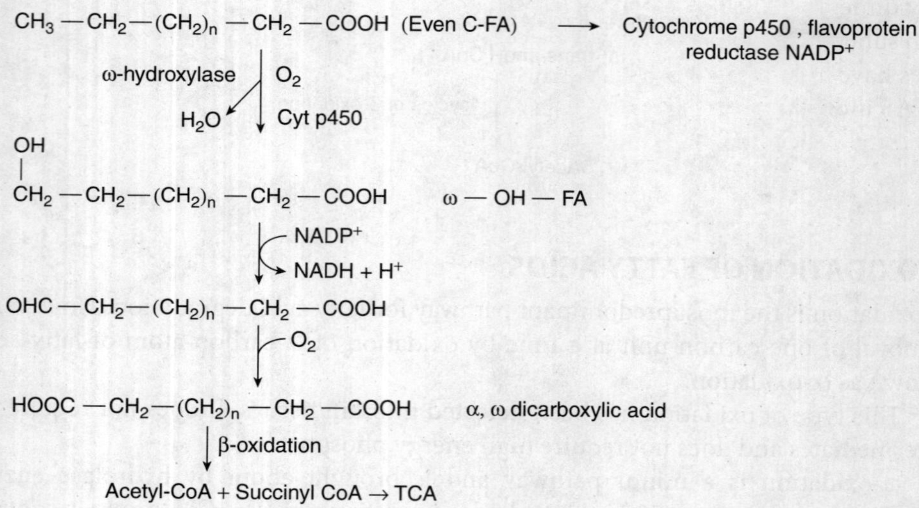